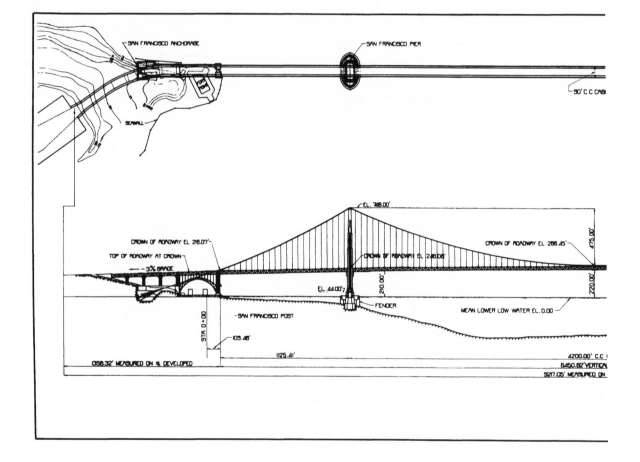

JOHN VAN DER ZEE

AN AUTHORS GUILD BACKINPRINT.COM EDITION

THE GATE

T H E

THE TRUE STORY OF THE DESIGN AND CONSTRUCTION OF THE GOLDEN GATE BRIDGE

The Gate:
The True Story of the Design and
Construction of The Golden Gate Bridge
All Rights Reserved © 1986, 2000 by John van der Zee

AN AUTHORS GUILD BACKINPRINT.COM EDITION
Published by iUniverse, Inc.

For information address:
iUniverse
2021 Pine Lake Road, Suite 100
Lincoln, NE 68512
www.iuniverse.com

Originally published by Simon & Schuster

Designed by Bonni Leon

ISBN: 0-595-09429-5

Printed in the United States of America

ACKNOWLEDGMENTS

This is many people's—not just one person's—book. It was Bob Asahina of Simon & Schuster who first suggested that the fiftieth anniversary of the completion of the Golden Gate Bridge might be a fit occasion for a new book about the bridge. The thought that there was a story here that hadn't been fully told elsewhere originated with Russ Cone, who had covered the bridge as a reporter and columnist for the San Francisco *Examiner* and whose father, Russell Cone, was resident engineer on the bridge during its construction. To Russ belongs the credit for whatever fresh perspective this book represents, as well as for pinpointing the existence of hitherto unused documents, correspondence, and photographs.

Mrs. Albert R. Mead—Eleanor Morrow Mead—of Tucson, who also felt there was a story yet to tell, generously shared letters, reports, and personal reminiscences of her father, Irving Morrow, while Herb Johnson, once a young architect on Morrow's staff, furnished a working associate's close-up view. Frank Stahl of Ammann & Whitney offered his time, thoughts, office, and the help of his staff while I plodded through the firm's twenty-eight volumes of bridge letters, reports, memoranda, and telegrams. The staff of the Water Resources Archives at the University of California at Berkeley patiently let me examine and copy the contents of Professor Charles Derleth's papers, while Hal Silverman and Robin McKenna of the San Francisco *Examiner* allowed me access to the paper's valuable library.

To Izetta Lucas Cone, Ruth Natusch, Professor Marion Scott, Ted Huggins, Frenchy Gales, Peanuts Coble, Harold McClain, Lefty Underkoffler, George Albin, Al Zampa, and Alfred Finnila, I am grateful for both the time spent in your company and for the parts of your lives you shared with me, moments past and present that I enjoyed and value.

Barbara Newcombe of the Chicago *Tribune,* Gladys Hansen of the San Francisco Main Library, Cecily Surace of the Los Angeles *Times,* and Harold L. Michael, head of the Purdue School of Civil Engineering, supplied source materials unavailable elsewhere, while Bob David of the Golden Gate Highway and Transportation District, along with Bruce Selby and Gene Rexrode, directed me to the people and pictures essential to bringing the construction of the bridge to life.

I am indebted once again to the staff of the Doe Library at the University of California, as well as to those of the Mechanics Institute Library of San Francisco, and to Stephen Fletcher and the rest of the staff of the California

Historical Society for the reassuring professionalism of their assistance and advice.

I would like to thank Meredith Greene of Simon & Schuster for her astute and insightful editorial suggestions—the best kind of compliment an editor can pay to a book.

Katy van der Zee took the picture of the Fourth Street Bridge and, along with her mother, Diane, and her brother Peter, served as a patient audience for, and sympathetic stimulus to, the writing of the book itself.

The book, then, is the sum of many people's interest, effort, imagination, and concern, whose congruence it is hoped might in some way reflect the great collaborative achievement of its subject.

To Bob Asahina

CONTENTS

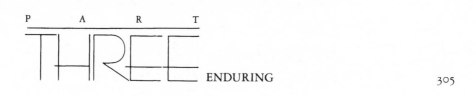

PART

THREE ENDURING

"a giant portal
that seems
like a
mighty door,
swinging wide
into a world
of wonders."

—Joseph Strauss,
   Dedication Speech,
   February 25, 1933

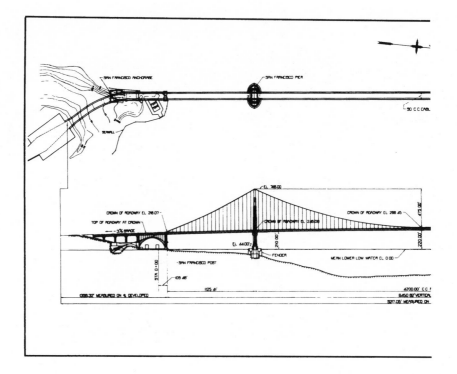

P A R T

ONE

DREAMING

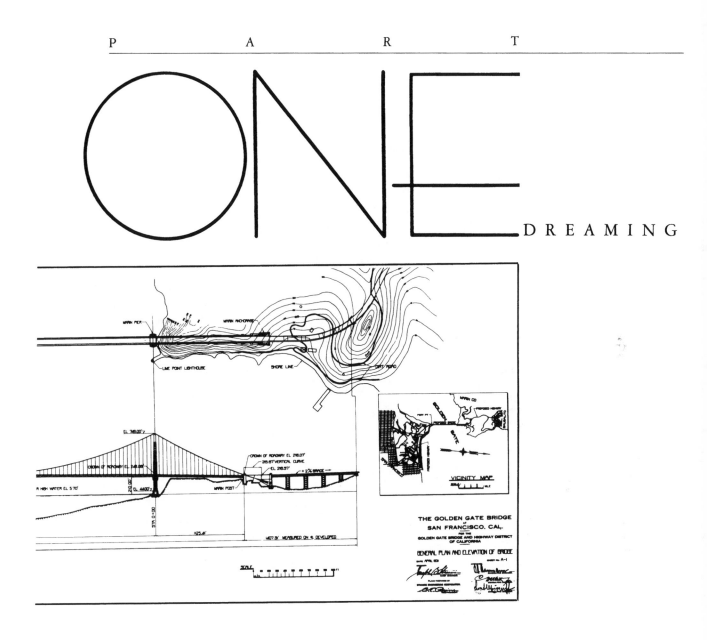

The first bridge ever built at the entrance to a major harbor, the Golden Gate Bridge stretches from the San Francisco Presidio (foreground) to the Marin County headlands (top). To the left is the Pacific Ocean and to the right, San Francisco Bay.

It is, along with the Statue of Liberty, one of the two structures that in and of themselves suggest the breadth and promise of America: the western half of the American ellipse. A work of everyday function, yet soaring inspiration, which, in its union of the organizational arts and skills of the industrial age with the enduring replenishments of water and land and sky, suggests some profound reconciliation between man and nature: perhaps the most successful combination of site and structure since the Parthenon.

It is at the same time a work of ego and politics, the end product of personal aggrandizement, promotional misrepresentation, deal cutting, the relentless pursuit of interest, the hidden use of influence, and broken health and careers; an engineering achievement whose intellectual sources have, for half a century, been obscured and denied.

Most significantly, it is a work conceived and carried to completion during the worst economic depression in American history, financed without either state or federal money, steered through the often bewildering shoals of American local politics at a time when that system itself was under severe ideological attack. It is interesting to note that the years of its construction, 1933–1937, coincide with the coming to, and consolidation of, power in Germany of the quintessential totalitarian of this century, a man whom Albert Speer has revealed to us as a frustrated architect, willing to destroy the world to rebuild it as his own gigantic mausoleum. Not one of his thousand-year works, with whatever degree of ruthless efficiency pursued, would have matched the simple daily usefulness and the stirring aspiration realized in this bridge, an idea filtered through all the compromising byways of the American political system, yet enduring as that rarest of all artistic achievements: a democratic masterpiece.

THE FIRST BRIDGE

I

Squat, bulky, unadorned, bluntly functional, and less than a hundred feet long, the Fourth Street Bridge is like the negative imprint of its famous counterpart on the other side of San Francisco. Instead of the entrance to one of the world's great harbors, it spans a contaminated backwater, a sluggish creek afloat with oil, garbage, Styrofoam, and condoms. The traffic beneath it—the bridge is a lifting drawbridge—has declined to a trickle of pleasure boats and small scows; at the mouth of the creek, two blocks away, one of the great liners of the Matson Pacific fleet sits rotting and neglected.

It is trucks that mainly traverse the Fourth Street or, more formally, the Peter J. Maloney Bridge, thundering in and out of San Francisco's declining industrial flatlands, a part of town earmarked for reincarnation as yet another visitors' attraction, perhaps including a sports complex with a retractable-domed stadium. But it is difficult to imagine the raw mechanism of this bridge in the midst of a genteel urban park-and-garden.

The bridge itself is so nondescript that one can easily pass over it without realizing it is there. Its most prominent feature, a 700-ton overhead concrete counterweight, which spans the roadway, makes it resemble an abandoned chunk of freeway overpass, like the blunted freeway arms one sees here and there about the city. At one time, San Francisco had been scheduled to be girdled by a freeway system linking both its great bridges and the freeways to the south. The network was completed everywhere but northward, in the direction of the Golden Gate Bridge, where its intrusions have been stopped, turned aside, diverted downward onto existing city streets.

No doubt it was the prospect of carving up what was left of the city's northern waterfront and the sundering of its magnificent Golden Gate Park that brought about the immediate civic outcry that opposed, and stopped, San Francisco's full freeway encirclement. But beyond this was something else. San Francisco, as we shall see, had been outrageously lucky in its most famous bridge, considering what some other communities have been forced to live with; the city had rarely been that lucky in civic enterprises before (the original City Hall took thirty years to complete, was a nightmare boondoggle of fraud and cost overruns, and collapsed in the first fifty seconds of the 1906 earthquake), and has never been as fortunate since (Candlestick Park, the Bay Area Rapid Transit District, the Hall of Justice, the Moscone Center). The Golden Gate Bridge remained, and remains, a stimulus to civic conscience, a reminder and a rebuke that the conventional dictates of the possible are not always inevitable.

That the Golden Gate Bridge could have ancestral ties with the ugly Fourth Street Bridge is one of those relationships that seem to challenge the whole known body of engineering and industrial genetics, like genius emerging from generations of unbroken idiocy. Yet that, in defiance of the conventional wisdom of several eras and of the laws of probability, is what happened. And the way that it happened, almost a cliché at first, fragmenting into a diffuse and complicated reality, emerges again as a truism of another, deeper sort, that breathes life into that oldest of New World wheezes, "only in America."

2

At three o'clock in the afternoon of February 20, 1915, President Woodrow Wilson pressed a gold key in the East Room of the White House; an instant later, across the continent in San Francisco, an elaborate "Fountain of Energy" leaped to life. Jets of high-pressure water were played high into the air before falling into an enormous pool. Resting on a pedestal in the center of the pool, supported by a circle of figures representing the oceans,

The Fourth Street Bridge—short, squat, ugly, bluntly functional, a typical Strauss lift bridge —it was the first Strauss bridge in San Francisco.

was the Earth, surmounted by a young man, the symbol of Energy, the force that dug the Panama Canal. From his shoulders, Fame and Victory blew their trumpets. While all the jets of water played, Energy, horsed, rode triumphant through the waters at either hand. The Panama-Pacific International Exposition was formally opened.

Set in a "City of Domes" spread over some three hundred landfill acres of what is today the prime real estate of the city's Marina District, the 1915 Panama-Pacific Exposition was the last big blowout of America's age of international innocence. While the nations of the Old World had already slipped into the world war that was to shape the history of the rest of the century, America had become, more than ever, the home of enlightenment and, especially, progress. "It was a happy time," Walter Lippmann recalled of the years before World War I. "The air was soft, and it was easy for a young man to believe in the inevitability of progress."

The exposition was a celebration of progress incarnate: the completion of the Panama Canal, the mythical linking of the oceans, made real by the greatest feat of engineering the world had yet seen, an ebulliently American enterprise at which the French had tried, and failed. It was engineering, above all, that people had come from all over the world to San Francisco to honor, and 257 different engineering societies were scheduled to meet at the exposition. For engineers generally, particularly sensitive now to their emergence from the level of artisans and craftsmen into the ranks of licensed and better-paid professionals, the exposition had an entire second level of significance: it was being held in a city that had been all but destroyed only nine years before and had now been restored, largely through the work of engineers, a fact that *Engineering News,* the bible of the civil engineering and construction industry, considered worth celebrating with a double-sized exposition issue featuring the "Notable Engineering Works of the City Engineering Department of San Francisco."

There was something in the reconstructed San Francisco with which an engineer of almost any specialty could identify: a new sewer system; a firefighting network that utilized both fresh and salt water reserves; a completely new municipally owned transit system that would eventually include tunnels, forty-eight miles of track, and "iron monster" railway cars built in the Municipal Railway's own shops; and, most significantly, the most sweeping and elaborate water supply system ever devised for an American city, the Hetch Hetchy system, which involved dams, lakes, and more than two hundred miles of aqueducts and conduits and which would require more than twenty years and over a $100 million to complete.

Especially poignant to the engineers was the fact that this enormous civic reconstruction was almost entirely under the direction of one man, a knock-about, practical civil and construction engineer not so different from themselves. And that, in San Francisco, he seemed to be the living embodiment of the engineering ideal of professional technical leadership showing the way to political and moral progress.

3

Born in County Limerick, Ireland, on May 28, 1864, Michael Maurice O'Shaughnessy was educated at the Queen's colleges of Cork and Galway, constituent colleges of the Royal University of Ireland, from which he received his bachelor's degree in engineering in 1884. The following year O'Shaughnessy arrived, "with brogue and parchment," in San Francisco, where he took a job in the engineering department of the Southern Pacific Railroad.

Michael O'Shaughnessy, the politically untouchable City Engineer of San Francisco, first proposed that Joseph Strauss design a bridge for the Golden Gate. An early Strauss ally, he later turned against Strauss and opposed the bond issue and the bridge, helping to put an end to his own career.

A strapping young man with a piercing, steady gaze, a thinning head of hair, a thick, trimmed mustache, and a taste for elaborately formal dress almost always including a boutonniere, O'Shaughnessy, as assistant engineer of the railroad and, later, civil engineer for townsites, rapidly learned the ins and outs of California construction and its politics. Bright, articulate, vigorous, and capable, O'Shaughnessy established himself rapidly in his adopted city. He married a young local woman and in 1893, while still in his twenties, was named chief engineer of San Francisco's first notable fair, the Midwinter International Exposition, designed to promote California's produce and her climate.

An ambitious young man who was gaining recognition in an expanding profession, O'Shaughnessy now went, for a time, wherever opportunity took him, although he was almost always either an independent consultant or else running the whole show: chief engineer of a copper company in Shasta County; construction and hydraulic engineer on sugar plantations in Hawaii. Returning to California in 1905, O'Shaughnessy became chief engineer of the Southern California Mountain Water Company at San Diego, for whom, between 1907 and 1912, he supervised the building of the Morena Dam, the largest rock-filled dam built to that time in the United States.

As O'Shaughnessy's reputation and career expanded, San Francisco, the city where he'd first lived and worked in America and his wife's hometown, endured what must still rank as the most severe natural convulsion, and among the worst politically, suffered by any major American metropolis.

Perhaps the best way to gauge the devastation of the San Francisco earthquake and fire of 1906 is to examine the photographs of the burned and shaken city. In the extent and depth of the damage—the heaps of rubble, the scarred, standing hunks of wall, the terrifying emptiness—the pictures are like a preview of the bombed-out European and Japanese cities of World War II. Nothing like the damage inflicted by nature on San Francisco was to exist for the world's cities for another thirty-five years, until the advent of long-range multiengine aircraft and the blockbuster bomb.

In addition to this physical devastation, San Francisco, in the years immediately preceding and following the earthquake, was probably the worst-governed city in America.

Under the administration of a stooge mayor, Eugene Schmitz, an ex-bandleader and local head of the musicians' union, and a wily political boss, Abe Ruef, San Francisco in the years after 1902 was like a municipality bearing a giant "For Sale" sign. Prostitution was open and rampant, and the dens and deadfalls of the Barbary Coast for years maintained an average of a murder a night. More serious, although less apparent, was the infiltration of the city's public utilities by graft-paying private interests. Patrick Calhoun, organizer of the United Railroads, the city's combined streetcar system, took advantage of the postearthquake confusion to string the city with the overhead electric power lines that had formerly been widely opposed, then sealed the deal by bribing members of the city's Board of Supervisors to pass an ordinance approving the unsightly lines. The city's water service, supplied by the private Spring Valley Water Company, which resisted public outcries for improvements, was so poor that in the Richmond District there was often insufficient pressure to produce more than a trickle out of householders' taps. The water company kept the entire Board of Supervisors on retainer.

Public indignation, and an effective reform movement led by former mayor James D. Phelan, industrialist Rudolph Spreckels, and Fremont Older, editor of the San Francisco *Call,* had led ultimately to the ouster and trial of both Schmitz and Ruef. As long as the city's water and transit utilities remained in the hands of the Spring Valley Water Company and the United Railroads, however, those hands could always line the official pocket and clutch the civic throat. By 1912, encouraged by the reform spirit of the Progressive movement, and particularly the California Progressives, led by Hiram Johnson, now governor of California and a man who had prosecuted Ruef, the people of San Francisco had voted to build their own municipal transit system, to run in competition with what was now the Market Street and United Railroads, and had decided either to build a new water system or to acquire the Spring Valley Water Company, or both.

What was required to keep these new systems too from falling into the hands of the grafters was a super-engineer, a man who could oversee the physical reconstruction of the city while remaining free of the pressures and plundering that had brought about its political decline.

Mayor James D. Rolph, who took office in 1912, was convinced he knew the man to handle both tasks.

4

"In the latter part of August, 1912," Michael O'Shaughnessy recalled years later, "Mayor Rolph wired me at San Diego . . . asking if I would be available for the position of City Engineer of San Francisco."

The task was enormous: the earthquake and fire had destroyed 2,300 acres of the city and left 100,000 people homeless; the entire public utilities system would have to be reconstructed. Also, O'Shaughnessy had been exposed to San Francisco politics before, an experience he remembered as "discouraging." And the salary, $15,000, was half the total of his income for the previous year.

Yet there was the challenge of undertaking the reconstruction of an American city—the chance to do it all as it really should be done, and the assurance from the mayor that, as city engineer, O'Shaughnessy would be politically autonomous. There was also something else, O'Shaughnessy admitted: "My wife being a native of the City, influenced my decision and favorable consideration of the Mayor's proposal."

Rolph, a shrewd campaigner who was to serve four terms as mayor of San Francisco, then one as governor of California, had the successful politician's instinct for pinpointing exactly what his audience most wants to hear:

"You must look on the City as your best girl," Rolph told O'Shaughnessy, "and treat her well. Do what you think is best for her interests."

When O'Shaughnessy expressed "my strong objections to political interference by elected officials," Rolph capped his sales pitch by offering the engineer carte blanche:

"Chief, you are in the saddle. You're it. You're in charge."

Riding the wave of a new, reform administration, responding to a public mandate to "get things done, (we don't care how)," O'Shaughnessy plunged into the problems of San Francisco with the bold, personal style that was to fill, and eventually overflow, the boundaries of his job.

"He has not waited for the reports of his subordinates," a San Francisco paper reported approvingly, not long after O'Shaughnessy took office, "nothing of the kind for this virile man. He investigates first personally, visits the spot where a correction is needed, makes a hasty, though no less accurate survey; arrives at his conclusions; tells what should be done . . . and the blueprints follow later."

The engineer as doer, the thinker as man of action, the theoretician combined with the practical man, all in the interest of public service. This was the O'Shaughnessy that O'Shaughnessy himself wished to be. And that, during these years at least, he was—to the administration, to the press, and to the overwhelming majority of the citizenry of San Francisco.

O'Shaughnessy delivered results. The city of vacant lots, newly built homes, and rising skeletal buildings bustled with the activity of his department's construction crews: track laying for the streetcars that were being welded together in the Muni's shops; tunnels under the Stockton Street hill and beneath Twin Peaks; pipelines and pumping stations for an auxiliary fire protection system; a sewer outfall at Baker's Beach—all working against the deadline of the 1915 fair. And, most impressively, Hetch Hetchy.

It was the development of the city's independent water supply that forced O'Shaughnessy's vision beyond the existing physical limits of San Francisco. Isolated on the tip of a peninsula, the city was both cut off by water—and in need of it.

To satisfy the city's thirst for fresh household water, San Francisco voters had agreed to reach some 245 miles into the Sierra Nevada Mountains and acquire rights to a valley called Hetch Hetchy, a neighbor of Yosemite Valley, and regarded by many people who had seen it, among them John Muir, as Yosemite's equal in beauty. In an act politically inconceivable today, and one opposed by Muir and the Sierra Club in a battle that dragged on for years and ultimately reached the Supreme Court, Hetch Hetchy was logged, dammed, flooded, and turned into the reservoir for the water supply of San Francisco. The project, which included the building of a railroad, tunnels, and access roads and the laying of a wilderness pipeline, cost eighty-nine lives and was not completed until 1934. It provided San Francisco, a city today of less than 750,000, with a water supply for 4 million people and is considered the model for urban water systems everywhere. That its design, construction, and completion represent the organizational ability and will of a single man can best be testified to by the fact that the dam at the mouth of the valley, the most significant structure of the project, was named at its completion, in 1923, after Michael O'Shaughnessy.

5

By the time of the 1915 exposition, O'Shaughnessy was a man of independent political power in San Francisco. Backed by the unquestioning support of a popular mayor, not answerable to the electorate himself, he wrote signed articles in the San Francisco newspapers on the state of the city and often advised readers how to vote on local bond issues. He served as a consultant on urban utilities to other cities and was retained in an advisory capacity by the boards of local municipalities' water districts. His reputation, expanding with the prestige of his profession, grew mightily on both the national and local level. In 1919 Rolph, in a speech at Hetch Hetchy, compared his city engineer to Hezekiah, the Old Testament king who "stoppered the watercourse of Gihon and brought it straight down . . . to the city of David."

A man of abundant physical and mental energies, O'Shaughnessy seemed not to be burdened by the breadth and depth of his responsibilities, but instead to be stretched by them. His imagination, he found, was like a muscle, which grew stronger with use, and the more local projects he took on and completed, the more O'Shaughnessy found his range of ideas expanding beyond San Francisco to the entire bay region. This city, this farthest west that he had come to, represented not only his own best possibilities but the country's. Here, where everything had begun over again, the future could be not only planned but engineered.

On Sundays, the tireless O'Shaughnessy took the ferry from San Francisco to the village of Sausalito in Marin County, where, accompanied by his two dogs, an Aberdeen Scotch terrier and a bulldog, he hiked through the territory he called Hobbylogland, the thickly wooded slopes and canyons of Mill Valley, often continuing up over the Coast Range into the cathedral redwood groves of Muir Woods.

"A chunky lump of a man," a reporter for the San Francisco *Call* described him, "with a chest broad and deep, bushy eyebrows behind which blue eyes twinkled, and he walked with the gait of a man accustomed to getting there." Led by his two leashed dogs, a cigar usually clutched in his mouth, O'Shaughnessy customarily devoted his day off to a hike of between twelve and fifteen miles.

His weekend hikes in the Marin woods stimulated O'Shaughnessy's active expansionist imagination. Here, around him, was water, plenty of it, tumbling down the fern-lined Marin creeks; and here also, more importantly, was room: space for the homesites that O'Shaughnessy could already see would be severely limited by the fixed dimensions of San Francisco. Here, in the most beautiful—and the least developed—of the counties bordering the bay, was the repository of the engineer-and-planner's hopes. O'Shaughnessy bought rental property in Marin County. And on his way back from his weekly hikes, riding on the ferry past the Golden Gate or, more memorably, waiting in his car in the long line that backed up from Sausalito almost every Sunday evening, he pondered the possibilities of engineering San Francisco's next great extension.

6

The 1915 exposition, while not planned itself by O'Shaughnessy, was nevertheless an impressive demonstration of his works. Visitors rode to the fairgrounds on streetcars built in O'Shaughnessy's Municipal Railway shops, on tracks laid by his crews, on streets paved by his gangs, with asphalt from his city asphalt plant. Indeed, the very arteries of the fair, from the water that first whooshed through the Fountain of Energy at Woodrow Wilson's touch to the sewage system that drained the fair's wastes, were appendages of O'Shaughnessy's reconstructed San Francisco. "The City That Knows How," President William Howard Taft had dubbed San Francisco at the groundbreaking ceremony for the fair in Golden Gate Park in 1910; but it was one man, above all, who'd overseen the job, in the city's greatest burst of civic construction, before or since. The shattered, corrupt San Francisco of 1906 had been replaced by another city, white and shining, full of hope, and as committed to progress as the imposing yet temporary domes and palaces of the fair.

Perhaps the most impressive view of this restored San Francisco, as well as of the exposition grounds, the bay, and the surrounding hills, was from the passenger car atop the Aeroscope, the largest amusement ride at the fair, where up to 120 passengers at a time, enjoying a "gentle spiral swirling motion, similar to that of an aeroplane," circled slowly some 260 feet above the ground. Although not a formal technical exhibit, and located in the "Zone" or midway of the fair, the Aeroscope made use of several advanced engineering devices and resembled a frame steel bridge set on end. An enormous concrete counterweight was held by trunnions, or fixed pivots, mounted on a swingbridge turntable. At the level of the trunnions, the turntable frame was 48 feet high and rolled by means of eight four-wheel "trucks" or carriages on a circular track 61 feet in diameter. The swinging frame arm, or bascule, at the end of which the passengers spiraled gently in an enclosed car, was 242 feet long. The bascule was raised and lowered and the tower rotated on its track by electric motors, with water ballast

The Aeroscope, essentially a Strauss bascule bridge set on end, was the most popular amusement ride at the 1915 Panama-Pacific International Exposition in San Francisco.

used to keep the structure balanced for loading. The Aeroscope was the most popular pleasure vehicle at the fair, the equivalent of the giant Ferris wheels at the 1893 Columbian Exposition at Chicago and the 1904 Louisiana Purchase Exposition at St. Louis. And it was the work of a man who was the most prolific, if not always the most insightful, bridge builder of his day.

7

Born in Cincinnati, Ohio, on January 7, 1870, Joseph Baermann Strauss was one of the most compelling and contentious figures of the dawning Age of American Engineering. An undersized man with a Napoleonic ego, yearning for a career in the arts, with only the most modest amount of formal engineering training, gifted with the ability to recognize in others the greater talent he longed for in himself, driven by a persistent energy that rivaled that of the lone, cold-call salesman, Strauss was a strange, at times almost self-canceling mixture of conflicting traits: promoter, mystic, tinkerer, dreamer, tenacious hustler, publicity seeker, recluse. The son of a father who made a living as a portrait painter and dabbled in literature and a mother who was considered gifted as a pianist, Strauss grew up intimate with the grandiose dreams and often grim financial realities of what his contemporary Ohioan, Sherwood Anderson, has tellingly described as "we little children of the arts":

> *How many there are of us! How many thousands. We are going to write world shaking novels, become intellectual leaders of a nation, paint a picture that will be at once recognized as a masterpiece, sing a song that will reverberate through the hearts of all men.*
> *"Oh, if I only had a little leisure. If I had money. If I had even a small income."*

This life of endless aspiration and frustration, with its rejection of, and obsession with, money, this existence suspended like a pendulum swinging between egotism and insecurity, was the given condition of Strauss's boyhood, the circumstance that shaped him, and the fundamental contradiction that he would try, all his life and with all his considerable energy and will, to resolve.

There was, in the Cincinnati of Joseph Strauss's youth, an example of how these fundamental conflicts might be settled, or at least externalized. The Cincinnati-Covington Bridge, spanning the Ohio River between Ohio and Kentucky, was the first successful long-span suspension bridge built in America. Designed by John August Roebling, begun in 1845 but not completed until 1866, a graceful structure with a suspension span of more than a thousand feet, the bridge was a constant and powerful suggestion of how an intense personal vision with its haunting ties to the subconscious might be linked to the practical world of everyday utility and getting ahead in one's career. John Roebling had gone on to design the Brooklyn Bridge, the most famous and the most beautiful bridge in the world. To design such a work, to see it finished, a kind of sculpture that not only enhanced but incorporated its setting, before it was even drawn, one had to be a dreamer, as Strauss knew himself to be; and to turn these dreams into monuments of concrete and steel and cable required not only the immense practical knowledge of the engineer, but the determined will to convince others of the correctness of one's vision, to win others over to one's dream though they might be instinctively opposed to it, or be unable to see it through a maze of practical obstacles, or even have dreams of their own that they secretly preferred. Strauss knew himself to possess this intensity of will. To do such work, creating the monuments of the age of engineering, was to be like Roebling, akin to the artists of the High Renaissance, like the Leonardo who planned fortifications and dreamed of machines in addition to his great frescos and canvases. And, more personally, it allowed the individual in whom both these often opposing currents ran strong to channel them in the direction of one great goal. And if Joseph Strauss dreamed more and greater dreams than most young men, then he compensated for that by being more determined than almost any young man.

At the University of Cincinnati, where Strauss studied engineering and commerce in the Charles McMicken College of Liberal Arts, he also wrote verse and plunged into the organization of campus activities. In 1889 Strauss founded the local chapter of Sigma Alpha Epsilon, the second national fraternity at the university. The following year Strauss, five-foot-three and slight of build, turned out for football and ended up in the university infirmary. It was not to be the last time his determination was to push him beyond the bounds of his physical endurance. For his class's graduation in 1892, Strauss, chosen class poet, composed, instead of the customary brief commemorative ode, a twenty-one-stanza rumination called "Reveries," full of the soaring allusions and elegant variation he was to favor in the verse he wrote all his life.

More impressive, and more memorable than this work, at least to one of Strauss's schoolmates, was Strauss's reading of his graduation thesis at the commencement ceremonies. "I recall distinctly the graduation exercises at the old Pike Opera House," Alfred K. Nippert, an admiring SAE fraternity brother, wrote years later, "when, before a crowded house, a bewildered faculty, and a distinguished group of visitors and speakers, this modest, soft-spoken young graduate unfolded his Utopian dream." Instead of the first tentative steps of a budding engineer, an apprentice work designed to display one's grasp of the practical realities of engineering and a mature sense of one's own limits, with perhaps an eye to attracting the interest of a potential employer, Strauss had instead taken this occasion to give his entire personality free rein, and out the aspiring poet in him had leaped: to an audience of "astounded listeners" Strauss calmly read a senior thesis proposing the bridging of the Bering Strait.

The modern re-creation of the prehistoric land bridge between Asia and America was an idea that had been proposed before, usually by amateurs or cranks. For a young man aspiring to be an engineer, at the start of his career, anxious to be taken seriously, speaking to an audience including senior members of his chosen profession, it was a startlingly bold stroke, a time at bat given over to one splurging swing from the heels. The proposal was, most of all, a revelation of the height of Joseph Strauss's imagination and the breadth of his ambition: he would surpass John Roebling in one great work. This "impossible and preposterous scheme which this young engineering graduate presented so seriously" had the desired effect of impressing the audience with the force of Strauss's imagination and will. People who were there, like Nippert, recalled it forty-five years afterward. However, it produced no job offers.

8

Although Strauss often implied that he was a civil engineer, and even claimed to have taught a course in this specialty at his university, he was in fact neither a member of the American Society of Civil Engineers nor a graduate of a formal college of engineering. The University of Cincinnati did not establish its engineering college until 1901 and did not graduate its first class until 1905, thirteen years after Strauss had received his degree. According to the university's records, Strauss never taught at Cincinnati in any capacity whatsoever. His degree was a liberal arts B.A., although he was awarded an honorary doctor of science degree in 1930, after he had achieved affluence and a certain fame as a bridge builder. He did, according to one undergraduate reminiscence, "like to tinker in the machine shop a lot."

Strauss's restless, dreamy nature, his eagerness to get on in the world, led him down less ordered paths. His was not the sort of temperament content to remain at the drafting table pondering the application of engineering theory; Strauss would, under almost any circumstances, prefer to be out and doing, thinking up and trying out mechanical improvements on existing structures and devices (he did become a member of the American Society of *Mechanical* Engineers). He longed to get things done, to make things work, to persuade and convince and make manifest in the world the works that some engineers seemed to prefer savoring as intellectual puzzles. Strauss *cared* about his ideas, committed himself to them, clung to them, enthusiastically promoted them to anyone who would listen. Yet with this intensity of interest there was a certain lack of staying power, a tendency to want to go on to something else, another idea, another invention to pursue, that probably sprang from personal insecurity over his lack of formal engineering credentials. He had an inclination, which increased as he grew older, to think and work in great spurts of effort that would be countered by prolonged periods of passive withdrawal.

In 1892, the same year he graduated, Strauss began his career in bridge engineering as a draftsman for the New Jersey Steel and Iron Company at Trenton, New Jersey. For the next ten years, Strauss served his apprenticeship learning the practical realities of his profession: inspector, detailer, and estimator in the bridge shops, then designer for the sanitary district of Chicago and other engineering organizations in that city. Here Strauss acquired an extensive practical knowledge of railway bridges and viaducts and first encountered the new art of bascule bridge design.

The bascule, or counterbalanced drawbridge, was an idea gaining momentum with the expansion of the American street and railway systems. Instead of swinging sideways in the turnstile manner common to most movable bridges, the counterweighted bascule bridge would simply tilt up, creating less of an obstruction while also saving time. These bridges, when Strauss first became interested in them, were relatively rare, limited in length, and expensive, largely because of the cast iron counterweights and the complicated mechanism required in their operation.

Strauss, who in 1902 had established a practice in Chicago as a consulting engineer for the general design of bridges, had been assigned to the job of revising several bascule bridges just introduced in the Chicago area. He looked at the existing idea, which he knew to be sound, and with his dreamer's and tinkerer's sense of imagination he set about modifying it, making it more practical, more salable. To replace the expensive cast iron counterweight, Strauss substituted a bulkier but lighter weight of poured concrete. Then, to allow larger-bulk counterweights to function without obstructing the bridge itself, Strauss developed a pinion-connected, parallel-link or "trunnion" system, a radical departure from conventional practice. By combining these modifications with a simplified operating mechanism, Strauss had produced a bascule bridge of dramatically reduced cost and a corresponding expansion of the size and weight of spans. Strauss patented his design as the Strauss Trunnion Bascule Bridge, and it became the basis for his own engineering firm and his fortune.

Within seven years Strauss, now president and chief engineer of the Strauss Bascule Bridge Company of Chicago, had developed four distinct types of bridge designs, which, although rarely aesthetically edifying, eventually reached just about everywhere in the world the automobile reached. By 1915, the year of the Panama-Pacific Exposition, Strauss had built forty bridges in the jungles of Panama alone, as well as a bridge across the Cuyahoga River in Ohio that was to revolutionize movable bridge design. He had also built a bridge—across the Neva River in Petrograd—that the Bolsheviks would storm over in October 1917 to seize the Tsar's Winter Palace. Of the more than 400 bridges that Strauss and his firm were associated with in his lifetime, probably 375 of them were of this basic bascule type.

As the world's preeminent designer and builder of bascule bridges, Strauss was the obvious choice to design the Panama-Pacific Exposition's Aeroscope, which was, essentially, a tilted bascule bridge, resting on its counterweight. Here on the newly filled flatlands overlooking the bay, Strauss could stand in a bridgelike frame structure of his own design, high above the water, looking out upon a bridge site that, in the drama of its setting and the challenge it presented to the bridge builder, was at least the equal of his bold undergraduate proposal to fling a bridge across the Bering Strait.

"Perhaps," Strauss recalled, in the grandiose prose he favored, yet that touched upon his parents' and his own deep feeling for the arts, "when the earth was young, the Master Artist who stretched the canvas for that vast picture, intended that, in time to come, presumptuous man might trace upon it his greatest etching in steel."

9

It is not recorded exactly where or when Joseph Strauss and Michael O'Shaughnessy first met, but it is almost certain they became acquainted during the preparations for the fair: their first formal collaboration began less than five months after the fair's opening.

There were qualities in each man, apart from professional credentials, that drew him to the other. As a Jew, Strauss found himself up against the tacit and sometimes overt anti-Semitism of American business and political life. He was the outsider who could become, at any moment, the object of whatever underlying hostility or suspicion might lurk in any man or group of men.

In July 1916 Strauss, attending a hearing of county commissioners in Jacksonville, Florida, objected so vehemently to the efforts of the board to rescind an agreement under which Strauss was to act as supervising engineer for a $750,000 bridge over the St. John's River that he was physically attacked. L. L. Meggs, the board chairman, forced Strauss into a corner and "pummelled him" until the two men were parted by spectators and board members. Strauss threatened a civil suit for damages, but the following day he quietly retired from the proposition.

Here, as in most bridge engineering jobs, Strauss was the interloper, the man who would be characterized as "pushy" or overly aggressive when, in his own view, he was simply meeting and contending with reality as he found it. He would have to be twice as good and work twice as hard in order even to be considered with someone more conventional, more familiar, more acceptable.

O'Shaughnessy, too, was an outsider, an immigrant Irishman whose speech was still seasoned with a brogue, yet an outsider who had made his way directly to the control room of local political power. As city engineer of a reconstructed and still expanding city, he was the operator, the man who did the doing when something had to be done and who, always thinking expansively, was feeling confined by the physical limits of the city whose future was now linked with his own. Strauss, on his part, had not built a successful Chicago-based construction company, dependent on contracts with public bodies, while remaining politically naive. A man whose connections with the city engineer of Chicago, Alexander Murdock, would result in Murdock's firing in 1923, on charges of using political influence in behalf of the Strauss Bascule Bridge Company, Strauss was acutely aware of how getting to and ingratiating oneself with the right people can dispel one's aura of strangeness in a city. And he seemed to possess qualities that intrigued the San Francisco city engineer. O'Shaughnessy had built or overseen the building of railways, tunnels, and sewer and firefighting systems and was building an immense water supply complex. Yet the one thing he was not, and could not convince his constituents he might become, was a bridge builder. Not yet, at any rate, and not alone.

On June 25, 1915, the Thomson Bridge Company of San Francisco was awarded the contract for removing an old swing drawbridge at Fourth and Channel streets in San Francisco and replacing it with a modern Strauss Trunnion Bascule Bridge, crossing Islais Creek. The foundation for the bridge, which was completed in 1916, was designed by O'Shaughnessy's Department of Public Works, while the superstructure was the work of the patentee, the J. B. Strauss Company of Chicago. It was a collaboration whose result, the squat, bulky, and ignored Fourth Street Bridge of today, was to link, with profound effects, the imaginations of two of the most ambitious and willful of men.

THE SECOND BRIDGE

I

It had become a kind of game with Michael O'Shaughnessy. He would meet a bridge builder—and in the course of his job as city engineer of San Francisco he was to meet just about every construction man of any real significance in America—and, as a sort of intellectual puzzle, he would pose the problem of bridging the Golden Gate.

That some sort of bridge system would eventually be built at San Francisco seemed inevitable: the city was the largest American metropolis still served primarily by ferry boats. Choked off at the tip of a peninsula, San Francisco faced a future of increasing congestion and economic strangulation. In 1920, local civic pride suffered its worst shock since the earthquake when the federal census showed that Los Angeles had replaced San Francisco as the largest city in California. It was a gap that the southern metropolis, with its unlimited access to adjacent space for homes and businesses, and its hunger for incorporation, was to widen throughout the next decade. Los Angeles had become the fastest-growing city in America, while growth rates in San Francisco for both population and industry had fallen below the national average. San Francisco's expansion had fallen behind that of the city's own suburbs, a fact that seemed to suggest even more strongly the need for a permanent link between the city and its communities around the bay. The city's future clearly depended on San Francisco's collective determination and resources and its willingness to commit them to the building of bridges.

The distance at the Golden Gate—a little over a mile—was the shortest of any point on the bay. A bridge here would link San Francisco with the resource-rich and underpopulated counties of all of Northern California. There was a natural symmetry to the site, two facing points of land, separated by a narrow strait; just looking at it, the Golden Gate seemed to present one of the most obvious places in the world to build a bridge. It also represented a variety of obstacles, the like of which had never existed in a single construction site before.

Geologically, the Golden Gate is a gap in a mountain range through which pours the outflow of seven different rivers. This gorge, carved out over the millennia by the action of the rivers, also fronts directly on the Pacific Ocean, producing a complex of tides and currents not duplicated anywhere in the world. At the center of the channel, the Gate is 335 feet deep. Any structure built here would have to be anchored in such a way that its base could absorb and withstand the force of both tidal currents and ocean waves while at the same time its superstructure would have to be firm enough and flexible enough to resist the battering of gale winds. Added to these hazards was the fact that the site was within twelve miles of a major earthquake fault that, within the last two decades, had caused an epic catastrophe.

These natural obstacles were compounded by other difficulties, mechanical, social, and political. The Golden Gate was the lone entrance to one of the world's great harbors, and no bridge had ever been built at a harbor entrance before. Not only would the bridge have to be anchored deep enough to survive tides and ocean waves, it would have to be tall enough for the largest ships to pass beneath its roadbed at high tide. There were naval bases inside the bay and army installations on either side of the Gate; any structure built here would require the approval of the departments of the Army, Navy, and War. How were the functions of a great seaport to be carried on while a construction project, which would undoubtedly take years, cluttered its mouth? In the event of war, couldn't a single well-placed bomb send a bridge roadway here crashing down to block the entire port? There were also aesthetic considerations. The Golden Gate represented one of the earth's most dramatic meetingplaces of land and water. Was a man-made structure, erected here, altering the landscape forever, a wise choice? Was expansion by this means really in San Francisco's best interests, or would it help destroy the fragile uniqueness of the city and make expansion alone its strongest characteristic, as in Los Angeles?

The men to whom O'Shaughnessy posed his problem were practical men, builders familiar with the realities of subcontracts and bids. Most of them didn't need to consider beyond the initial questions of time, design, and the cost of materials and money. Bridge building, after all, followed a certain logic, one theorem leading to another, with the solution to a particular problem usually being an extension or an adaptation of some other solution. The larger and more complicated the problem was, the larger and more complicated, usually, was the solution. A bridge could conceivably be built anywhere, providing someone was determined enough to build it and, more crucially, to pay for it. The typical rough estimate that O'Shaughnessy received for the cost of constructing a bridge over the Golden Gate was $250 million, a sum equal at the time to two-thirds of the assessed value of all the property in San Francisco.

2

At the time of his initial collaboration with O'Shaughnessy on the Fourth Street Bridge, Joseph Strauss was forty-five years old, affluent, and unfulfilled. His patented bascule bridge design, easily adapted to streets, highways, and rail systems, had made him independent and kept his firm busy on jobs all over the United States and in many places abroad; these works, however effective, were merely functional devices, pieces of other systems that would never be remembered in and of themselves. The great lasting work, the indisputable achievement that Strauss had always longed for, had, as yet, eluded him.

It was not for want of trying. Strauss was constantly generating ideas. In addition to developing the trunnion bascule bridge, Strauss also designed a rack-and-pinion mechanism for lift bridges that replaced the less-safe cable structure. He designed a number of portable searchlights that were used by both the United States and Russia during World War I, and constructed a disappearing tower at Fort Hancock, New York, with an elevating arm, raised and lowered like a bascule bridge, carrying a sixty-inch searchlight on a swinging platform. He patented the first yielding traffic barrier, a cable device designed to stop automobiles at railroad crossings, an idea that, although it failed to conquer America's highways, did inspire the arresting cables eventually used on the decks of aircraft carriers. And he wrote: monographs, verses, and odes, some of which were published and even anthologized.

Strauss's most ambitious and industrious endeavors, however, always carried a desire for acceptance, which was somehow always denied him. Although there were hundreds of Strauss bridges and numerous Strauss patents, there was no Strauss paper read before a meeting of the American Society of Civil Engineers, no publication of Strauss engineering theory bearing any sort of official imprint, no body of work by Strauss that was the subject of study by aspiring student engineers. He was the promoter, the salesman, at best, the tinkerer and gadgeteer, the entrepreneur, but never the poet-engineer, the John Roebling he so clearly longed to be.

The Aeroscope at the 1915 fair was a good example of the niche into which Strauss had fallen. Engineers visiting the fair, perhaps to attend one of the conferences there, might spend a few idle minutes at the midway, waiting, as their wives or children rode Strauss's gadget, and they might themselves enjoy the fetching ingenuity of it, but the thing was, after all, an amusement ride, not the work of a man they would seek out, a powerful and original thinker at whose feet any engineer would wish to sit. It was, on balance, just another functional device, from a man who longed for grandeur in his life. And thus far had failed to find it.

How accurate a measure Michael O'Shaughnessy had taken of Joseph Strauss we can only guess. As one of the nation's leading city engineers, O'Shaughnessy surely knew that Strauss had never supervised the building of any bridge on a scale comparable to that required at the Golden Gate; he was essentially a builder of highway bridges, the lip-to-lip movable structures dictated by America's vast river system. He was not a profound or original thinker, and his success as a builder was based on easily reproduced adaptations of a single idea. Still he was a bridge builder, and he had an obvious yearning, typical in men of a certain age, for an achievement that would catch the eye of posterity. That Strauss was not a licensed civil engineer was probably not of any great concern to O'Shaughnessy. If Strauss couldn't actually do the job, he might, with his promoter's enthusiasm and entrepreneurial energy, serve as a stalking-horse for the man who actually would do it; and that man might well be the engineer who, by that time, would have completed the most ambitious water project ever constructed for an American city, the Hezekiah of Hetch Hetchy.

When O'Shaughnessy suggested to Strauss that he address the problem of bridging the Gate, he included the reservations that other builders had expressed about the project, as well as the fact that the standard estimate for the job was a quarter of a billion dollars. To some men this authoritative negative opinion would have been a deterrent; but to Strauss, an outsider, short, middle-aged, Jewish, a man who had hungered all his life for the chance for some great achievement, which he realized might elude him forever if he didn't leap at it now, the unfavorable opinions of other builders was a goad. Strauss replied that he thought he could design and build the bridge and that he could bring the job in for under $25 million, providing O'Shaughnessy authorized an underwater contour survey. O'Shaughnessy agreed, and the partnership that had begun with the Fourth Street Bridge entered its next, and more grandiose, stage.

3

The collaboration was to be an extension of that on the Fourth Street Bridge, with O'Shaughnessy's department furnishing location surveys, soundings, grade line profiles, and maps and the Strauss Bascule Bridge Company providing the design. Beginning in 1919, and while they and their staffs continued other design and construction work, Strauss and O'Shaughnessy undertook the first serious examination of the problems of building a bridge over the Golden Gate.

In February of 1920, soundings were made by the U.S. Coast and Geodetic Survey, using the steamer *Natoma,* between Fort Point on the San Francisco side of the Gate and Lime Point on the Marin side. The findings, according to the San Francisco *Chronicle,* were not encouraging. "Federal experts believe it will be impossible to put piers at this point owing to strong currents and great depth." This information, along with substantial additional details, O'Shaughnessy dutifully transmitted to Strauss. He also sent it to two other bridge builders: Frank McMath, of the Canadian Bridge and Iron Company of Detroit, and Gustav Lindenthal, design engineer of the Hells Gate Bridge in New York.

Strauss was not aware of this unacknowledged competition. Even if he had been, it probably wouldn't have deterred him; he would work himself and his people harder, perform better, come in lower, and win the job. Strauss's people would spend a good part of the next year at the task.

Like most successful men, Strauss had a tendency to repeat what had worked for him in the past. He was a designer and builder of a certain kind of bridge, and he approached a new job with a mind that was the product of these earlier tasks, successes that encouraged him to recast a new situation, if need be, to fit a solution that was comfortable to him. What Strauss and his firm built were movable lift bridges, and what Strauss saw the Golden Gate as was the site for a kind of super-drawbridge, only with the lifting part replaced by something else.

The distance from the San Francisco to the Marin shore at the Golden Gate is some 6,700 feet. According to the proposal prepared jointly in 1921 by Strauss and O'Shaughnessy, "at 1,345 feet from the shore on each side there are rock ledges with a depth of only fifty feet of water." Although this neatly symmetrical conclusion was not borne out by another survey made of the same site less than ten years later, it was, nevertheless, literally the foundation upon which Strauss and O'Shaughnessy rested their bridge design.

Long-span bridges had traditionally been of two types, cantilever and suspension. The cantilever was a fixed-frame bridge in which two trusses, frameworks of metal bars arranged in several triangles, supported a flat span extending out between them. The design offered the advantages of stiffness and rigidity at a cost of great weight, which limited the cantilever style of bridge to spans, between piers, of less than 2,000 feet. The suspension bridge, consisting of spun steel cables suspended from towers, supporting a roadbed, allowed great reductions in weight and so could extend over a greater length, but at a cost of much less rigidity. Strauss proposed to achieve, at San Francisco, the best of both worlds by combining the two different types of design in a single bridge.

The bridge would rest on two concrete, rock-faced piers, 200 feet from base to top, set on the matching rock ledges found by Strauss and O'Shaughnessy to be in 50 feet of water. From each pier would rise a crisscross steel beam tower, 747 feet high, for a total height of some 950 feet, or 34 feet less than the Eiffel Tower. Reaching out toward the towers from each shore would be a cagelike cantilever frame span 1,320 feet long. From each tower, an additional cantilever framework would extend out over the water. From either tip of this framework, cables would be suspended 2,640 feet over the main channel, supporting a center span that stretched 4,000 feet from tower to tower, the longest suspended span ever built. The cross-section width of the bridge would be 80 feet, allowing room on the bridge for two lines of trolley cars, four lanes of automobile traffic, and two 7-foot sidewalks. There would be but one level to the roadway.

By combining the best features of both types of bridge, Strauss contended, he had invented a new kind of structure, superior to either the pure cantilever or pure suspension bridge in its combination of strength, lightness, and flexibility. He was so proud of it that he applied for a patent for the design, in his name.

What Strauss had designed, in effect, was a bascule bridge, only more massive than any ever built, with the movable arm or lifting drawbridge replaced by a suspension span. It was, like most of his highway bridges, bluntly functional, with little or no aesthetic concessions to its site.

With its immense, spidery framework and thick, crosshatched girder towers, Strauss's original design suggests an enormous railroad bridge, dark and ponderous, a soulless mechanism to be traversed by other mechanisms, something out of the early industrial era of soot and smoke and pounding repetitive noise rather than a structure intrinsically pleasing, a link to the natural world of air and water and light, where people might like to promenade and talk and think.

Strauss submitted his design to O'Shaughnessy in June 1921. What the city engineer's first reaction was, we do not know. He did, however, take a great deal of time to ponder Strauss's work: the design was not made public for another year and a half.

VIEW OF THE PROPOSED GOLDEN GATE BRIDGE

Although not an aesthete, O'Shaughnessy probably had reservations about Strauss's design from the start. As an engineer, he must have questioned the wisdom of combining two different, and in some ways opposed, types of bridge design, and of attempting this combination for the first time in a place as challenging as the Golden Gate. According to Strauss's plan, the cables of the suspension span, instead of being firmly anchored at either shore, were to be fixed to the extensions of the cantilever frame, an obvious compromise that would make the suspension center of the bridge extremely vulnerable to the high winds that periodically tore through the Gate. O'Shaughnessy almost certainly realized that the locations indicated for the piers had been chosen perfunctorily, and the action of riptides and ocean waves at the Gate, which were strong enough at times to swing a ship around, would require more complicated solutions than were indicated in either Strauss's design proposal or his estimate. Undoubtedly there would be opposition to this structure from the kind of people who had attacked the damming and flooding of Hetch Hetchy. Still, O'Shaughnessy was convinced that San Francisco's future depended on expansion to the north, and of all the builders to whom he had proposed the idea of a bridge at the Golden Gate, Strauss was the one who was willing to give it a serious try. Strauss's design was not necessarily final, after all; no contracts had been signed, and everything was subject to modification or replacement. Whatever its actual merit, Strauss's design could be useful as a test, to help stimulate public interest in building a bridge and to determine what financial and political support really existed for such a project.

Strauss's original design for the Golden Gate Bridge was a ponderous, ugly structure of mixed parentage, based on erroneous survey information and precious little actual engineering. Strauss proudly patented his design and campaigned eight years for it.

Most importantly, whatever O'Shaughnessy's personal reservations, Strauss had planned a bridge for a price that O'Shaughnessy could live with: Strauss's estimate for the total construction costs of his bridge had come in at $17,250,000. The bid represented, at least, a bird in hand. O'Shaughnessy's other two bridge designers, McMath and Lindenthal, had not yet even submitted estimates. McMath never did, and it would be two years before Lindenthal would furnish his estimates: for a bridge at the Gate, a minimum of $60 million, with a probable eventual cost of $77 million. Perhaps, at Strauss's $17 million figure, an unaesthetic bridge would not be so hard to accustom oneself to after all. In time, O'Shaughnessy knew from experience, the wounds caused by civic improvements healed: people grew used to the necessary intrusions on their lives. And an ugly bridge would serve San Francisco's interests better than no bridge at all.

O'Shaughnessy had failed to take into account a crucial element in all this: the degree of promotional energy that the prospect of bridging the Golden Gate had released in Strauss. Here was the great thing, the justifying work he had been stretching toward all his life. Strauss was, if anything, stronger at marketing and promoting ideas than he was at conceiving them. He seemed to know instinctively whom to reach, whom to get to and persuade, who were the decision makers, the people who mattered in any given situation. And he was not a man to sit by, passively waiting for a decision to be made.

In 1922, while Strauss's design still lay somewhere in O'Shaughnessy's office, Strauss turned up unexpectedly at a City Council meeting in Sausalito, in Marin County, the nearest town across the Gate from San Francisco. The local mayor, James Madden, had made a place for Strauss on the agenda, but the appointment had been all but forgotten. Only the previous day, the city clerk had reminded the mayor that a man who had come all the way out from Chicago was scheduled to appear to talk about a bridge at the Golden Gate.

Before a small audience, in a town of less than 2,000 people, Strauss, whom Mayor Madden remembered as "the world's worst speaker," first publicly outlined his proposal for a bridge at the Golden Gate. Strauss had chosen his spot carefully, and the people he addressed that day were to prove influential. As an experienced bridge builder, Strauss knew that the richest advantages of development lay not in the crowded cities pushing for expansion but in the adjacent open spaces awaiting the touch of residential and commercial development: his bridge was the wand that would work the magic in Marin County: from San Francisco to the north, the bridge would boost property values, encourage developers to lay out homesites, stimulate construction, settlement, and commerce of all sorts, and encourage tourism all the way north to the Oregon border. The day the bridge opened, every person who owned property in Marin County would automatically be wealthier. This was the kind of speech that didn't require eloquence to be convincing. Earnestness and the appropriate credentials were sufficient, and Strauss had both. By the time he left Sausalito, his listeners were sold on Strauss and on his bridge.

This performance, with local variations, Strauss was to repeat throughout the "cow counties" of Northern California during the remainder of 1922, and at frequent and often unexpected intervals afterward. To assist him in these efforts, and to serve as a sort of cultural interpreter between the Jewish Chicagoan Strauss and the rural Northern California locals, Strauss retained, as his "public relations counsel," Charles Duncan, an executive of Foster and Kleiser, the West's largest outdoor advertising firm. Duncan, who was the eloquent speaker that Strauss, as yet, was not, and who, from riding locations as an outdoor salesman, knew every village and byway north of San Francisco, was to be another valuable wedge in Strauss's efforts to make his way into what remained in many respects a closed community.

Strauss's stumping of the northern counties was to stimulate instant demand for his bridge when the design was publicly announced. Upon publication of Strauss's proposal in December 1922, the mayor of San Rafael, Marin's county seat, proclaimed that the bridge would "open the gateway to thousands of homeseekers now held back by lack of proper transportation facilities."

Farther north, in Sonoma County, Frank Doyle, a local banker and president of the chamber of commerce, called a mass meeting in the county seat of Santa Rosa in support of the bridge. At the meeting, held on January 13, 1923, and attended by representatives from twenty-one counties, it was agreed to form a Bridging the Golden Gate Association for the purpose of encouraging construction of a bridge at the earliest date possible. An Executive Committee, headed by W. J. Hotchkiss, was appointed to study the means of getting the project under way.

The committee appointed George Harlan, a Marin County assistant district attorney who had attended Strauss's maiden Sausalito speech and who was a specialist in local enabling acts, to draft legislation authorizing the establishment of a Golden Gate Bridge and Highway District, the first legal body of its kind ever proposed: never before had a group of property owners joined in a local tax district for the purpose of building a bridge. The proposal was introduced in the State Assembly by Frank L. Coombs, assemblyman from Napa, another of the North Bay counties.

While this bill was being considered in the Assembly, members of the association canvassed the counties north of San Francisco, talking with boards of supervisors and city and county officials, urging them to pressure their representatives to pass the enabling legislation. The most effective campaigner, whatever his limits as a speaker, was Joseph Strauss. The appearances of the "famous designer" were events, reported in the local press as major news items, the occasion of speeches where he was able to seize upon and address his audiences' most urgent concerns.

The Golden Gate Bridge was to be one of the wonders of the world, Strauss told a meeting of the Lions Club in San Rafael in March 1923. "It would be two and a half times larger than any similar bridge in the world. The towers would be ten feet higher than the Eiffel Tower. There is no better place for the eighth wonder of the world than Northern California."

Asked whether the War Department would not object to construction of a bridge at the entrance to a major harbor, Strauss declared: "I have conferred with the Secretary of War and several members of Congress, [and] I am in a position to inform you that there will be no objection from Congress or the War Department." Strauss's authority, like a gas, could expand to fill any space.

Two days later, to a gathering of automobile dealers in San Francisco, Strauss pointed to the bottleneck that, due to the present ferry system, existed at Sausalito, where, "at the height of the touring season, thousands of automobiles are lined up for miles every Sunday," while on the San Francisco side an equally frustrating line developed every Saturday. All this would be history with the opening of the bridge, Strauss promised, when "San Francisco motor car owners could step into their cars, travel to the Presidio or some point near there, and cross the Golden Gate above the water, and then drive clear to the Oregon border without using a ferry boat." The bridge would build a great community on the north side of the straits, would make the vacation land to the north easily and quickly accessible for "hundred of thousands of motorists," who would pay bridge tolls less than the current ferry boat fares. Nothing could have been more pleasing to the car dealers' ears than this glowing, and eventually accurate, forecast.

The Golden Gate could be bridged by 1927, Strauss assured a San Francisco *Chronicle* reporter, "if the people of San Francisco and other communities of the Bay region are willing to spend $20,000,000 and if they are successful in obtaining the sanction of the War Department." The bridge would be paid for, Strauss explained, out of income from tolls, which, based on existing ferry traffic, would be in excess of $1,290,000 a year.

In his speeches, in the timing of his arrivals and departures, in his skillful use of the press, Strauss possessed what can only be described as star quality. His sure sense of drama had him always in motion, always arriving on a train from Chicago or en route to serve as a consultant to the Port of Los Angeles, imbuing whatever the present occasion might be with an added weight of national importance. Added to this was an actor's sense of audience, of calculating or sensing what his listeners yearned most to hear, and of addressing it with his aspiring poet's sense of trenchant phrase. Most impressive of all was Strauss's ability to respond to, or parry, questions. Never an eloquent speaker, least so when he tried most, Strauss was a telling counterpuncher, probably most effective when questioned or opposed.

His campaign proved tremendously effective politically, as he addressed local interests as a worldwide authority on bridge building and received newspaper coverage of his appearances, usually including the endorsement of one or more local officials. During this time, officials of twenty-one Northern California counties went on record as favoring passage of the bill.

In March 1923, Strauss testified before the Assembly's Committee on Roads and Highways on formation of the Bridge District. "The engineering plans are fully solved," he announced to the legislators. "It is up to California to go ahead."

Presented with this authoritative endorsement, backed by what seemed to be an endless prairie of grass-roots support, the legislators yielded to the challenge of inevitable progress. The Coombs Bill, enabling the formation of a tax district for the purpose of building a bridge at the Golden Gate, was signed into law on May 25, 1923.

4

The whirlwind acceptance of the bridge proposal could only have aroused profoundly conflicting feelings in Michael O'Shaughnessy. While the enthusiastic response to an idea that he had been nurturing for more than four years was gratifying, as was the mention of his name in the press in association with the bridge, Strauss's leap to the forefront of the project must have caused him a certain alarm. The idea of a bridge, which O'Shaughnessy had initiated, had been seized by the terrier Strauss, who had run away with it. Campaigning mostly in the northern counties, beyond O'Shaughnessy's territory of San Francisco, Strauss, the outsider, the "famous designer," as the papers described him, had made the bridge *his* bridge. O'Shaughnessy was included in some of the meetings and conferences, and his name mentioned in descriptions of the project, but only as a sort of junior partner, someone who had been responsible for some useful preliminary survey work but who would now presumably step aside while the Master Bridge Builder saw the job to its completion. The attention given Strauss personally, and the sweeping nature of his assurances about Congress and the War Department, financing and construction, only reinforced this impression. This was not at all what O'Shaughnessy, who normally insisted on controlling all his jobs, had in mind. The city engineer had underestimated Strauss's promotional ability and his political skills, with the result that O'Shaughnessy's professional pride, already of a sensitivity that, in a few years, would lead him to threaten resignation rather than accept criticism of his performance as city engineer, had been wounded. Strauss and O'Shaughnessy remained cordial and continued, for the present, to cooperate on the bridge project, but it was a wound that would continue to fester over the next few years.

There was surprisingly little public objection to Strauss's design, either on engineering or aesthetic grounds. In *Harper's* of June 1924, Katherine Fullerton Gerould objected to the idea of a bridge, rather than the proposed design, urging, "Dear San Francisco . . . in the interest of your own uniqueness, do not bridge the Golden Gate." The project proved that "San Franciscans, too, are prone to some of the worst American faults." Expansionist gestures of this kind, she suggested, would be better left to Los Angeles, "which, if it had a Golden Gate, would most certainly bridge it, and sink oil wells into bay and ocean on either side of the bridge."

For most people, however, in the decade of America's first great automobile boom, the idea of a bridge at the Golden Gate appeared to be simple human destiny, with the convenience of being able to cross the Gate by car offsetting any reservations people might have about the looks of the bridge itself. The public officials and booster groups involved appear to have been gratified to have an actual engineer's design of any kind available for promotion.

At the same time there seem to have existed, even at this early date, some private reservations, among responsible parties, about the appropriateness and rapid acceptance of Strauss's design. The most prominent local authority on bridges was Professor Charles Derleth, Jr., chairman of the University of California's Department of Engineering at Berkeley. Derleth, who was chief engineer of the cantilever Carquinez Straits Bridge then under construction over the eastern San Pablo Bay arm of San Francisco Bay and who had worked with O'Shaughnessy on certain aspects of the Hetch Hetchy project, had followed news of the Golden Gate Bridge since its earliest announcements and was known to have expressed the opinion that he considered an all-suspension bridge at the Golden Gate "more feasible" than Strauss's combination design. Derleth was a man well connected in academic, social, and political circles, and there were already feelings within the Executive Committee and the association that Derleth should be brought into the Golden Gate project in some capacity, feelings that Derleth did nothing to discourage.

A man with something of the look and a good deal of the wiliness of a Chinese Mandarin, Derleth was born in New York City in 1877, graduated from Columbia, and taught for two years at the University of Colorado before arriving at Berkeley in 1903. He was made a full professor and dean of the College of Civil Engineering in 1908 and was appointed chairman of the department in 1924. Derleth was a joiner, a member of more than a dozen fraternal, professional, and social organizations, and active in the Bohemian Club, whose Isle of Aves camp at the club's annual summer encampment at the Bohemian Grove was to prove extremely useful over the years to the University of California's administrators and department heads for recruitment, fund raising, and the advancement of the university's prestige. Derleth was known by the national civil engineering figures who had never wholeheartedly accepted Strauss. He also seemed well acquainted with just about everyone of consequence in Bay Area construction circles, and for any one or more of these reasons his services were sought as a consultant on nearly every major construction project in Northern California. There would be increasing pressure in the next few years to include Derleth in any bridge construction proposed for the Golden Gate.

There was also now a certain skepticism toward Strauss's design among the members of the San Francisco Board of Supervisors. Richard J. Welch, who in November 1918, in a burst of post-Armistice enthusiasm, had introduced the initial resolution in favor of building a bridge at the Gate, remained an enthusiastic booster, but Michael O'Shaughnessy had a number of increasingly critical rivals on the board, men acutely sensitive to the city engineer's penchant for mighty projects and his ability to use these works to assure and expand his own authority; as the decade wore on, there would

Allan Rush, a Los Angeles engineer, suggested a graceful suspension span for the Golden Gate in 1924. Though the navigational clearances and wind stresses were not entirely thought out, it suggested esthetic possibilities beyond Strauss's eyesore.

be growing sentiment among board members to bring O'Shaughnessy to heel. The situation put a certain political edge on every issue that involved O'Shaughnessy, and assured that any proposal with which the city engineer was associated would be subject to intense scrutiny by at least some members of the board. In addition, there was the complication of an alternate design.

On May 14, 1924, a more adventurous bridge proposal was submitted to the San Francisco supervisors. Conceived by Allan C. Rush, a Los Angeles engineer who had been fascinated for years by the idea of bridging the Golden Gate, the design was for an all-suspension span extending from great piers at either shore. A slender, graceful structure with the look of a giant expansion watchband, stretched taut, the Rush bridge may not have been thoroughly worked out in terms of construction details, ship clearances, and the effect of winds, but its bold, clean, futuristic look makes the original Strauss bridge seem a relic from an earlier century. Anyone seeing both designs could not have remained convinced that Strauss's proposal had exhausted the aesthetic possibilities at the Gate.

Even Joseph Strauss himself, for all his promoter's confident air, appears to have had certain private reservations aboat whether he could really deliver on the design of what would be the largest bridge of its kind in the world. In 1922, Strauss had added to his staff Charles Alton Ellis, professor of structural and bridge engineering at the University of Illinois, and before that design engineer for the Dominion Bridge Company of Chicago.

It would be difficult to find two men in the same professional specialty more unlike each other physically, emotionally, and intellectually than Joseph Strauss and Charles Ellis. Born in Parkman, Maine, in 1876, Ellis appears to have been an almost classic "downeaster," tall, slender, reserved, scholarly, with a fine dry wit and a lifelong interest in intellectual pursuits. Although not a graduate of an engineering school, Ellis took four years of mathematics and courses in higher mechanics at Wesleyan University, where he received his A.B. in mathematics and Greek in 1900. Discovering that there was not much of a living in doing Greek translations, Ellis went to work for the American Bridge Company, where his abilities at calculus won him the responsibility of figuring out the stresses of the subway tubes under the Hudson River. From 1902 to 1908, Ellis branched out into structural engineering, a field not yet specialized, and worked himself into a new profession. He took additional courses, studied theory, and eventually wrote a textbook, *Essentials in Theory of Framed Structures,* which became a standard work on the subject.

In 1908, Ellis joined the faculty at the University of Michigan as an instructor in civil engineering, later moving to the University of Illinois, where he became part of a famous engineering school headed by Dean Hardy Cross. He was an active member of the American Society of Civil Engineers and contributed articles and papers to its journal and its meetings. He had been teaching for most of the past fourteen years and was considered an authority in his field when Strauss hired him.

A man whose hobbies were doing Greek translations and finding structural engineering problems he couldn't solve, and who maintained all his life that "happiness cannot be found by merely seeking it. It is far more satisfactory than mere pleasure," the austere Ellis seems to have had little in common personally with the flamboyant, mercurial Strauss. As a professor at the University of Illinois, Ellis was part of one of the nation's most respected engineering faculties. He was, according to his former students, an inspiring and profoundly influential teacher. A man with little desire for the limelight, Ellis would seem to have had the ideal temperament for academic life, with its deferred satisfactions, peer appreciation, and sense of private rather than public achievement.

Yet even in a man as self-disciplined as Charles Ellis there was the urge to make one's way in the world, to be a part of some monumental task, a desire, arising perhaps from his very reticence, to lose oneself in some great enterprise that would mark forever one's passage on the earth. This fundamental desire, approached from an entirely opposite direction, is what attracted Ellis to his boss. Strauss would be the front-runner, the face-man, the chieftain who negotiated and bartered with the various local tribes who controlled so much of the savage politics of American construction. He would handle the wheedling and cajoling and hassling, Ellis would see to the staff work that the restless Strauss so abhorred; Ellis would be the man at the desk, patiently working out the equations, pondering the application of theory or evaluating the work of the draftsmen and sending them back to revise their drawings again, accumulating satisfaction, detail by detail.

For Strauss, adding Ellis to his staff was like acquiring the professsional and academic credentials that, until now, he had lacked. Here, as he was to demonstrate again later, Strauss had an unusual ability to locate and draw to him men of greater abilities than his own, men who would accept his leadership. In his own eyes, Strauss was an impartial employer: he didn't care who came up with an idea, or how good it was, just as long as Joseph Strauss got credit for it. In this one-sided division of recognition, Strauss did not represent an exception to American management practice generally so much as he typified it.

Strauss made the most of Ellis's professional background and accomplishments. In his firm's engineering proposals, in his prospectuses and presentations, he dwelled on Ellis's academic background, and cited with a certain savoring detail his book and his articles. Strauss relished the use of Ellis's academic title, referring to him not only in formal proposals but in letters and even in telegrams as "Doctor" and "Professor" Ellis. Strauss might not possess these credentials himself, but during working hours at least he possessed a man who did, and that, for most purposes, was just as good. Nor was Ellis some drab clerical figure confined to a back room full of ledgers or drafting tables. Although reserved, occasionally to the point of remoteness, Ellis was experienced and knowledgeable at both the theoretical and practical aspects of structural engineering. As the relationship wore on, he became, more and more, Strauss's emissary to the professional and academic world, the licensed professional talking to other licensed professionals, the architects and engineers with whom the Strauss Engineering Corporation, as it was now called, regularly collaborated. In addition, Ellis was a surprisingly effective on-site engineer, going out in his wire-rimmed glasses and dark three-piece suit to discuss with contractors and foremen and supervising engineers the job and the materials and specifications. Thorough, exacting, tough when he had to be, Ellis could make himself understood and respected, just as he had in the classroom. "He was what he was," recalls a former student, whom Ellis influenced to become a professor of engineering himself. "Competent, but never flouted it. Hard to convince, but could be won over."

With Ellis on his staff, Strauss now had freedom to dream up and promote new projects, the opportunity to be out and doing and above all traveling, making the dramatic appearances, arrivals, and departures that appealed so to his romantic nature. With Ellis working out the design details, Strauss could concentrate on the big picture: advanced ideas, great bridges, monorails, inventions, poems, songs.

It was during these years, with Ellis serving as the firm's design engineer, that the Strauss Engineering Corporation advanced into the construction of long-span fixed bridges—such as the 3,511-foot Quincy Memorial Bridge over the Mississippi River, completed in 1927, and the 5,478-foot Longview Bridge over the Columbia River, completed in 1929—in addition to the classic Strauss lift spans, which continued to be the firm's bread and butter.

A subtle change was taking place in both men, and in the relationship between them. While the reserved academic Charles Ellis was becoming more and more at home in the practical world of meetings and site visits and estimates, Strauss, while not becoming any more familiar with the complexities of engineering theory, was indulging more and more in what could be described as flights of fancy. He had placed a tremendous burden of personal hope and ambition on the successful construction of the Golden Gate Bridge. It was a work that would establish him not only as a great engineer but as a great man. With this work, he would surpass Roebling. Yet he hadn't even officially got the job yet, and there were some doubts surfacing about his proudly patented design. Strauss knew it would take determination to win and complete the job, and determination was one quality he possessed in abundance. Yet at times determination can edge over into desperation, promoting people to do things they might not otherwise. And as the delays of the approval process wore on, an element of desperation entered into the character of Joseph Strauss.

As for Charles Ellis, somewhere in his relationship with Joseph Strauss he made a miscalculation in the equation between personal accomplishment and public recognition. The bargain struck with a man like Strauss, the acceptance of shelter from the world in order to do one's work in peace, carries a heavy price. The differences in rewards—money, fame, personal prestige—are not arithmetical, but geometric. The world has not the time or the interest to sort out the individual sources of achievement; people will, in most instances, accept the proponent of an idea as its progenitor. Indeed, this is so common that it is widely accepted as not only just but natural. If someone who works under you presents an idea to you that you approve, advises a contemporary book on American business management technique, the idea is yours. Both Strauss and Ellis underestimated the prevalence of this frame of mind, and the intensity of feeling it can provoke in people—to what were eventually drastic effects.

3

"WE WILL GET THIS BRIDGE, IN THE END."

I

On May 16, 1924, a hearing was convened in San Francisco at the direction of the War Department on the subject of the construction of a bridge at the Golden Gate. The hearing was to be conducted by Colonel Herbert Deakyne, head of the local board of army engineers, who would be presented with evidence in support of the project by representatives of civic and commercial organizations in Northern California. Despite the fact that this was to be a federal hearing and was being held in the San Francisco City Hall, it was clear in advance that the star of the proceedings was to be a man who, as yet, held no official position with any of the participating communities, the Bridging the Golden Gate Association, or the Executive Committee. It was Joseph Strauss whose picture appeared beside the article announcing the hearing in one San Francisco newspaper, under the heading "Engineer of Gate Bridge," Strauss whom another paper described as the "bridge builder of Chicago," who had arrived in the city for the hearing and who would "explain in detail the feasibility of erecting what will be the largest bridge of its kind in the world." It was mentioned, almost in passing, that "City Engineer M. M. O'Shaughnessy will assist Strauss."

This take-charge tone accurately reflected the stance taken by Strauss at City Hall. Although the purpose of the hearing, according to Colonel Deakyne, was to determine, first, whether the idea of a bridge was in compliance with the requirements of navigation and mechanical safety and, second, whether there was a preponderance of evidence that public sentiment was in favor of a bridge, Strauss used these issues as a mere runway toward a soaring flight of promotional bombast.

"San Francisco has often done the impossible," Strauss informed Colonel Deakyne and his board in an imposing opening statement, and now his bridge could "make her the great city she is destined to be." The bridge would be, in Strauss's opinion, "the greatest feat of construction ever developed." Wrapped in his "famed designer" mantle of international authority, Strauss assured the board: "San Francisco is one of the few cities that has all the energy, all the wealth, all the courage and all the ability that is needed to undertake and carry this project to success."

Questioned by Deakyne on the possibility of ships colliding with the bridge piers, Strauss replied that improvements in radio guidance made such accidents extremely unlikely. What about bombing, Deakyne wanted to know. If the bridge was hit dead center, Strauss maintained, and both cables snapped, the roadbed would fall into 300 feet of water and leave the channel freely navigable. Having made his point, Strauss again took wing. "During the late War," he informed the military engineers, "not a single bridge was destroyed. London and Paris were bombed, but none of the bridges was damaged. Nor were any of the bridges in this country or Canada. If the enemy got close enough to this city as to be able to bomb the bridge, I am afraid that there would be very little left of the city." The military safety of the bridge, in other words, was the responsibility of the military.

Although a number of other speakers addressed the board, only Strauss's testimony was reported in the press. In fact, probably the most telling piece of evidence offered on the navigational hazard of the bridge piers was a bit of personal history by Michael O'Shaughnessy.

Some years earlier, O'Shaughnessy stated, he had been aboard the steamer *Alameda* when she had been wrecked near Fort Point, on the San Francisco side of the strait, almost exactly where one of the piers of the proposed bridge would be built. "With the lights and bells that could be established there," the city engineer concluded, "I am of the firm belief that it could be made of great assistance to navigation."

Buoyed by this degree of personal conviction, convinced by what seemed broad and deep support for the bridge among the community at large, swept up a bit perhaps in Strauss's whirlwind promotional fervor for his particular design, Deakyne concluded his hearings on a note of confidence, assuring the members of the Executive Committee that a final, official decision would be coming from the War Department within two months. Thereupon Strauss, who was already adding the Golden Gate Bridge to his resume of accomplished works, went off to promote other projects, while O'Shaughnessy returned to the responsibilities of running his department and building Hetch Hetchy. At both tasks, the city engineer was encountering a greater degree of opposition than he'd been used to.

O'Shaughnessy's career, a mountain range of peaks so far, was now showing its inevitable valleys. In 1920, the city engineer had arranged to make a triumphant return visit to his mother's home in Limerick, Ireland. On the eve of his departure, he was presented, on behalf of members of the Board of Supervisors and his friends at City Hall, with a diamond stickpin in recognition of his services to the city. Even more gratifying was the ceremony in May 1923, christening the 344-foot-high dam at the mouth of Hetch Hetchy Valley O'Shaughnessy Dam, an unusual act of hagiography for a live and sitting city engineer. O'Shaughnessy's prestige would never be as great, or his authority as unquestioned, again. Already there had been rumblings of resentment over the city engineer's exemption from accountability to the usual public bodies. There were cutting references to him in the press and among the city supervisors as "The King." O'Shaughnessy's reputation and prestige had attracted a certain critical attention in themselves: was it really appropriate for an appointed city official to possess unquestioned power?

Among the critical eyes that O'Shaughnessy's string of successes had caught were those of William Randolph Hearst, who maintained a certain watch over San Francisco politics from the bridge of his flagship San Francisco *Examiner.* Hearst had now decided to start firing a few shells in the city engineer's direction.

In a page-long bold-type "Open Letter to Our City Engineer" on December 10, 1923, the *Examiner,* which described itself as "always your friend in the past," had taken O'Shaughnessy to task for failing to enforce the Raker Act of 1913, which had required that the electric power generated by the Hetch Hetchy system be distributed from public, not private, bodies. Instead, the letter charged, "There are indications that you and your assistants obstinately oppose municipal electricity, and openly espouse turning over the city's power output to the Pacific Gas and Electric Company." Although professing to admire O'Shaughnessy's engineering skills, the *Examiner* suggested he leave matters of public polity to elected representatives and the mayor's advisory committee on Hetch Hetchy. The city engineer, warned *Examiner* editorialists, was faced with two choices: "Either to forget your opinions on matters of polity and finance; or to acknowledge frankly that your differences of viewpoint are too great to warrant your remaining in office."

O'Shaughnessy was defended by the other San Francisco papers, who accused the *Examiner* of trying to turn the city engineer into a Hearst stooge, but the issue was a live one, one that remained unresolved and has been periodically revived over more than sixty years. More immediately, the letter demonstrated that the city engineer, long regarded as a local fixture beyond politics, was no longer immune to public criticism. O'Shaughnessy's unchallenged honeymoon with his wife's hometown was over. By November 1924, a measure had been introduced before the Board of Supervisors that would "shear" O'Shaughnessy of his "carte blanche" to spend city bond issue money on Hetch Hetchy. O'Shaughnessy would continue to defend his hard-won autonomy and prestige vigorously. He would also grow increasingly touchy about what he considered attempts to diminish it.

2

The issue of War Department approval was crucial to construction of any bridge at the Golden Gate, and not only for reasons of defense. The land on either shore of the strait was military property that, upon approval, could be acquired without the tedious, often agonizing process of purchase and condemnation. There would be no angry property owners to negotiate with, no painful resettlement of families, no razing of cherished neighborhoods or historical landmarks. At the same time, the military, as landlord for either of the two anchorage sites of the bridge, could exact a stiff price for its approval and assure that what might have become a military obstruction would be turned into a military asset.

This complicated approval process, involving the Department of the Army and the Department of the Navy, as well as the War Department and the Corps of Engineers, dragged out the permit application through the summer and fall of 1924 and, with it, the earliest possible start-work date for bridge construction. It was now apparent, even to the professionally enthusiastic Strauss, that his bridge would not be completed by 1927, as he'd promised. And there was the growing possibility that, with further delay, there could be a waning of enthusiasm for the bridge among the public and a reappraisal of the likelihood of their being taxed to support it.

During the wait for the War Department's verdict, Strauss, perhaps sensing reservations about the appropriateness of his original design, made certain modifications in it. To appease concerns about navigational clearances, the height of the bridge roadway was raised considerably. The height of the steel-beam girder towers was also increased so they would now be taller than the Eiffel Tower. This Parisian motif was extended to the toll plaza, where Strauss now proposed to build an ornate entryway modeled on the Arc de Triomphe. The bridge, already a crossbreed of two different engineering theories, was now further complicated by two conflicting architectural styles, one dark, massive, and bluntly functional, the other wedding cake-like, decorative, and ornate. This piling on of borrowed styles and compromise solutions seemed to weigh the whole design down so that here, at one of the world's most exhilarating meetings of land and water, the bridge proposed lacked a single breath of originality.

At last, on December 20, 1924, the idea of a bridge received its first nod of approval from Washington. "The general project for construction of the Golden Gate Bridge is approved," Secretary of War John W. Weeks announced, in a wire to W. J. Hotchkiss of the Executive Committee, "subject to conditions which follow by letter." The permission to build was greeted in San Francisco as the government's Christmas gift to the city, but the War Department and the army and navy were, in fact, exacting a substantial price in return. The letter spelled out the terms. The military objections to the bridge could be eliminated providing the city of San Francisco and the other interested counties bore all expenses connected with "moving, rebuilding and replacing of elements of the defensive and other military installations damaged by such construction." They must also bear the costs of building and maintaining approaches to the bridge. They would cede to the federal goverment, in time of war, all control over the bridge. They must permit government traffic on the bridge free of charge at all times and would be required to subject the construction of the bridge to the consent of the secretary of war and his representative, the district engineer.

Even with these conditions, the approval was not final. What the War Department had granted was permission to proceed. Further permits would have to be applied for when detailed plans for the bridge were ready to be submitted.

The War Department's conditions were accepted. The bridge proponents, the Executive Committee, and the San Francisco supervisors had no choice. Plans were immediately announced to start raising finances for the bridge, and, Chairman Hotchkiss stated, Joseph Strauss, "the engineer who drew the plans and under whose supervision the bridge will be erected," would be summoned to San Francisco as soon as the financial campaign was under way.

Joseph Strauss needed little encouragement to plunge into what he regarded as the crowning task of his life. Within a month of the War Department announcement he was in San Francisco, where he'd set up headquarters in the Palace Hotel for the purpose of overseeing the formation of a district to finance construction of the bridge. Strauss had so committed himself to the project that he put his Chicago home up for sale, with the expressed intention of moving to the West Coast for the duration of the job. In announcing the sale of the house, in March 1925, the Chicago

Tribune described Strauss as having been "awarded the contract for the $26,000,000 new bascule bridge in San Francisco." In fact, the district that could let bids for such a job had not yet even been formed.

Strauss had also taken steps to buttress his credentials as a bridge designer. Working through his academically respected structural design engineer, Charles Ellis, Strauss had arranged for Professor George F. Swain of Harvard University and Leon S. Moisseiff, designer of the Manhattan Bridge over the East River in New York, to serve on a "board of consultants" on the project of bridging the Golden Gate. Both men, it was announced, had examined the Strauss plans and found them to be practical from an engineering standpoint and capable of being built within the estimated budget, which was now some $21 million. Professor Swain, in a letter to the Bridging the Golden Gate Association, said that he considered the proposed cantilever suspension design "perfectly practical." Moisseiff, an independent consultant who was perhaps the leading living theoretician of bridge design, confined his endorsement to a rather lukewarm approval of the cost figure, which he found "is about correct and may be exceeded by not more than $2,000,000."

Strauss also requested Moisseiff to prepare a report on a comparative design for a suspension bridge based on identical specifications and prices as Strauss's cantilever-suspension combination. Moisseiff's report, submitted in November 1925, suggests a bridge similar in many ways to the one that was eventually built. The bridge would consist of two wire cables spanned over two towers and anchored on each side into rock. Wire ropes, suspended from the cables, would support stiffening trusses and a floor system extending from tower to tower. The distance between towers would be 4,000 feet. The towers, rising from piers set on the floor of the strait, would be made of cellular steel, in the form of two columns braced into one unit. The columns would have a trapezoid form, tapering from 40 feet at the base to 20 feet at the top. The towers would be 735 feet high, as compared to the completed bridge's 746 feet. Moisseiff estimated that this bridge could be built for a total cost of "Say—$19,400,000."

Moisseiff's report seems to have been regarded by Strauss as a sort of straw man, to be set up only to be knocked down in the process of selling his own cantilever-suspension idea. The center suspended span, some 4,000 feet long, would be more than twice the length of any yet built, and the cables to support it, "about 35½ inches in diameter," thicker than any yet strung. Since it was Strauss's sales approach, at this time, to present his bridge as composed of elements, each within the realm of "things that had

been done," the all-suspension bridge may well have been used as an example of a radical design that could be used to counter the sweeping
modernistic suspension bridge recommended by Allan Rush. There is no
record of its seriously being considered at this time, although the memory
of the Moisseiff design would surface, along with its designer, significantly,
later.

Strauss's summoning of outside experts was the beginning of a dependence upon consultants that was to grow throughout his association with
the planning and building of the bridge but that would come to be increasingly at odds with the projected image of himself as sole mastermind of a
great work. Eventually he would speak of conceptual ideas beyond his
capabilities as if they were his own, and accept, without correction, praise
for work that he knew he had not done. He became, in time, an accumula-
tor of credit and praise, never sated, enlarging his role guiltlessly, as if he
were being compensated for the agonizing delays, the rebuffs and frustrations, professional and personal, that he had earlier endured.

Under the terms of the 1923 enabling act, the Northern California
counties that agreed to form a bridge district would be able to float bond
issues, take out loans, and charge bridge tolls. If these sources of funds
proved insufficient, they were legally empowered to levy taxes on property
owners in all the counties of the district. If the bridge job turned out to be
a boondoggle, with construction dragging out over a decade, or halted
completely, every property owner in the counties involved could turn out
to be financially liable. The time element was crucial. Delays in approval
and a protracted construction period would drive up the cost of money and
increase the likelihood of taxpayers having to shoulder the costs of the
bridge.

The legal process of forming a Bridge District was patterned after that
for putting initiatives on an election ballot. In each interested county,
petitions would be circulated in favor of formation of a district for the
purposes of building the bridge. When 10 percent of a county's population
had signed, the petition would be submitted to the county's board of
supervisors, who could then vote either for, or against, joining. This meant
that the approval process for the bridge would now be turned into an
election campaign, in twenty-one Northern California counties, complete
with committee work, publicity, canvassing, speechmaking, and intense
electioneering among the members of the various county boards of supervisors.

The wheelhorse of the campaign was Joseph Strauss. He had made himself the living embodiment of the bridge, and he was available to go on the stump, to address the chamber of commerce and Lions and Rotary luncheons, to meet with the smalltown press, and to offer expert testimony to the local political bodies. It was a part he assumed with the combination of conviction and rediscovered ebullience and vigor of a man in his middle fifties, confirmed at last in the chosen role of his life.

"If you had followed me today, you would be tired," Strauss boasted to a San Francisco reporter during this period, who described Strauss as "enthusiastic" after a day of bouncing between various conferences, meetings, and discussions with the chairman of the association, Mayor Rolph, and O'Shaughnessy. The details to be discussed were endless, and the hours, extended by socializing and travel, were grueling. Strauss was to have dinner that evening with O'Shaughnessy, then travel to San Rafael the following day for a luncheon speech at a local hotel, leave that afternoon for Los Angeles to make an appearance before the Harbor Commission, then return for several more days with the association in San Francisco. Yet always, Strauss retained energy enough for selling, opportunism enough to plant one more reminder of the significance of his bridge. "This bridging of the Golden Gate will make a wonderful city of San Francisco," he told the reporter. "You are bottled up now, with a wonderful, undeveloped country lying to the North."

It was a very undeveloped county indeed, lying considerably to the north of the Golden Gate, that cast the first vote in favor of the district. The supervisors of Mendocino County, halfway to the Oregon border and stretching from a thickly forested Pacific Coast to inland farming valleys, voted to join the Bridge District on January 7, 1925. An underpopulated county, with an economy split among farming, timber, and redwood tourism, parts of Mendocino County were so remote that, on the coast at least, until well into the twentieth century, transportation between communities had been easiest by sea. In was an encouraging sign that one of the counties more remote from the bridge's immediate benefits would be so quick to join.

In Marin County, directly across the Gate from San Francisco, the issue was never in doubt. Strauss gave a speech in Sausalito before a crowd estimated at approximately 25 percent of the town's population, the petition was quickly circulated, and the Marin supervisors voted to come aboard on January 23. The two adjoining counties, Sonoma, a farm county devoted to orchards, vineyards, and hops, and Napa, even in Prohibition the nation's most prominent wine-growing area, followed soon afterward.

In these counties, which proceed in a tier directly north from San Francisco, the advantages of a bridge were both public and private. It was the property owners in these counties whose land values would increase almost as soon as the district was approved by the supervisors, thus offsetting any individual worries about being taxed to support some long, wasteful bridge project. Beyond this hot core of obvious commuter and commercial benefit, however, support for the idea of a bridge had begun to cool. Humboldt County, just north of Mendocino, was dominated by the lumber industry, companies that depended on large tracts of lightly taxed forest land and that would have little to gain from an influx of tourists or residents—and much to lose from an increase in property values and taxes. Responding to the most emphatic, if not the most numerous, voices of their constituents, the Humboldt County Board of Supervisors declined to join the district, going on record as believing that the question of whether or not a bridge should be built properly belonged to San Francisco.

It had begun to dawn on people, in the counties away from the Gate, that they had little to lose by voting against the district. If the bridge were not built, or went bankrupt, or operated at a loss, they couldn't be taxed to support it. But if the district were formed, and the bridge were built successfully and operated efficiently without them, their property would not increase all that much because of it, and they would enjoy whatever benefits it offered anyway. The opportunity, as some local people and many local politicians began to see it, was that of securing an inevitable, if moderate, gain at a choice of considerable risk or no risk at all. This, plus the political fact that few individuals are ever voted out of office for saying no to anything, began to lend opposition to the bridge a certain air of astuteness.

This is what appears to have happened in Lake County, immediately adjacent to Napa and Sonoma, an area that would eventually enjoy a recreation-based boom directly attributable to the opening of the bridge. When the Lake County Board of Supervisors received the petitions signed by 10 percent of the county's voters in favor of the district, the board members promptly voted against it. Their justification was the power given to the Bridge District's Board of Directors to meet any deficit in construction costs, maintenance costs, operating losses, or payments of principal and interest on bonds, by taxing all property within any of the counties of the district.

These officially articulated second thoughts prompted doubts in other quarters, some of which had been assumed to be automatically in support of the bridge. In San Francisco the Board of Supervisors, where the initial resolution in behalf of building a bridge at the Golden Gate had been passed in 1918, now put the district-favoring ordinance back in committee, where it languished for more than two months. On March 26, 1925, with Mayor Rolph out of town, either by necessity or design, the issue was finally introduced, but in the form of a substitute resolution proposed by the acting mayor, a supervisor by the name of Ralph McLeran.

As we have seen, there was already openly expressed opposition to Michael O'Shaughnessy among the supervisors, and O'Shaughnessy was associated with the bridge as it was now proposed. His department was, in particular, responsible for the survey work: the borings, soundings, and foundations. McLeran's resolution called for a new engineering survey of the Golden Gate. There would be new borings, new foundation tests, additional plans, and more hearings. It was proposed that the city allocate $150,000 for this new study.

Richard J. Welch, the supervisor who had introduced the original pro-bridge measure in 1918 and who had built a considerable political reputation as "the father of the Golden Gate Bridge," exploded at hearing the resolution. Addressing an audience that included representatives from the northern counties, come to see what they thought would be San Francisco's keystone joining of the district, Welch told the people that they had been "slapped in the face and told to go home" by the city's acting mayor. The people may have to go without their bridge now, Welch said, but, he vowed, "we will get this bridge in the end."

Supervisor McLeran had made a miscalculation of a degree that causes the death of political careers. The San Francisco *Examiner,* which was certainly not friendly to O'Shaughnessy at this time, was firmly in favor of the bridge, and McLeran found himself, following his day in the sun as acting mayor, attacked editorially in the *Examiner* as an obstructionist and accused of favoring interests that intended to replace the proposed publicly owned bridge with a more expensive private one.

McLeran immediately started backpedaling, insisting that he had not really wanted a new engineering study so much as he wanted a change in the composition of the district board. Since San Francisco, which had the most valuable property in the district, would be bearing the overwhelming share of the cost if the bridge didn't pay its own way, then the city should have a dominant majority of members on the board.

Expressions of this sort did not sit at all well with people who had previously committed themselves to the building of a north-reaching bridge over San Francisco Bay. Frank L. Coombs, the California state assemblyman from Napa County who had introduced the enabling legislation, now warned the San Francisco supervisors that if they did not agree soon to make their city a member of the district, then the member northern counties would reconsider building their bridge to Oakland.

At a meeting in the city of Napa on April 5, 1925, a compromise was ironed out. The San Francisco supervisors agreed to drop their demands for a new engineering survey and for guaranteed outright control of the district board. Instead, a formula was developed for representation based on population. It included a provision that all counties with populations of more than 500,000 would have representation equal to that of all the other counties combined. San Francisco was the only such county. A week later, Supervisor Welch introduced legislation stating San Francisco's intention to become a member of the Bridge District; the San Francisco supervisors passed it unanimously.

The dust kicked up by the Board of Supervisors, in the city where the idea of a bridge at the Golden Gate originated, gathered into a lingering cloud of doubt over the whole project. There were, it seemed, some legitimate doubts about the thoroughness of the original engineering survey, as well as discrepancies in some of the overall budget figures quoted. Property owners *were* being asked to cede to this new body an unlimited power to tax. These feelings might have been resolved if only there had been some romance or grandeur about the bridge itself, some soaring sense of human possibility or the expression of great ideals inherent in its design. This inspirational quality the present bridge did not possess, except perhaps in the mind of Joseph Strauss.

When, on August 24, tiny Del Norte County, all the way north on the Oregon border, voted in favor of the district, it was by a three-to-two margin of the county board of supervisors. And it would be the last county to join. The campaign could be regarded, at best, as a marginal success. Of twenty-one Northern California counties invited to participate in forming the district, only six had voted to commit themselves. And even within some of those counties there were now serious misgivings.

In Mendocino County, the original county electing to join the district, there was now an intense reaction against the idea on the part of the local timberland owners. Should their lands be included in the district, a number of the timbermen testified before the county grand jury, they could be taxed out of all proportion to any benefit they might gain from the construction of a bridge. Unspoken, yet equally strong among the lumbermen, was the fear that the bridge would bring an influx of tourists, and with them increased public pressure to reduce or eliminate logging among the *Sequoia sempervirens,* the giant redwoods whose groves tend to be clustered in the foggy areas of the coast. Panicked by the prospect not only of confiscated timberlands but of lost jobs, the grand jurors recommended that the board of supervisors withdraw Mendocino County from the Bridge District, which the supervisors voted, on September 15, 1925, to do. The panic in Mendocino now spread to the individual signers of the original petition, some 180 of whom now signed a counterpetition stating that they had changed their minds and did not want their properties liable for the bridge's cost and possible losses. Although small in number, the petitioners represented enough of the county's population to bring the original petition below the required 10 percent. And they inspired similar counterpetitions among disgruntled citizens of the other counties that had joined the district. People in the rural counties were told that the bridge would cost more than $100 million, that the cables it would require would be so enormous that no bridge tower could possibly support them. A San Francisco attorney retained by the Mendocino timber interests prepared a printed protest form; more than 2,000 of these were signed by district property owners.

The Mendocino counterpetition, when it was filed in January 1926, caused the licensing secretary of state of California, Frank Jordan, to refuse to certify that the Bridge District had, in fact, been organized. The number of voter signatures had now dropped below the minimum.

In response, Supervisor Welch and Frank Doyle, the Sonoma County banker who had called the first mass meeting in behalf of the Golden Gate Bridge at Santa Rosa in 1923, turned to the California Supreme Court. Here, they argued that support for the district, once given in the form of one's signature on a petition, was like the casting of a ballot and could not be withdrawn. The court agreed, upholding the original petitions and on December 30, 1926, ordered the secretary of state to certify that "portions of six counties" had formed the Golden Gate Bridge and Highway District.

The district formed was, and is, one of the oddest-shaped of geographical and political bodies. There is a core cluster of counties immediately adjacent to either the Golden Gate or each other: San Francisco, Marin, Sonoma, Napa. Then a rump county, consisting of a strip adjoining the Redwood Highway as it passes through Mendocino County—and excluding most of the county's timberlands. Then a leap of empty space to the northern state line and remote Del Norte County. It is a district of extremes, including some of California's most densely and least populated areas, affluent and poor, rustic and sophisticated; an organization created for a single and temporary purpose, which has somehow managed to survive for more than sixty years.

3

It seems inevitable that progress must claim its victims. The giant dam that will bring flood control and drought relief to millions is constructed at the expense of the farmers near the site whose lands will be submerged; the freeway alleviating traffic congestion for the thousands entering and leaving a city will also cause the sundering of neighborhoods and slumlike darkness for the residents who remain. Yet without a dam's abundant water on tap or a freeway's speed and ease, we identify, during the process of construction, with the dislocated, the uprooted: the lonely widow, barricading herself in her house against the sheriff, the farmer with a shotgun ready to take on a giant Caterpillar bulldozer. They could be us, were time and circumstances different.

At the Golden Gate, however, there were no immediate victims. With military property at both ends of the bridge, the only relocation would be of military installations. One official body would be displacing another official body, at practically no cost in individual pain or inconvenience. Yet the feeling, the residue of injury that had accompanied every change since the coming of the railroad, remained, an instinctive sense that this bridge, like every other advance of the industrial age, would exact its inevitable human price.

The inheritors of this feeling turned out in this case to be the taxpayers, especially the isolated dissenting voices in the counties that had formed the district, who now faced the possibility of having all they owned forfeit to an idea that they had opposed. Some of these individuals—sheep ranchers, dairymen, smalltime loggers—represented some of the deepest Western feelings about personal autonomy. Their misgivings about the district and the bridge prompted some fellow citizens—engineers, lawyers, accountants —to ask some more pointed and detailed questions about the bridge and its financing. And the more questions that were asked, the more uncertainty the present proposal seemed to raise. There were many things being attempted for the first time here, and the moment to challenge them was at the beginning, before everything became set in cement.

In March 1926, while the petitions in behalf of the district were still circulating and the movement to withdraw was under way in Mendocino, and just after Lake and Humboldt counties had refused to join, the Joint

Council of Engineering Societies of San Francisco held a general discussion
of the Golden Gate Bridge project. The engineers, many of whom owned
property or had worked in the northern counties, or had friends or relatives
who did, discussed the proposed district and how it would go about raising
funds for construction. There was particular concern expressed about the
district's power to tax, "and that this taxation and expenditure could go on
year by year indefinitely unless stopped by action of the state legislature."

At the conclusion of the meeting, the engineers' council formally resolved: "1. That the project for bridging the Golden Gate has not as yet
been adequately investigated," and "2. That council does not approve of
the methods by which plans for the bridge district have thus far been
advanced."

The resolution, from a joint body of engineering professionals, criticizing
not only the idea of the bridge but the nature of the district's promotion,
was a direct slap at Joseph Strauss, the outsider, the promoter-engineer.

To make this point more emphatically, the council recommended that
"any project for bridging the Golden Gate should first have a proper investigation by a commission of not less than three competent bridge engineers." This investigation and the necessary studies, including borings, the
council advised, would probably cost no less than $500,000.

Once these studies were made, it would be possible to determine what
type of bridge would be most feasible and to prepare plans and estimate
costs. Only then, the council maintained, would it be possible to make an
economic study as to the justifiability of such a project's costs. "These data,"
the resolution concluded, "are unquestionably necessary preliminaries to the
formation of a bridge district."

Eight years after the passage of the initial resolution in favor of a bridge,
its proponents were being advised to start over.

Unquestionably, injured regional pride had a certain influence on the
engineers' disapproval. Experts from outside were being brought in to do a
job for which local professionals hadn't even been considered. Also, certain
private bodies, in particular the Southern Pacific Company, which owned
the ferry line operating between San Francisco and Marin, had a strong
financial interest in discouraging any competition for auto traffic across the
Gate. Yet even to laymen, the approval process for the bridge design and
the governing district had a certain fishy quality to it. Wasn't the idea of a
bridge the thing to be approved first, and then an engineer brought in to
build it? Instead, here was Strauss, the bridge builder from Chicago, campaigning all around Northern California, speaking often to public bodies

about his bridge and the wonderful things it would do, as if the contracts had already been signed. Wasn't there at least the possibility of some better, more aesthetically pleasing, less costly design? Whatever happened to competitive bidding? How competitive could any alternative bid now be when one engineer obviously had the inside track?

Although the council had no legal voice in the formation of the district or the approval of the bridge, its disapproval, reported nationally in *Engineering News-Record,* raised questions to which the proponents of the bridge would have to respond. In the six counties that had agreed to form the district, hearings were now called for to determine whether "the benefits accruing from the construction of the Golden Gate Bridge will be commensurate with the cost as financed under the assessment district plan."

At the first of these hearings, in November 1927, held in Santa Rosa, the seat of Sonoma County, Strauss testified personally in behalf of the proposed district and the proposed bridge. His estimate of the total cost was now $27 million, of which $21 million was for the foundation and superstructure, $4.5 million for highways and structural approaches, and $1.5 million for preliminary and administrative expenses.

In opposition to Strauss's figures, a group of Sonoma County taxpayers presented another report, prepared by J. B. Pope, Charles B. Wing, and W.J.H. Fogestrom, consulting engineers. This team had been retained to estimate the cost of the bridge and approaches, the revenue that could be expected from tolls, and the amount of money that would have to be raised by taxes to carry out the project. Their report ran for some sixty-six pages and included charts, diagrams, and tables.

In preparing their report, the engineers had evidently worked from original sources, Coast Guard soundings and tide tables, because their conclusions concerning the piers and foundations particularly were both considerably more detailed than the original Strauss–O'Shaughnessy proposal and extremely critical of it. The proposed center span of the bridge, the three engineers concluded, which was now designed to stretch 4,010 feet, "is probably not practicable" because of the realities of the foundation conditions. To support such a span, the pier at the San Francisco side of the bridge would have to be 940 feet out from shore, while the pier at the Marin side would have to be 400 feet from the opposite shore. Both piers would be in water 75 feet deep, and the foundations would probably have to be carried down to a depth of at least 115 feet below high tide level to assure a firm footing. Both pier locations would be exposed to the heavy seas that roll in directly from the ocean and to tidal currents that run as much as 9½ feet per second. These circumstances and pressures made conditions at the pier sites unstable.

Changes in the bottom, the report stated, "have been noted at both pier sites, at one of which the depth has been increased 40 feet in seven years." To seat the pier foundations under these circumstances would require the sinking of enormous and expensive concrete caissons—the giant compressed-air chambers in which men could walk the ocean floor—and a quantity of poured concrete seven times that allotted in the proposed estimate. "Under these conditions," the engineers maintained, "even if the borings show favorable rock formation, the cost of piers and anchorage would be $28,800,500."

The report also disputed the specifications proposed for the bridge's steel superstructure. Despite the problems of an exposed location, 950-foot towers, and the fact that the structure would be parallel to the San Andreas earthquake fault some twelve miles west, proponents of the Strauss bridge design were suggesting a bridge width "40 percent less than that required by standard practice for lateral stability." Allowances for "dead load"—the permanent weight of the bridge itself—were less than one-third those used in the George Washington bridge, then under construction, whose total length was only 3,500 feet. As for Strauss's patented cantilever-suspension combination, the engineers concluded "that it would afford no saving in cost and would probably increase the total weight on foundations."

The actual cost of the bridge, assuming the proposed 4,010-foot span could be built, the engineers estimated at $112,344,778. With the additional 600–1,000 feet that they predicted would be necessary for the center span, the figure would be substantially higher. On the basis of only the $112,000,000 total, however, the engineers' report concluded that revenue from bridge traffic would not pay off the interest on the construction costs until twenty years after the bridge had opened. At this rate, the total cost of building the bridge and paying off the money borrowed in the form of bonds to do it, and compensating for the anticipated losses caused by its operating costs, would amount, forty years after the bridge's opening, to $396,700,000. "Such a bridge," warned the engineers, "would start its long deficit period by causing an increase of 20 to 25 percent in the taxes of all counties in the district."

4

It would be difficult to imagine a more thoroughgoing condemnation of a proposed engineering project by a group of fellow professionals. Second opinions were usually more guarded, less inclusive than this. Like other professionals, engineers usually treat one another with kid gloves; one never know when one's own work might be subject to the unexpected judgment of peers. Yet the report seemed to justify every reservation anyone had ever expressed about the bridge—except the most obvious one: its looks. The conclusions were so sweeping that they threatened to destroy not only Strauss's plans for the bridge and the operation of the still-unformed Bridge District but to eliminate also the possibility of anyone attempting a bridge at the Golden Gate in the foreseeable future.

Not surprisingly, the report went all but unmentioned in the San Francisco newspapers, most of which had gone on record as in favor of the present bridge, as proposed. Nor was the subject raised by the Bridging the Golden Gate Association or the members of the various official bodies, whose members were now jockeying for appointments to the board of the district. Despite the unfavorable report, it seemed as if this were still the West, where a discouraging word was seldom heard, or at least not made public.

Within the engineering community, there was less reticence. "No competent engineer has said that the Golden Gate cannot be bridged," *Engineering News-Record* conceded in an editorial in December 1927. "However it is a far cry from that which is possible, regardless of cost, to that which is feasible from the financial viewpoint."

A bridge of the sort now proposed across the Golden Gate, the editorial continued, would require a span of more than 4,000 feet between supports —more than double the length of any bridge yet built—and would incorporate two towers almost as high as the Eiffel Tower; but the Eiffel Tower is a light structure resting on widely spaced footings and, most importantly, carries no load at its top. The Golden Gate Bridge towers would have to sustain an enormous top load, while resting on a relatively narrow base. The steel members would have to be of very great size simply to support such a structure, thus giving the bridge, overall, a tremendous dead load.

"It is natural, therefore," *Engineering News-Record* said, "that such a bridge scheme, unsupported by data about the various problems involved, should be opposed by the engineers of San Francisco."

As to the question of finance, the magazine pointed out that proponents had placed the cost of the Golden Gate Bridge lower than the cost of the Philadelphia-Camden Bridge, even though that bridge had a span of only 1,750 feet and occupied a much more favorable site.

The editorial concluded with a plea in behalf of the reputation of the engineering profession. To preserve that reputation, which had been won only during the last quarter century, the financial side of great engineering projects must be kept on a sound basis. Were the Golden Gate to be marred by a partially completed and abandoned structure, resented by every taxpayer assessed to support it, the work would not only be a "mortification to the proud Golden State" but "a misfortune to the engineering profession."

No Golden Gate Bridge project should be launched, *Engineering News-Record* recommended, until there was assurance from the most competent and experienced engineers that the project had been studied in detail and was proved feasible; that it had been planned and financed so as to assure its completion within a reasonable time; and that its total cost would be within a reasonable amount commensurate with the service the bridge would render.

"No scheme for bridging the Golden Gate," the editors advised, "that meets all or any of these requirements has yet been proposed."

These attacks on his patented design and on the ambition that had sustained him over at least the previous six years can only have come as body blows to Joseph Strauss. Once again, the engineering profession, resentful of his broad success, his notoriety, his wealth, was ganging up against him. He was being condemned for proposing something that had never been done, judged by theorists, academics, politicians, and dilettantes, engineers of the draftsman mentality, people to whom "it has never been done" was synonymous with "it can't be done."

Yet Strauss knew that *any* structure built at the Golden Gate would have to be in the nature of an experiment. The extraordinary conditions of the site precluded an unadventurous solution. Realizing in advance that he would encounter resistance of this nature, he had tried, as much as possible, to keep his plans for the bridge within the bounds of "things that had been done." At the same time, he had kept an eye on recent developments in bridge building that had extended the limits of what had traditionally been considered possible, and he had tried to incorporate these elements in his proposal for the bridge.

At the time Strauss had first proposed his design, the largest cable ever constructed was for the Manhattan Bridge over the East River in New York, a span 1,470 feet long suspended from four cables, each twenty-one inches in diameter. The largest cantilever that had ever been built was the Quebec Bridge, which contained structural members that had 1,941 square inches of metal in their cross-sections. If a bridge could be designed that would have a cable no larger than the ones at the Manhattan Bridge and that used structural members no larger than those in the Quebec Bridge, Strauss reasoned, then nobody could honestly say that the structure was not feasible, or that it could not be built. That was the bridge that Strauss had proposed, and that was the bridge that opposing engineers were now calling physically unsound and financially unfeasible. By vying for the approval of these nay-saying minds, Strauss had only further exposed himself to their opinions and given them the opportunity, once again, to reject him.

At times like these, Strauss could and did console himself with the thought that other men, too, men who dreamed great dreams and aspired to do great things, had, traditionally encountered fierce opposition. Indeed, one sure mark of the power inherent in an idea was the intensity of the resistance to it. A proposal that provoked no emotional response in anyone was one that most likely held little potential anyway. Besides, what was life, at best, but battle, and achievement, the overcoming of obstacles, and with it the sense of satisfaction? Was it not better to be at the center of things, on the field of life, engaged and involved, rather than sitting on the sidelines or up in the stands? These feelings, which ran deep in Strauss, he eventually committed to verse, in celebration of his bridge,

> LAUNCHED MIDST A THOUSAND HOPES AND FEARS,
> DAMNED BY A THOUSAND HOSTILE SEERS.

At the same time, he acknowledged the pain it had cost him:

> BUT ASK OF THOSE WHO MET THE FOE,
> WHO STOOD ALONE WHEN FAITH WAS LOW.
> ASK THEM THE PRICE THEY PAID.

Strauss was now fifty-seven years old. Uncertain about his health, he had begun consulting a physician regularly and would require increasing degrees of medical attention. His marriage, strained by his absences and travel, was in trouble and would soon end in divorce. Even if construction were started now, he would be in his sixties by the time his bridge was completed. Where again would he ever find a project whose possibilities spoke so deeply to him? How would he gather the energies required to will such a work to completion? Where again would he be presented with the great, confirming opportunity of a lifetime?

Strauss battled back. At the next feasibility hearing, held in Napa County Superior Court on February 15, 1928, Strauss brought in his own support troops: Charles Ellis, his staff structural design theorist; and Leon Moisseiff, who had designed the Philadelphia-Camden Bridge to which the opposition kept referring. Moisseiff, pressed for time, considered the process tedious and inconclusive; there was no way to prepare properly for these public bridge hearings, where one was always confronted with something unexpected.

The unpleasant surprises continued. On February 22, an official-sounding firm calling itself the San Francisco Bureau of Governmental Research issued a forty-page report that found, among other things, that the existing data on the Golden Gate Bridge was too incomplete to enable anyone to tell either what the bridge would cost or what its revenues would be. The report noted that estimates of the bridge's cost now ran from $25 million to $112 million, that each estimate altered the estimated tax rate accordingly, from zero, to four cents, to ten cents, to thirty-eight cents in the first year, and varied the length of time required for the bridge to pay its expenses from one year to twenty-four years.

The report recommended no other commitments be made until a complete study was made of the underwater foundations, the cost of approaches, and all engineering features. Until this was done, at an estimated cost of $500,000, "it will not even be known whether the bridge is possible or whether it can be built at any cost within reason."

While Strauss tried to cope with legal and engineering challenges to the bridge itself, the Bridge District, which was still under question in the courts, had already become the setting for political infighting.

Originally intended as a nonpolitical body, the Board of Directors of the Bridge District was viewed by the San Francisco supervisors as a potential breeding ground for mavericks with an insufficiently developed sense of political realities. To ensure that the city's, not to mention their own, interests were adequately represented, the supervisors appointed three of their own members to represent San Francisco on the bridge board. Their fourth appointee was Richard J. Welch, the former supervisor who was now in Congress, his career rising, even though the bridge that had stimulated it was, as yet, not. The supervisors' self-appointment provoked a storm of resentment among the private citizens who had been active in the Bridging the Golden Gate Association. The ideal of creating a great utility entirely without public funds, and without politicians and political problems, had been honored in all the other counties of the district. Now the San Francisco

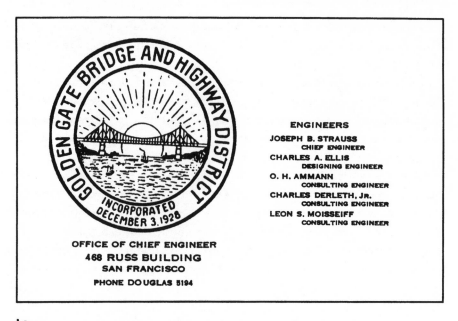

ENGINEERS
JOSEPH B. STRAUSS
CHIEF ENGINEER
CHARLES A. ELLIS
DESIGNING ENGINEER
O. H. AMMANN
CONSULTING ENGINEER
CHARLES DERLETH, JR.
CONSULTING ENGINEER
LEON S. MOISSEIFF
CONSULTING ENGINEER

OFFICE OF CHIEF ENGINEER
468 RUSS BUILDING
SAN FRANCISCO
PHONE DOUGLAS 5194

Strauss's bridge design, incorporated into the Bridge District logo, was still there as late as 1929.

supervisors had compromised the whole process by appointing politicians as their representatives on the board. What voter wouldn't be gun-shy, W. J. Hotchkiss, chairman of the association, pointed out, of a bond issue proposed by such a body?

Indeed, there were rumors that some of the supervisors appointed to the board had already made deals with lumber and railroad interests who were opposed to the idea of any bridge at all. According to one of the supervisors, Franck Havenner, selection of the the supervisors was being influenced, if not masterminded, by a near-legendary San Francisco influence peddler and bagman, Abraham "Murphy" Hirschberg.

Born in Poland, Hirschberg had emigrated to America at fourteen and arrived in San Francisco, where he sold newspapers in the streets, shortly before the earthquake. During the 1920s and 1930s, under the tutelage of Tom Finn, an old-line political boss whose career dated back to the labor-backed administrations of Schmitz and P. H. (PinHead) McCarthy, Hirschberg had come to wield enormous influence at San Francisco's City Hall, where he was such a familiar figure in the corridors that he was known as the "Twelfth Supervisor." Hirschberg, who was considered the go-between for Finn and the board and was accumulating a fortune for himself in real estate, was rumored to be picking and choosing the San Francisco members of the Bridge District Board himself. Havenner, an anti-Hirschberg reform figure who had been one of the appointed supervisors, resigned from the bridge board and introduced a resolution, endorsed by Major Rolph, calling on the other supervisors to do the same. After some fits and starts and a certain amount of huffing about impugned integrity, the other three supervisors either declined to follow Havenner's example and resign or, in one case, withdrew a resignation already submitted.

The unseemly mess caused by the San Francisco supervisors had given the entire district a bad odor. On January 16, 1929, an irked Mendocino County state assemblyman named R. R. Ingels introduced a bill repealing the 1923 enabling act and abolishing the Golden Gate Bridge and Highway District altogether. Ingels cited both the sorry political spectacle the San Francisco supervisors had made of themselves and the danger inherent in the district's unlimited power to tax.

The district board members, fighting for their organizational lives now, retained George Harlan, the Marin County lawyer who had drafted the original enabling legislation, to oppose Ingels's bill. Harlan, who had been nursing the project along ever since Strauss's first shaky speech at Sausalito in 1923 and who knew all the ins and outs of the efforts that had been made in its behalf, made "a spectacularly successful presentation" in support of the bridge. On March 5, the Assembly Roads and Highways Committee tabled the Ingels proposal. The committee then sent to the Assembly a bill introduced by a Marin County assemblyman validating all legal measures that had been taken toward the formation of the Bridge District. This bill passed into law, and the Golden Gate Bridge and Highway District became an official body. The board members promptly chose, as the logotype on all the district's correspondence and documents, Joseph Strauss's ugly, maligned, and probably unfeasible cantilever-suspension bridge.

"I CAN HAVE PROFESSOR ELLIS ADJUST MATTERS."

I

To Michael O'Shaughnessy, the fire that the proposed bridge design and, in particular, the orginal engineering survey and foundation plan, had come under was simply the latest volley in what had become a continuing barrage of politically motivated criticism aimed at him and all his works. At first he had been surprised and wounded by the intensity of such attacks and by the readiness of elected officials to go public with them. When, in September 1926, a pair of San Francisco supervisors, Warren Shannon and James McSheehy, accused the city engineer of low-balling (deliberately underestimating) bids on a portion of the Hetch Hetchy water line and considering only one bid for the building of the dam that had been named after him, O'Shaughnessy was so stung that he confided to Mayor Rolph that he was contemplating resigning. To O'Shaughnessy's dismay, the mayor announced to the press that the city engineer so resented the statements made by the supervisors that his resignation was "probable." Rolph said that he was "very much afraid that we shall lose O'Shaughnessy" and that "it is well-known that he has several outside offers carrying a much larger salary."

For O'Shaughnessy, any advantage he might have gained by having his importance to the city endorsed by its mayor was offset by the public admission that the supervisors had got his goat. "I thought it was a private conversation between the Mayor and myself," O'Shaughnessy responded. He admitted that he resented the continuing attacks and that they had become "almost unbearable." Yet he now announced, "I will never quit under fire."

In part, O'Shaughnessy was now paying the price of the extraordinarily favorable press he had received earlier in his career. The making and unmaking of public personalities has always been a reliable stock in trade for certain elements of the press, and it had a particularly strong history in San Francisco, where William Hearst had begun his newspaper career, as well as Fremont Older, the respected San Francisco *Call* editor who had campaigned so long for the trial of Abe Ruef and, upon his conviction, had immediately launched a campaign to get him out. The making and unmaking of celebrities had news value, and any individual involved had to develop a thick skin and a strong survival instinct. If this were the way the game was played, O'Shaughnessy concluded, then he would learn to play it with the best of them.

When, in October 1927, Mayor Rolph vetoed an attempt by the supervisors to appoint an outside engineer to make a street railway transportation survey, O'Shaughnessy made a public statement that he had not been consulted in the choice of engineer and that his own proposal to conduct the survey himself at a cost less than the outside engineer's fee had been submitted to the supervisors, who had tabled it. This growing instinct to go public in matters that he would previously have dealt with privately or remained aloof from, O'Shaughnessy now extended to the city's Board of Public Works. In an interview with an engineering trade magazine, O'Shaughnessy described this civic watchdog group as being composed of "a green grocer, a boilermaker and a retired military officer," men so technically unsophisticated that they were political weathervanes. The temperature of invective between O'Shaughnessy and his enemies would continue rising.

In 1929, however, O'Shaughnessy was still a man of enormous power and prestige in San Francisco. Mayor Rolph, his patron, still reigned at City Hall. The Hetch Hetchy project, still uncompleted and requiring periodic transfusions of capital in the form of new bond issues, still represented the city engineer's great vision made manifest. A new road, linking the city's Glen Park District with the thoroughfare around Twin Peaks, was designated to be named O'Shaughnessy Boulevard. Even his severest critics among the Hearst press and on the Board of Supervisors did not dispute the value of his services to the city. It was his power and autonomy they feared, and they let slip no opportunity that offered the possibility of curtailing them.

In the matter of the bridge, O'Shaughnessy, with his heightened political sensitivity, could feel himself being pushed out of the picture entirely. The call for a new engineering survey at the foundation sites was a rejection of the survey previously done under O'Shaughnessy's authority; now, neither he nor his department were even being considered for the new study. Should discrepancies be found between a new survey and the old one, it would undoubtedly be O'Shaughnessy who would be targeted for the blame by the Bridge District Board of Directors: Warren Shannon, who along with

McSheehy represented the most vehement anti-O'Shaughnessy faction on the Board of Supervisors, had been one of the San Francisco appointees to the Bridge District board.

For the most part, all these recent political jolts represented the normal ups and downs of political life that O'Shaughnessy had learned to endure with a certain amount of equanimity; but beyond them were certain foreseeable long-term changes that O'Shaughnessy must have found profoundly disturbing. If the bridge project went forward without him, it would represent the first major work undertaken in San Francisco since 1912 that did not bear his stamp. In a sense, it would mark the limits of his professional life: he'd been good enough to go this far, but no farther. It suggested that all his earlier efforts had somehow been flawed and that later reconsideration might find them wanting. There was also the fact that O'Shaughnessy's works, no matter how respected among civil engineers, were, to the people of San Francisco, all but invisible. Everyone in San Francisco might one day use the water from Hetch Hetchy, but hardly anyone ever went there, stood on the dam, traveled the railroad, marveled at the size of the culverts and tunnels and pipelines and the mighty effort that had been required to build them. It was the same with the city's streets, sewers, tunnels, rail lines, and streetcars. People used all these things, but they did not marvel at them. With a great bridge it was different. The works of it were there, for all to see, an enabling power that, like a giant sculpture, incorporated into its vision even the people beholding it. Almost everyone who crossed a great bridge must be aware, if only for an instant, of his dependence upon its engineer. O'Shaughnessy had earned the right to be part of such a work, and now it was about to be denied him.

What must have been most infuriating of all was the fact that O'Shaughnessy, the engineer who in many ways had originated the bridge project, was being excluded from it, while Joseph Strauss, to whom he had first suggested bridging the Gate, was not only still part of the proposal but among the engineers being considered to be put in charge. O'Shaughnessy, well informed about local politics, knew how hard Strauss was campaigning for the chief engineer's job. He had heard the rumors that Strauss had imported an advance agent, one H. H. Meyers, a Los Angeles oil promoter, and that Meyers had rented a hotel suite, where he was wining and dining supervisors and other public officials who would have a voice in choosing a chief engineer for the bridge. There were also rumors that, if Strauss got the job, he would give Meyers a percentage of his fee. No doubt O'Shaughnessy reflected on the irony that some of the supervisors most critical of his performance as city engineer were among the recipients of Meyers's hospitality.

As city engineer, O'Shaughnessy appeared at many of the same gatherings as did the various board members and promoters of the bridge. He sat alongside them at the dignitaries' tables of the civic club luncheons and local association dinners. He sometimes addressed the same groups or responded to questions from the audience or the press. And as the appointments of the Bridge District board were being made and the hierarchy chosen that would oversee the construction of the bridge, O'Shaughnessy could not but feel an increasing sense of having been left out and, with it, a growing urge, conditioned by his new habit of going public with his feelings, to speak his mind.

On April 10, 1929, Alan MacDonald, senior partner of the San Francisco contracting firm of MacDonald and Kahn, was appointed Manager of the Golden Gate Bridge and Highway District. MacDonald's name had been proposed to the district's board by San Francisco Supervisor William P. Stanton, and the nomination seconded by Congressman Richard Welch. MacDonald's first task, it was announced, would be to direct the work of making surveys, soundings, and plans for the bridge. Congressman Welch, along with William P. Filmer, a San Francisco businessman and member of the bridge board, would be traveling to New York the following week, where they would consult with "prominent engineering and legal firms in that city." There was no mention of O'Shaughnessy or the previous survey, or for that matter of the Strauss-designed bridge. The Bridge District, formed to build the bridge that Strauss had been campaigning for since 1923, and which had incorporated his design into its own logo, was, at the same time, quietly abandoning it. The project, which still had to be proposed to the district's voters as a bond issue, was starting over.

To people who had followed the fortunes of the Golden Gate Bridge since its earliest proposal and who could read between the lines of the official announcements it was clear that, where the bridge was concerned, all previous bets were now off. Among these observers was the influential dean of engineering at Berkeley, Charles Derleth. The morning after the announcement of MacDonald's new job, Derleth wrote him a formal letter congratulating him on his appointment and, in the same breath, applying for a job on MacDonald's engineering staff, either as chief engineer or as one of the project's consultants. Derleth then listed his credits, which dated back to 1896 and included the job as chief engineer of the Carquinez Straits Bridge between 1922 and 1927. Derleth, who sent a copy of his letter to Robert Gordon Sproul, president of the university, had no doubt concluded that, with new surveys and soundings being proposed, the Strauss design was being scrapped. And if the Chicago bridge builder's design was out, then Strauss could no longer have the chief engineer's job locked up.

This was also apparent to a number of other bridge builders who were approached by the district during the spring and summer of 1929. Among them were George Swain and Leon Moisseiff, Strauss's consultants who had earlier endorsed, with varying temperatures of enthusiasm, his design for the bridge; O. H. Ammann, the Swiss-born chief engineer of the Port of New York Authority and designer of the George Washington Bridge; Ralph Modjeski; Gustav Lindenthal; Charles Evan Fowler; and the bridge-building firms of Jacos & Davies and Waddell & Hardesty. Several of these prominent engineers made a point of coming to San Francisco to present themselves, their credentials, and their proposals personally.

That Strauss was not a certainty for the job was obvious most of all to Joseph Strauss. He, too, surely realized that the call for new surveys and soundings meant that his cantilever-suspension design might well be rejected. According to Congressman Welch, the new survey would determine whether the two rock shoals near the shorelines extended far enough out into the strait to justify erecting the heavy towers required to support the bridge. If the shoals were there, Welch said, the bridge was an assured fact, because the central span would not have to be more than 4,000 feet long, which would make it feasible from both an engineering and financial viewpoint. If, however, the rock shoals were incapable of bearing the weight of the towers, then the towers would have to be built at either shore, as in the bridge proposed by Allan Rush, which, said Welch, would make the cost of the bridge prohibitive.

For Strauss, this was the crucial turning point of his career, the decision that would determine how he would be remembered, as an industrious engineer-promoter who dabbled in the arts—or as something more, a conceiver of a monument that functioned, a poet in steel, a Roebling. His qualifications, he knew, were in question. The survey that had been part of his original proposal was being retaken, his design itself was in doubt. The foundation tests might demonstrate that the only bridge that could be affordably built at the Golden Gate would be an all-suspension type. And Strauss had never built a suspension bridge.

Even if he got the job, he would never have the autonomy he craved. He would have to accept the advice and submit to the approval of a consulting board of engineers, chosen by the bridge directors, not by him. Probably he knew through his sources, Charles Duncan and the promoter Meyers, that Derleth, critical of Strauss's original design, was being considered as one of the consultants on the bridge. Strauss had run his own firm, controlled his working environment, for more than twenty-five years. Could he really yield gracefully now to the opinions of others?

In a way, it would have been understandable if Strauss, resenting the criticism of his original design, or the compromises required in the chief engineer's authority, had withdrawn his name from consideration. Yet he had made a tremendous investment in this bridge over the years, not just in hopes and ambition, but in time, energy, and money. Not only was Strauss out of pocket for the revisions and expenses and trips he'd made over the years to sell the idea of the bridge, but he had also passed up other jobs to press forward on this one. How else would he ever recoup?

There was also something else. More than most men, Strauss understood the discrepancy that often exists, particularly in this century, between work and its recognition. In his early jobs, in the bridge shops in New Jersey and with the sanitary district of Chicago, he had found his own ideas summarily rejected or appropriated by and attributed to others. That was the way things were when you worked for someone else, and it was one of the reasons why Strauss had set up on his own. Strauss now controlled the environment of his work, and if someone objected to that particular climate and terrain, then let him do what Strauss himself had done: let him set up on his own, rely on his personal salesmanship and force of personality, and see how far even the most intellectually brilliant engineer or designer might get in the rough-and-tumble real world.

Years later, on the eve of the beginning of construction on the bridge, Strauss gave a speech in which he described his long struggle for approval and financing as "a 13 years' war . . . a long and tortuous march" at the end of which "the worn sandals of weary crusaders sink to rest at the foot of a long-receding goal."

Strauss certainly knew that although many soldiers contribute to a victory, history attributes it to the commanding general; he must be a man of mechanical ingenuity, careful, persevering, sagacious, kind and yet severe, open yet crafty, careful of his own but ready to steal from others, profuse yet rapacious, cautious yet enterprising. Joseph Strauss would qualify on every count. And he was determined to seize his next objective: the post of chief engineer.

2

Welch and Filmer returned from New York riding a tide of lifted optimism and lowered estimates. Among the engineers they had met with in the East were men handling "the great New York tube and bridge projects," including the George Washington Bridge under construction over the Hudson River. These engineers undoubtedly included Moisseiff and Ammann, or members of their staffs. Based on the judgment of these engineers, Welch told the Golden Gate Bridge Board of Directors, meeting at San Francisco's City Hall, that the cost of their bridge, far from reaching the $100 million that the skeptics had been quoting, should not go higher than $35 million.

"The new bridge over the Hudson River," Welch elaborated at the board's July meeting, "with far greater problems to overcome, will cost only $42,000,000 with a total, including foundations, of $60,000,000." The foundation detail on the Hudson had been a huge problem, Welch explained, and both bridge and foundation costs at the Gate should be considerably less.

Exactly what type of bridge and foundation this estimate was based upon Welch did not make clear. The George Washington is an all-suspension bridge for which the Golden Gate, at present, had no comparable plan. To the relieved and encouraged bridge board, however, this didn't seem to matter.

Considering the "marked enthusiasm" displayed at this meeting, observed Filmer, president of the Bridge District, there was no reason why the bond issue for actual construction could not be submitted to the voters before the end of the year. The directors approved a budget of $300,000 for preliminary work, including engineering and legal costs, the manager's costs, and directors' expenses, which would be paid by a three-cent tax assessment applied on property throughout the district.

At the time this tax was being proposed, angry property owners in four of the counties to be assessed were still contesting the formation of the district. In February 1929, lawyers representing nearly one thousand taxpayers had filed a petition against the district before the California State Supreme Court. The court announced in June that, in its view, the district had been legally organized, whereupon the taxpayers' lawyer announced he would appeal to the U.S. Supreme Court. It was becoming apparent that, where this bridge was concerned, nothing would come easily.

The political maneuvering in the choice of a chief engineer had intensified. Since the day of Alan MacDonald's appointment, Derleth had been in contact with the Bridge District manager at least two or three times a week, advising him how to proceed in setting up an engineering organization, suggesting how best to approach some of the individuals being considered as consultants, while at the same time lobbying for one of the top jobs himself. Derleth had asked President Sproul to speak to the manager on behalf of his dean of engineering, and Sproul had agreed to help.

The two ideal engineering choices would be Moisseiff and Ammann. Moisseiff was perhaps the world's outstanding theoretician of suspension bridge design, and his most recent work, on the Philadelphia-Camden and George Washington bridges, was expanding the entire possibilities of span length and lightness. O. H. Ammann, Swiss-born, was to the practical implementation of engineering design what Moisseiff was to theory. A civil engineer and bridge designer, he had become chief engineer of bridges for the New York Port Authority, responsible for the construction and maintenance of all bridges in New York harbor. A gifted administrator as well as a conceptualist, Ammann was especially adept at seeing a bridge project whole, breaking a task down into its components, and bringing to them exactly the talent most suited to each job. Along with Moisseiff, Ammann was still committed to the construction of the George Washington Bridge. In addition, Ammann had also been named chief engineer for all Port Authority projects, in charge of the plans, construction, and maintenance of everything from bridges to tunnels. There would be serious question of his availability.

Derleth, while not a bridge engineer of world rank like these two, represented the required local figure, necessary by now to avoid alienating the Western engineering fraternity. Also, he was a man of excellent academic credentials with an impressive network of personal and social connections. And he was willing to work as an engineer on the bridge in just about any capacity.

The crucial question now was that of Joseph Strauss. No one doubted his promotional abilities; indeed, through his "representative," Meyers, the bridge manager and board members were undoubtedly feeling a certain amount of pro-Strauss promotional pressure right now. Strauss wanted the chief engineer's job—he wanted it desperately. But what would happen if the board appointed as chief a man whose design, which he had been promoting now for eight years, was subsequently rejected? They would be putting a man in charge of building a kind of bridge that he had never built before. Within months, the bridge district would be going before the voters with a bond issue to raise the money to build the bridge. There would be opposition to it, broad-based and well-financed. These opponents would surely point out that the whole idea of a bridge district had originated with a man from Chicago whom the district had afterward appointed chief engineer. The whole thing could be made to seem a political setup, at the taxpayers' expense.

On the other hand, Strauss's long campaign in behalf of the bridge had given him at least some sort of moral equity in it. The district itself had been born not only of his promotional energies but also of his now-disputed design. He had worked on behalf of the bridge longer than anybody but Welch, who was now on the board, and O'Shaughnessy, who was apparently out of the picture entirely. Strauss was feisty. He could be litigious: he had once sued the city of Chicago for infringing on one of his bridge patents and had battled the local authorities for four years before accepting a settlement. If he chose to bring suit for compensation for his years of effort in behalf of the district, it might tie up construction of the bridge for years and cast yet another cloud over the entire proposal. Whatever choice the directors made, there was going to be opposition and criticism. Looked at this way, it might well be easier to award the chief engineer's job to Strauss than not.

Whatever were Strauss's technical shortcomings as an engineer, he did seem to be able to lead men and organize them. He was known by Moisseiff and Ammann, and he had the capacity to attract men of exceptional ability, like Ellis, to work for him. With the manager and the bridge directors firmly in control, the consulting engineers could, like pillows, be used both to bolster Strauss and to muffle him. They could furnish Strauss with the expertise and credentials he needed, but at the same time, they could prevent him from dictating policy.

Strauss realized that the only way he could be appointed chief engineer was if he could guarantee, in advance, that other bridge engineers of the highest reputation were willing to serve with him as advisors. In March of 1929, working through Ellis, Strauss approached Moisseiff and Ammann, inquiring whether they would accept appointments as advisory engineers. Moisseiff, who made his living as a bridge consultant, expressed immediate interest. Ammann was more reluctant, because of his responsibilities with the New York Port Authority and because of certain misgivings about the way the new bridge's engineering organization was to function. Any appointment of Ammann would have to be approved by the New York Port Authority commissioners, and Ammann was personally determined not to be put in the position of competing with other engineers for a job.

On March 21, Ammann received a letter from Filmer, as president of the district, inviting him to apply for a position as one of three consulting engineers retained by the district to ascertain the feasibility of building the bridge. Filmer also requested Ammann, if selected as one of the consulting engineers, to name two or more engineers with whom he would like to work.

This was not at all what Ammann had in mind. As an engineer occupying an official position, he did not feel he could consistently offer his services in competition with engineers in private practice. This situation was not helped by a letter from Strauss, a week later, informing Ammann that Moisseiff had agreed to serve as one of the advisory engineers and requesting Ammann to "confirm our verbal agreement that you will act in an advisory capacity to me on this project in consideration of a compensation of $15,000 for the first year's services and $10,000 for the second year's services and $10,000 for the third year's services."

Ammann was not aware that he had agreed to anything, verbally or otherwise, and he promptly wired Strauss that he was unable to accept. For one thing, he had not received approval from the commissioners of the Port Authority. And for another, there was the letter from Filmer, Ammann wrote, "requesting offer to serve as one of three consultants for preliminary work. This is a new situation not before realized. My offer even if made through you would put me in the embarrassing situation of competing with other engineers which I desire to avoid." Under the present circumstances, Ammann felt he could accept the position only if his appointment were made upon request by the other two engineers engaged by the Board of Directors.

Ammann's refusal had pulled the rug out from under Strauss. If he couldn't deliver consulting engineers of unquestionable credentials, chances were he wouldn't get the chief's job at all. Strauss wired Ammann at the New York Port Authority: "AS I UNDERSTAND IT YOU HAD ALREADY AC-CEPTED OFFER AND MY LETTER WAS MERELY WRITTEN CONFIRMATION OF VERBAL AGREEMENT REACHED AFTER THOROUGH DISCUSSION AND AP-PROVED AT YOUR REQUEST BY MOISSEIFF." Strauss claimed Ammann had also previously indicated his willingness to serve by both wire and letter. "BASED ON THESE FACTS," Strauss continued, "I HAVE ALREADY NAMED YOU AND MOISSEIFF AS ADVISORY WOULD THEREFORE GREATLY EMBARRASS ALL CONCERNED WERE IT NECESSARY TO CHANGE."

Strauss assured Ammann that he would not be at a disadvantage and "kindly reminded" him that he had taken up this matter with him long before. If Ammann desired to discuss some modification of the confirming letter, Strauss wired, "I CAN HAVE PROFESSOR ELLIS COME ON AND ADJUST MATTER WITH YOU. PLEASE WIRE QUICK. JOSEPH B. STRAUSS."

Three days later, Ammann wrote to Strauss in San Francisco, explaining in detail his position. When the two men had first discussed the job, Ammann had objected to the chief engineer's employing two advisory engineers, believing "that in the best interest of all concerned the advisory engineers should be employed and paid by the Board with the consent of the engineer in charge." Strauss had told him that this arrangement could not be changed. Ammann had reluctantly accepted it.

Also, Ammann had understood from Strauss that there was to be no competition for engineering services and that the appointment would be in connection with both preliminary work and the execution of the project when financed. Filmer's letter seemed to say otherwise.

Under these circumstances, Ammann wrote Filmer that he could accept only if requested by the board, and with the consent of the other consultants. He would also consider an amount of $25,000 to be "fair and proper compensation for advisory services," and he recommended a number of engineers with whom he had worked and would not hesitate to collaborate, among them Moisseiff and Strauss. In part, Ammann had wanted to be courted.

He got his wish. On August 13, Ammann received a telegram from Filmer, on behalf of the board, asking Ammann to work with "Joseph B. Strauss as Chief and Leon Moisseiff, Waddell and Hardesty [the engineering firm, who later withdrew] and Charles Derleth as Consultant Engineers." Ammann accepted. Strauss now had both his unquestioned experts in bridge engineering. And a lock on the job of chief.

On July 24, MacDonald had made a tentative offer in a telegram to Derleth. The Berkeley engineering dean would be appointed a consulting engineer, pending approval by the board and the chief engineer. The following week, at the Bohemian Grove, Derleth visited with Sproul and left a letter for MacDonald at his camp. On August 5, Derleth visited both MacDonald and Filmer. There was some sensitivity as to how Strauss would work with Derleth, the local academic who had criticized Strauss's earlier design. Also, there were reservations concerning the appointment of Derleth, expressed by board member Warren Shannon, O'Shaughnessy's old foil from the San Francisco Board of Supervisors, who seemed to view any local engineer with a certain suspicion. What was established either now, or in practice later on, was that Derleth would be the directors' man on the

Engineering Board, an unofficial eye within the supposedly closed sessions of the engineers, and keeping MacDonald and the directors informed about what was really going on. This was something that Derleth, an experienced faculty politician, understood thoroughly and apparently had no problem accepting. On August 7, Filmer made another "tentative" offer to the Berkeley dean, who promptly accepted.

As if to underscore the fact that the directors intended to be running the show, the announcement of the appointment of all four engineers was made at the board meeting on August 15. Strauss was not even given a day to bask in the sunlight alone, but presented along with Moisseiff, Ammann, and Derleth as one of "four engineers of national and international reputation" appointed to the bridge's Board of Engineers, with Strauss as chief engineer. In only a few regional papers was the announcement accompanied by a picture of Strauss's proposed cantilever-suspension bridge.

The engineers would begin work almost immediately, the directors enthusiastically announced. Strauss was already in San Francisco and Derleth in Berkeley, and Moisseiff and Ammann would be arriving the following week "for a preliminary investigation of the conditions . . . and to arrange for the surveys necessary." Derleth, the day after the announcement, wrote a diplomatic letter to Strauss congratulating him on his selection as chief engineer and saluting him for "your pioneer work on this project [which] justifies the confidence which the board has placed in you." Derleth went on to express the honor he felt at being associated with Strauss, Ammann, and Moisseiff and assured Strauss of his loyal cooperation.

In fact, Strauss had paid a stiff price to get the chief engineer's job. According to Richard Welch, who, as the only man associated with the bridge longer than Strauss, probably made him the offer, the terms of Strauss's contract were "the lowest ever written for a bridge job of such magnitude." For a fee of 4 percent, as compared to the standard 7 percent suggested by the American Society of Civil Engineers, Strauss's firm would be expected to absorb all engineering expenses, and there would be no royalties beyond the fixed construction percentage. On a total construction estimate of $30 million, for example, Strauss would only have $1.2 million to cover salaries, plans, travel expenses, testing, and consulting fees. Strauss, who had campaigned long and hard for the job, had to swallow his financial as well as his professional pride to get it.

The professional blows had not ended. On August 20 Moisseiff, realizing apparently for the first time that the arrangement for his employment was between Strauss and himself, wired Strauss that he was withdrawing his acceptance. Like Ammann, Moisseiff wanted to assure his independence by being hired and paid by the Bridge District directors, not the chief engineer. Now there was a danger of his jumping out of the boat. On behalf of the Bridge District, Filmer wired both Moisseiff and Ammann, urging them to come to San Francisco anyway, assuring them that "arrangements can be made satisfactory to all parties."

At Moisseiff's and Ammann's insistence, their contracts were modified to specify that they were being appointed by the Bridge District to serve on an engineering board of which the chief engineer was also a member. They would be working with Strauss, but not for him. They were also to be paid well for their services: Ammann's contract guaranteed him a fee of $7,000 to be paid out of the proceeds of the district's first tax levy, with an additional $43,500 to be paid in six semi-annual installments if the bond issue passed and construction on the bridge itself began. It was an enormous fee for the time, and in painful contrast to that of Strauss, who had been required to cut his own percentage to get the job.

The announcement of Strauss's appointment produced surprisingly little criticism, alarm, or skepticism, considering that it was the appointment of a chief engineer who had never built a bridge approaching this scale and whose earlier design could not have been built within his budget, if at all. Strauss had been associated for so long with the idea of bridging the Golden Gate that his appointment to most observers must have seemed to happen years before. There was, however, one emphatically dissenting voice.

For Michael O'Shaughnessy, the selection of Strauss, in so many ways his own creature, as chief engineer was the final, unendurable affront of his ten-year association with the bridge. Almost immediately following Strauss's appointment, O'Shaughnessy, the most prominent local engineering figure not included in the new board, was quoted as saying that he was opposed to building the bridge because in his opinion, at least for the record, its costs would be in excess of $100 million.

At a Public Spirit Club luncheon on August 31, Congressman Welch responded with heat when questioned about O'Shaughnessy's statement. "Knowing as I do the disposition of my friend, M. M. O'Shaughnessy," Welch began disarmingly, "I urge you to disregard all opposition which springs from professional jealousy and pique." Welch pointed out that the engineers now consulting with the board, who included the foremost bridge engineers in the country and who had been employed by the district to build it, estimated that the bridge would cost no more than $27 million.

In Marin County, the body that probably felt most immediately threatened by O'Shaughnessy's statement, the local county real estate board, passed a resolution calling for the discharge of O'Shaughnessy as consulting engineer to the Marin County Municipal Water District. Michael O'Shaughnessy's association with the Golden Gate Bridge, like that of an estranged parent and child, was turning into an injury- and pride-ridden feud.

First meeting of the engineering board, Alta Mira Hotel, Sausalito, California, August 27, 1929. Front row, Joseph Strauss (left) and William L. Filmer. Second row, Charles E. Derleth, Jr., O. H. Ammann, Army officer, Leon Moisseiff, Charles Ellis. Rear, Supervisor Stanton, Congressman Welch. Strauss was so sensitive about his height that he once asked his secretary to guide him to high ground whenever a picture was about to be taken.

THE THIRD BRIDGE

I

In one of those rare photographs that in its subtlety of human posture and arrangement not only reveals the individual personalities involved but suggests the texture of the entire relationship, the principal figures concerned with the design and construction of the Golden Gate Bridge have gathered on the steps of the Alta Mira, the largest hotel in Sausalito, California. The date is August 27, 1929, less than two weeks after the engineering staff has been announced. On the top step, in the back row, stand two of the directors, William P. Stanton and Representative Richard Welch, the "daddy of the bridge," posed now like fathers at a graduation, overseeing with mixed anxiety and pride the symbolic release of their offspring upon the world.

On the step below, along with an army representative, are the thinkers, the men of ideas, Charles Ellis and the three consulting engineers. Charles Derleth, on the left, is bulky, round-faced, enigmatic. O. H. Ammann, a step behind him, has his hands clasped behind his back, as if in reticence. Next to him stands the army officer, and beside him Leon Moisseiff, feet apart, chest out, with the dark goatee and flashing eye of a road company Svengali. Next to him, at the extreme edge of the picture, wearing a dark suit and standing a step below the other men in his row so he will not tower over them, is Charles Ellis, reserved, scholarly, professional, diffident. In the first row, a step in front of the rest, are two men: William Filmer, the district president, a bankerly man with white hair, in a dark suit with a watch chain across his vest; and Joseph Strauss. Of all the men in the picture, Strauss is the one who immediately claims your eye. He stands hands on hips with one foot forward, elbows thrusting his coat back, facing directly into the camera, his head cocked as if to take on any challenger, while, at the same time, announcing to the world that, in matters of significance concerned here, he is the man in charge.

While standing still for the photographer, Strauss has somehow managed to strut. It is, perhaps too deliberately, not the obvious stance of a man who has just made important concessions in order to get a desperately sought after job.

Strauss had wasted no time in asserting his authority as chief engineer. The day after the announcement of his appointment, when he was joined by a board of consultants—one of whom had criticized his earlier bridge, and two of whom had resisted working under him—and with the legality of the district still in doubt, he announced that he was summoning his consultants to San Francisco for a conference. Now, some eleven days later, the three consultants, along with Ellis, designated by Strauss as his personal representative in San Francisco, were making a three-day reconnaissance of the two bridgeheads at the Golden Gate, examining the land approaches and making a cruise on the choppy waters above the proposed pier. They were also, undoubtedly, as men do in such situations, taking the measure of one another.

It is easy to imagine Strauss, with his ten-year familiarity with the site, assuming the role of tour guide, pointing out to the others conclusions drawn from O'Shaughnessy's earlier survey, standing on the pitching deck of a Corps of Engineers vessel, describing the treacherous tides and currents at the Gate. With a little further stretch, perhaps we can even imagine O. H. Ammann, charged with paying special attention to the study and solution of foundation problems, looking at the rocky outcrops and swelling choppy water here; or Derleth, comparing conditions here to those at the Carquinez Straits a dozen miles or so farther up the bay. We might imagine Moisseiff and Ellis, however, looking up, away from the sites of piers and foundations, gazing high overhead to where the cables and suspenders and tower tops would be, their minds wandering off in calculations of winds and stresses, dead and live loads. The other men would most likely be seeking to apply what they had already done to a new situation, but Moisseiff and Ellis, comfortable with theory, would be listening to the Gate itself, waiting for it to speak to them, to tell them something new.

In Charles Ellis, Strauss had what he had not possessed when he had made his first attempt at designing a bridge for the Golden Gate: an academic, a university professor capable of discussing and elaborating upon the most abstract and complex engineering theory. By appointing Ellis as his personal representative on the bridge, Strauss was dispensing with two potentially troublesome problems at once. He was delegating to someone else the technical complexities of dealing and debating with the engineers on the board. And by designating Ellis, his employee, to meet with the senior engineers who accepted Ellis as their peer, Strauss was suggesting that he was their superior.

In a concentrated period, the men spent a great deal of time together. There would be a full day out on the site, with a break for lunch at someplace like the Alta Mira, followed by dinner at the Palace Hotel in San Francisco. For Ellis, this experience must have been in the nature of a waking dream. He was actually working and walking, talking and eating with Leon Moisseiff, a man whom the restrained Ellis openly declared a genius and whose ideas he freely admitted being influenced by, elaborating upon, and borrowing.

In the twenty years between 1915 and 1935, the science and art of bridge design in America underwent greater change than in the entire previous century. It is not an unwarranted statement to suggest that the most significant of these changes originated within the head of one man. Born in Riga, Latvia, in 1872, Leon Solomon Moisseiff received his early training at the local polytechnic institute and emigrated to the United States in 1891. He received his degree in civil engineering from Columbia University in 1895 and became a U.S. citizen the following year. An enthusiastic American, Moisseiff named one of his daughters Liberty.

From an early job with the New York City Department of Bridges, he acquired the interest in bridge design that was to absorb him the rest of his life. Moisseiff was an original, a seminal thinker, with an ability to cut through complex and abstract problems to fundamental and influential conclusions. It was a gift that Moisseiff strengthened through application over the years, propagating his ideas in papers that made him the nation's leading theoretician of suspension bridge design, and in a series of bold bridge projects that altered forever the American idea of what a bridge might be.

These bridges, the Philadelphia-Camden (Benjamin Franklin) Bridge, the George Washington and Bayonne bridges at New York, the Ambassador Bridge at Detroit, and the Maumee River Bridge at Toledo, represented a step-by-step extension of the boundaries of bridge design and were based on a profound grasp of theoretical knowledge that incorporated technological advances in metallurgy and the drawing and spinning of cables. Gradually, largely through Moisseiff's ideas, America was evolving from a nation of ponderous, rigid, resolutely functional bridges to structures equally sturdy but of a soaring, almost airborne lightness and flexibility.

To Ellis, Moisseiff's most profoundly influential discoveries concerned the measurement and distribution of the stresses caused by winds upon bridges. In the course of his work on the Philadelphia-Camden and George Washington bridges, Moisseiff had developed a theory to distribute wind stresses on suspension bridges by balancing the sideways movements—the horizontal deflections—in the cable with those in the bridge's stiffening truss—the beam framework that supports the roadbed.

Moisseiff would begin by guessing what portion of the total wind load was carried by the truss; the remaining portion was obviously borne by the cable. This gave him an assumed load line—a stress limit—for both the stiffening truss and the cable. By using the principles of integral calculus, Moisseiff was able to use the assumed load line of the stiffening truss to estimate its shear curve—the extreme extent of its bending under wind stress. And, using the shear curve, he could calculate the truss's moment curve—its average deflection under all conditions. From these integrations, Moisseiff was able to determine the slope of the stiffening truss's elastic curve—what degree of movement would be necessary for the truss to return to a stable position.

While four different integrations were required to find the slope of the elastic curve for the stiffening truss, only two were required for the cable, thus giving Moisseiff the horizontal deflections at various points both in the truss and in the cable. Since the relationship of the vertical dead load— the weight of the bridge itself—to the horizontal wind load at any point in the stiffening truss determines the slope of the suspender—one of the hundreds of smaller cables that hang vertically from the main cable and connect it to the roadbed and the frame—once this was known, the difference between the deflection of the truss and that of the cable was known. Using the slope of the hanging suspenders as his guide, Moisseiff could measure his estimate of the balance between the two at various points along the structure. If the two checked, his original estimate was accurate. If they didn't, it was necessary to repeat the operation perhaps half a dozen times, each sequence requiring perhaps two days of close application, until the balancing was accurate.

According to this theory, developed by Moisseiff in collaboration with Frederick Lienhard, an engineer for the New York Port Authority, as much as half of the pressures caused by winds could be absorbed by the main cables themselves in a long suspension bridge and transmitted to the bridge towers and abutments. At the same time, the deflection of the truss and of the cable would tend to offset one another, thus working to restore the bridge to equilibrium. If this were true, then in a properly balanced bridge, with the suspenders appropriately adjusted, the highest probable winds would not damage it if the bridge were flexible enough to bend and sway with the wind. Bridges in high wind areas like the Golden Gate would not have to be heavy, ponderous, and rigid to withstand gales. The spans could be lighter, longer, and narrower than had ever been thought possible. And they could be quicker, and less costly, to build.

For Ellis, who fully grasped the implications of these discoveries, the Golden Gate represented an opportunity to implement some of the most profound bridge-building ideas ever conceived, all of them the product of recent years, when America had moved to the forefront among the bridge-building nations of the world.

2

Even though he did not entirely grasp the mathematical subtleties of these new theories as Ellis and Moisseiff did, Joseph Strauss was now willing to concede the desirability of building a suspension bridge at the Golden Gate. At the first deliberations of the consulting board in San Francisco, Strauss went at the problem through the process of elimination. There were, as Strauss saw it, three types of bridge construction possible at the Gate. One was his own type, the cantilever-suspension hybrid. This, Strauss now finally suggested, should be eliminated, since it was clear to see that the erection of this type of bridge, with its complex, heavy metalwork and ponderous piers, would take a year longer than the other two possible kinds, with a correspondingly larger cost in interest on the money used for construction. Strauss's concession on this issue must have produced an audible sigh of relief among the other engineers.

The second possible choice, a suspension span with straight backstays or support members, Strauss eliminated for two reasons. First, it was a less aesthetically pleasing structure—this was definitely a new tack for Strauss —than one with curved or sloping backstays; and second, the side span on the San Francisco end of the bridge would have to be supported by additional piers in the bay, in violation of the original permit plans as submitted to the War Department in 1924. If the design now called for more piers, it would not conform to the original application, and the approval process would have to begin all over again. In this manner, Strauss and the board arrived at their recommendation: a suspension-type bridge, with curved backstays.

The engineers also brought a fresh perspective to the location of the bridge foundations and piers. In the original Strauss–O'Shaughnessy design, the proposed length of the center span had been 4,000 feet. This placed the San Francisco pier 1,100 feet from the shore and the Marin pier about 200 feet from the opposite shore. Both piers would stand in about 60 feet of water, but would have to rest on foundations considerably deeper to withstand the Gate's extreme currents and tides. If, however, as Moisseiff's new theories suggested, a suspension span of much greater length could safely be built at this site, it would only require an extension of 200 more feet to rest the Marin end of the bridge on a pier at the Marin shoreline. Building a pier at the shore would cost nearly $1 million less than sinking one in the tide-rip and ocean swells of the strait. The board recommended that test drilling, to determine the extent of the rock shelf from the Marin shore at Lime Point, be undertaken as soon as possible.

Even in these early meetings, it is apparent, the question of the bridge was receiving a quality of analytical thought it had not been given before. The minds being applied to its problems were probably the most capable, on these particular matters, in America, if not the world. They constituted the cutting edge of the most recent developments in bridge-building technology, and most of them were at the peak of their physical and intellectual powers. Perhaps this is why Strauss, realizing how fortunate he was to have men like this on his team, so obligingly abandoned the design he had originated and patented and campaigned for these last eight years. Or, it may simply have been a condition of his appointment, as laid down by Welch and the other directors: Strauss may have been told that he could serve as chief engineer only if he did not interfere with the decisions of the consulting board. Yet in the account of these meetings left by two of the men who took part in them, Strauss seems a willing and active participant in the dismantling of his proudest idea.

Probably Strauss knew he was onto a good thing, which, as the deliberations progressed, kept getting better and better. Not only was Strauss going to be chief engineer of a suspension bridge, a type that he had never built, it was also to be the most advanced design of its kind, the longest single-span suspension bridge in the world. And the board was suggesting designs that were actually going to save money. It was like the early days of the Strauss bascule bridge all over again, when, with his combination of trunnions and poured concrete, Strauss had been in the forefront of his profession with an idea that was not only dramatically new but more economical than the structures it replaced. For Strauss, now approaching sixty, to be remembered as commander-in-chief of such a mighty project could be heady stuff, even though he might have a pang at times, knowing at heart it was not really his own design. But that too might change, among a public known for its distractable eye and faulty memory.

3

On August 28, Strauss, along with Moisseiff and Ammann, departed for the East. Before leaving, Strauss announced that he would be returning in about eight weeks. Meanwhile, Ellis would be his personal representative in San Francisco, while the "engineers"—presumably Ellis and Derleth—prepared specifications for the bids on the test borings for the foundations. It was hoped, Strauss said, that the specifications would be ready to submit to the Bridge District directors' meeting in September.

Then Strauss was gone, and Charles Ellis was on his own in San Francisco, alone with the dreams and theories of a bridge, confronting the practical realities of how to build it.

For Ellis, a self-disciplined man of the most restrained emotions, this must have been at the same time the most euphoric and the most intimidating task of his life. He was to be the key, integrating part of a great venture, which seemed to call upon both everything he knew and everything he was. Ellis was, at least in part, a mathematician, a man who loved the challenge of analytical problems to the point of indulging in them as a hobby. Now he would be working out the mathematical computations for a theoretical problem that would be transformed into a great work, perhaps the greatest of its kind. It was like academic scholarship at its best, yet at a professional engineer's salary, and with a project that was not a theoretical problem but a work that would be made manifest and endure perhaps for centuries.

There was another side of Charles Ellis that this project touched: the Greek scholar, the translator and classicist. The Golden Gate, one of the world's most inspiring junctures of land and water, must have suggested to Ellis the stories and legends that surrounded similar sites in the ancient world: the Dardanelles, bridged by Xerxes the Persian king, using a string of boats in 481 BC, and by Alexander the Great, using the same method 150 years later; and the Bosporus, named for the crossing of the ox that had been the outcome of Zeus's love for Io. At the western entrance to the Hellespont, a site not that different from the one San Francisco now occupied, the city of Troy had flourished and fallen. Even the name *Golden Gate* came from the Greek *chrysopylae:* John C. Fremont had coined the name in homage to that given the ancient harbor of Byzantium, Chrysopceras—the Golden Horn. Standing here, at the far edge of a recently settled continent, bearer of the advanced ideas of a technological society, Ellis was at the same time back near the sources of Western civilization.

Conferring regularly with Derleth, the Berkeley dean, Ellis was in touch with the academic environment he had come from and to which one day he would return. At the same time, he was engaged in the practical work of talking to surveyors, traffic engineers, and drilling contractors and overseeing work crews. On September 7, it was announced that on the following Tuesday two work crews under Ellis would begin preparatory surveys to determine the lines for the test borings that would decide the final site of the bridge. In order that the bids for the boring contracts might be called at the earliest possible moment, two surveying crews would be working simultaneously on both sides of the Golden Gate.

This work had barely gotten under way when, on September 11, Ellis, appearing before the Bridge District Board of Directors for the first time without Strauss, asked the directors for authorization to call for bids for soundings and borings. Ellis, now serving the functions of both staff planner and line officer, did not mince words with the directors. He stressed the fact that the entire future of the bridge depended on these borings and soundings. If no firm foundation for the piers supporting the structure could be located, then the bridge plan would have to be abandoned. Preliminary plans and specifications necessary for advertising bids would be presented to the directors later that month. The directors, responding to the urgency of Ellis' appeal, gave permission for the calling of bids and resolved that the Engineering Board could ask for them when ready. The preliminary work was completed, and the bids were formally advertised on October 8.

While Ellis, Derleth, MacDonald, and Harlan, the district's lawyer, worked out the details of the contracts to be let for the test borings, the bridge project came under attack from an entirely unexpected source. Major General Charles Jadwin, chief of the Army Corps of Engineers, a man who had approved the initial proposal for a bridge at the Golden Gate and had reaffirmed this approval to Welch as recently as 1927, stated emphatically, in making his final recommendations to his superiors on the day of his retirement, that he was opposed to the construction of the bridge for military reasons.

This statement was first reported by San Francisco Mayor Rolph, who had been visiting in Washington at the time and who, concluding that Jadwin's feelings represented those of the War Department, apparently panicked. The army chiefs had altered their opinions about the Golden Gate Bridge, Rolph reported in a letter to the directors, and would "fight any attempt to connect San Francisco and Marin County."

Welch, who was in San Francisco and who knew something of Jadwin from negotiating with him over the bridge before, sent a telegram to Filmer, who, along with Strauss, was in New York, urging them both to go to Washington and determine whether Jadwin's opinion was indeed the War Department's opinion. This Filmer proceeded to do. Calling upon California Senator Samuel J. Shortridge on October 15, Filmer asked the senator to arrange an appointment with the secretary of war for the following day. The appointment was granted: it was indeed a simpler time.

On October 16, Filmer, accompanied by Shortridge and Strauss, called on Secretary of War Goode and General Deakyne, the Corps of Engineers officer who had held the original bridge hearings in 1924. Referring to the permit that had been granted following those hearings, Filmer wired Welch the results that afternoon: "SECRETARY GOODE SAID THAT GOVERNMENT REGARDED THAT INSTRUMENT AS A CONTRACT AND WOULD CARRY OUT ITS PART IN GOOD FAITH AS LONG AS WE COMPLIED WITH THE PROVISIONS THEREIN CONTAINED STOP GOODE TOLD US TO GO AHEAD WITH THE WORK AND DEAKYNE LIKEWISE."

Time, as Ellis had been telling the directors, was of the essence. Eleven years had now passed since Welch's initial bridge resolution had passed the San Francisco Board of Supervisors. Any delay now would only make the idea of a bridge seem more remote and unlikely and lend further encouragement to the opposition.

The opposing forces who needed the least encouragement were the dissenting taxpayers from four of the Bridge District counties. On October 23 their attorneys filed an appeal before the U.S. Supreme Court, asking that body to pass upon the decision of the California Supreme Court holding that the district was legally organized. The taxpayers contended that they had been included in the district without their consent and that they had not been given a chance to vote on the question of taxes on their property. The court agreed to take the matter under study.

In this atmosphere of haste and uncertainty, with the plunge of the stock market to new lows daily dominating the national news, the prospect of beginning physical work at the bridge site itself must have seemed a relief, an end to delay and frustration, and the beginning of tangible accomplishment. The bids, opened on October 28, included a low proposal from the E. J. Longyear Company of Minneapolis. Longyear's bid was accepted, the contracts were signed, and the preliminary arrangements were made. On November 25, 1929, the first diamond drill bit into the rocky soil on the San Francisco side of the strait. Work on the Golden Gate Bridge had finally begun.

"MR. STRAUSS GAVE ME SOME PENCILS . . ."

I

With its mile-wide stretch of moving, swelling water, backed by the sudden upthrust of the Marin hills, its intermittent rivers of fog, and the broad sweep of the Pacific opening out on one side, and the bay on the other, the Golden Gate at San Francisco is one of the world's great natural stage settings. It lends an aura of drama to anything happening there. This in part was why, on December 9, the day the test drillings for the water foundations began, more than a thousand spectators appeared at either shore of the strait.

The land tests, begun two weeks earlier, were in the nature of a warmup. Behind Fort Point, a red brick Civil War–era army blockhouse which was patterned on Fort Sumter but had never fired a shot in anger, a small crew of workmen in Depression-uniform cloth caps and overalls had set up a ladder-sided rig, like that used by oil wildcatters, and had quietly gone to work. It had attracted only modest attention. Now, however, all had changed.

Out in the strait some one thousand yards from the San Francisco shore, a large dredge was anchored in place; powerful motors were ready to be started, great shafts about to be sunk beneath the waters of the strait. The first steps were being taken toward the mastery of the Gate itself.

On the shore at Fort Point, the San Francisco Municipal Band was playing, while on both the Marin and San Francisco sides of the bridge a full program of speakers made addresses that were carried by radio throughout California and up to the Oregon border. Joseph Strauss, as chief engineer, was the principal speaker at both sides of the strait. An original song entitled "The Bridge Across the Bay," written by Mrs. Sigfrid Millhauser, was sung by Mrs. Millhauser, accompanied by the San Francisco Municipal Band. Supervisor Dan Gallagher, presiding at the San Francisco shore, pronounced the song "a happy inspiration." When the starting switch was thrown, operating the motors in the diamond drill aboard the dredge, and the band struck up its most stirring air, while radio listeners strained their ears to hear through the static, the crowd at Fort Point let up a mighty cheer. It was as if years of accumulated anticipation and frustration had come to an end.

Strauss, returned from the East, had plunged into his pet project with renewed vigor: the stimulus of travel always seemed to restore him. It seemed he was involved in everything: chairing meetings, making speeches, meeting the press, conferring with experts. Strauss seemed to be everywhere, but then, abruptly, he was gone. His presence was intermittent. A few days in one place and he was off, promoting another project or overseeing one in progress. It seems that, as much as winning the chief engineer's job had meant to Strauss, his real interest was more in promoting the bridge than in building it. Indeed, throughout this period, the man who was there on the job every day was Ellis. He was at the test site, not just on dedication day but every day. He oversaw the borings as they were made, making sure the contractors adhered to the guidelines established by the surveyors. He met not only with Strauss when Strauss was in town but with MacDonald and Lawson and Captain Savage, the army engineer, and the concrete man who wanted to talk about aggregates. He was the man on the line, the man people came to each day with questions, expecting to leave with answers. He was also, presumably in his free time, supposed to be thinking about the mathematical formulas involved in developing the design for the bridge.

Ellis's relationship to Strauss had become like one common on ocean liners, where, because of the press of social functions, there are actually two captains. One, the ceremonial captain, hosts different passengers at the captain's table at dinner, takes people on tours of the ship, and poses for photographs. The other, the working captain, oversees the day-to-day running of the ship and the performance of the crew. Charles Ellis had become very much the working captain on the Golden Gate Bridge, and if he felt at all neglected by the lack of official notice, it was something he was too busy to devote much thought to; more than that, it was fundamentally contrary to his nature to complain. Besides, he was enjoying himself enormously.

One day, while the borings were still going on, Ellis was meeting at Fort Point with Professor Andrew Lawson, the Berkeley geologist who would be analyzing the samples from the test borings, when the question of earthquakes arose. Professor Lawson, who, with his snow-white hair, walrus mustache, and horn-rimmed spectacles, looked like a cartoon absentminded professor, had made a more thorough study of earthquake questions than any other man in America, according to Ellis. Now, as the two men watched the drillers at work on the rise behind the fort where the giant cables of the bridge were to be anchored, Lawson asked Ellis what would happen if one of the anchorages should instantly slip six inches.

Ellis, a man who was at home with questions of structural engineering and who had been devoting concentrated study to the behavior of cables, replied that as far as the vibrations of the bridge were concerned, a slip of six inches instantly would be the same as if the anchorage slipped twice as much gradually.

They were two professors talking now, illustrating points of theory by using simple anecdotes, working preferably with the material at hand, as one would in a classroom. "Supposing I have a fish," Ellis improvised in explanation, inspired no doubt by the fishermen who invariably line the seawall near the fort, "and I lay my fish on a scale pan, letting go of it gradually as the pan sinks. When I have completely released it, and the pan is carrying the fish, the recorded weight is, say, five pounds. Now, if instead of letting go of the fish gradually, I hold it down very close, practically touching the pan, and release it instantly, the scale sinks twice as far and registers momentarily ten pounds. The scale vibrates back higher than its original position and down again, back and forth, until it comes to rest again at five pounds." The question, in other words, becomes "what would be the effect if one of the anchorages slipped a foot gradually." The answer, said Ellis, "is that nothing would happen—the practical effect would be a lengthening of the cable by one foot"—something that could happen anyway with a simple rise in air temperature.

Lawson, clearly enjoying this interplay of theorizing and pedagogy, next asked Ellis what would happen if one of the main piers dropped instantly six inches. The answer again, said Ellis, "is that nothing would happen that would seriously affect the bridge. It would simply increase the length of the cable slightly. The structure is a limber structure and is susceptible to considerable variations in deflections."

This was for both men like faculty club conversation at its best, an interchange of disciplines between peers, yet more invigorating than that because it was real, because they stood twelve miles from a major earthquake fault, discussing its possible effect on a structure their ideas were to help create.

Professor Lawson proposed a more difficult question: What would happen if one of the piers were to be moved sideways, to slide transversely? According to studies of earthquake situations in Japan and San Francisco, scientists had arrived at the conclusion that for structures built on any kind of soil and made up of heterogeneous material, such as steel, terra cotta, brick, and concrete, the effect of a severe earthquake shock is equivalent to a force equal to 10 percent of the acceleration of gravity. What would happen to the tower above the pier in such an event?

Ellis replied that this bridge would be built of homogeneous material and on rock, and therefore 5 percent would give an equivalent force of a severe earthquake. Say the total dead load at the top of one tower made up of two posts was 100 million pounds. Applying a transverse force of 5 million pounds at the top of the tower would not exceed the working stresses at nearly all points in the tower. Double the transverse force to 10 million pounds and the working stresses might be exceeded, but they would not exceed or even equal the yield point of the material.

Ellis, who was to refer to this conversation later in the year, in an address before the National Academy of Scientists, summarized his feelings regarding the bridge and earthquakes to the scientists as this:

> *If I knew that there was to be an earthquake in San Francisco tomorrow and I couldn't get into an airplane and had to remain in the city, I think I should get a piece of clothesline about 1,000 or 2,000 feet long, and a hammock, and I would string it from the tops of two of the tallest redwoods I could find, get into the hammock and feel reasonably safe. If this bridge were built at that time, I would hie me to the center of it, and while watching the sun sink into China across the Pacific, I would feel content with the thought that in case of an earthquake, I had chosen the safest spot in which to be.*

2

On February 3, 1930, Joseph Strauss proudly announced to the press that a rock formation so hard that it had worn away more than three carats of diamond drills had been discovered at Lime Point, the proposed Marin County pierhead of the Golden Gate Bridge. The rock in this leg extending out from the shoreline was so hard that it had kept an expert diamond setter working continuously. Its discovery fully vindicated earlier reports made by the bridge geologists and, said Strauss, "clears the way for work to commence within the next two weeks on final plans and specifications."

The Bridge District directors were elated. The findings of the drillers had refuted the testimony of engineers employed by opponents of the bridge, who had repeatedly asserted that no suitable foundations for so heavy a structure existed on either side of the Gate.

Privately, the conclusions drawn from the test drillings were somewhat less enthusiastic, particularly those of Professor Lawson, the consulting geologist. The rock on the Marin shore, Lawson informed Strauss, consisting partly of sandstone and partly of greenstone, was indeed strong enough and amply sufficient to support the bridge pier. The south pier, to be set in the strait some one thousand feet off the San Francisco shore, would be founded on serpentine, which is not uniform in strength. Because of the effects of water, the rock had suffered internal shearing, so it now "consists of an aggregate of ellipsoids or spheroids of strong rock embedded in a matrix of sheared rock of very small strength."

While the drilling was going on, Ellis had conducted a pressure test at Fort Point on a typical stretch of the same kind of serpentine. He had found and demonstrated to Lawson that the rock could sustain a load of 32 tons per square foot, or 450 pounds per square inch. Since the supporting strength of the same kind of rock below the pier, confined by concrete, would be in excess of this, Lawson concluded that the foundation rock would be adequate, providing "the excavation for the foundation be not less than 25 feet below the lowest point of the rock surface."

It was the south *anchorage* of the bridge, on the hill behind Fort Point, that represented the greatest problem to Lawson. Here the rock was made up of "serpentine in a rather badly sheared condition with feeble coherence and very low tensile strength." Because of the structural weaknesses in the rock, the south anchorage of the bridge, binding its two great supporting cables to the San Francisco shore, would, in Professor Lawson's view, "have to be designed to depend upon dead load (the weight of the concrete base anchoring the cable) rather than upon the tensile strength of the rock, or the frictional resistance of the latter to the pull of the anchor."

Professor Lawson also had some chillingly fatalistic conclusions about earthquakes, pointing out that any earthquake so violent that it would destroy the bridge would also destroy San Francisco. Yet, "tho it faces possible destruction, San Francisco does not stop growing and that growth necessarily involves the erection of large and expensive structures."

More surely destructive, Lawson pointed out, would be the fact that the life of a steel bridge near salt water is limited and that, from an economic view, the fact would have to be faced that the bridge would have to be replaced once or twice a century owing to deterioration by rust.

Lawson's conclusions, while favorable, did not provide an airtight argument in support of the bridge. People who wished to take issue with the conclusions could conceivably find openings here, and with a bond issue election on the horizon, Strauss and the directors could be sure that the opposition would exploit any weakness. O'Shaughnessy in particular, it now appeared, would be aligning himself with the anti-bridge forces. Under these circumstances, Lawson undoubtedly felt himself under pressure to produce the most conclusively favorable geological report possible. At the same time, he was expected to adhere to the academic standards of his profession, yet he was on the faculty of a university whose president and engineering dean were firmly in support of the project. On March 7, before his report had been completed or submitted to the bridge directors, Lawson abruptly resigned.

Derleth, apparently the first to get the news, spent two hours talking to the geologist on the telephone. Then he called Ellis, who called Strauss in Los Angeles. Strauss called Derleth. The following day, Derleth met with Lawson at eight-thirty in the morning; the two men talked for two hours. When the meeting was over, Derleth telephoned Ellis: the consulting geologist was back on board. In Lawson's geology report, submitted on April 7, the critical remarks and conclusions were included.

Like a military invasion, with Strauss its flamboyant, peripatetic commander-in-chief, the bridge project was now proceeding on several fronts at once. On March 3, the Supreme Court dismissed the appeal of the four counties' taxpayers challenging the legality of the bridge district, ruling that, in this matter, no federal question was involved. The way was now cleared to call a special election on the bond issue. Strauss, who summarized Lawson's as yet unsubmitted report as stating that the underwater foundations "are the finest kind of rock" for both the pier and the anchorages, said that he expected that final plans and specifications for the bridge would be ready within a week or two of the receipt of the Lawson report. "Strauss," the newspaper report concluded, "is out of the city and will not return until Monday."

Within the Engineering Board, the emphasis had shifted from making theoretical decisions to designing the structure itself. Nine years had been spent fine-tuning and deliberating Strauss's original design for the bridge. Now a new design would have to be prepared, from scratch, in a matter of months. In their deliberations, extending through several sessions and amplified by correspondence, Strauss, Ellis, and the consulting engineers had arrived at certain overall conclusions about the type and style of the bridge. These guidelines, reached amid the deepening shadow of the Great Depression, show an increasing obsession with economy, and many of them were dictated by necessity.

It was to be an all-suspension bridge, since that was the most economical kind to build and, according to Moisseiff's recent discoveries, also the most appropriate. There would be two piers, one in the strait, 1,100 feet off the San Francisco shore, and the other at the Marin shoreline. Again, this positioning was justified by its saving of a million dollars in construction costs. For the sake of symmetry, it was decided to make the two side spans —the parts of the bridge stretching from shore to tower—the same length: that of the stretch required between the San Francisco tower and the shore at Fort Point. Since the topography on the two sides of the strait differed, there would be no savings in using straight backstays in the side spans. Therefore, Strauss was able to recommend the more aesthetically satisfying curved backstays.

In the interests of lightness and economy, the rail lines proposed for the original roadbed would be dropped. This was justified by the conclusion that the day of the trolley as the most widely accepted means of mass transit was ending and that rail vehicles everywhere would be replaced by cars and buses. It may have been one of the few instances when the engineers' projections would later be proved wrong, although the jury on rail mass transit is certainly still out.

The roadbed would be made of concrete slabs, seven inches thick, weighing eighty-five pounds per square foot. If, at some later date, a slab of sufficient strength could be designed of lighter aggregate, the concrete could be replaced and the live-load capacity of the bridge increased. The floor beams would be designed to carry six 24-ton trucks abreast, with the heavy axle loads directly over the floor beam. There would be wide, ample pedestrian sidewalks. The versine, or sag of the cable, which can vary in suspension spans from one-eighth to one-eleventh of span length, could in this case vary from 380 to 525 feet. The smaller the cable, or the higher the tower it was suspended from, the greater the sag. Again, in the interest of economy—the thicker the cable, the more expensive it would be—and also appearance, it was decided to compromise on a versine of 475 feet. The towers, it was agreed, should be made of steel.

"All these decisions," Charles Ellis recalled a few months later, "were the result of deliberations of the board through several sessions. All the major questions requiring experience and sound judgment having been settled, there was little left to do except to design the structure. At this point, Mr. Strauss gave me some pencils and a pad of paper and told me to go to work."

"SOME VERY MYSTERIOUS THINGS . . ."

7

I

Charles Ellis's dry sense of humor was definitely at work when he made his statement, in a speech to a group of scientists, about the simplicity of designing the bridge. It was a simple problem, but only once you understood it and had it thoroughly defined. This was something new. Moisseiff understood it, and Ellis, and that was about all. Derleth, who admitted it, did not understand it; nor did Strauss, who never would have admitted anything of the kind. Ellis, working now mostly in Chicago with the Strauss staff, and in consultations with Moisseiff in New York, was now the man completely on the line.

An engineer, any engineer, is a man always working within the state of the art—the sum of the achieved knowledge and accrued experience of his time. Ellis was now going to jump across it—go beyond the state of the art. Assumptions had been made about the way certain things must be done. Now Ellis was challenging those assumptions.

According to Ellis, any young engineer who has mastered the fundamental theories of moments, shears, and equilibrium or coplanar forces should have little difficulty in computing the stresses in a plate girder or simple truss. With a few years of experience, he can design a girder or truss himself. "He will find that it is but a step further to the analysis and design of a cantilever bridge." If he advances to the study of an arch, the engineer meets one or two interesting new problems, but nothing for which his earlier work has not prepared him. When he comes to the stress analysis of a suspension span, however, the engineer "will find himself in a totally undiscovered country."

In undergraduate courses in bridge analysis, applications of fundamental principles are almost always based on the beam, plate girder, and simple truss types of bridges. The reason for this is that an engineering student, in order to develop suspension bridge theory, must of necessity have had more mathematics than is given in a four-year undergraduate program. "Of course, the young engineer can take the formulas as given in treatises on suspension bridges, but his success will be very doubtful, because any person who uses a formula which he cannot develop is on dangerous ground."

Even an engineer who has had sufficient math to master suspension bridge theory and understand it thoroughly will find, on his first application of it to a real problem, "some very mysterious things."

To begin with, a suspension bridge does not behave like any other type of bridge. It behaves like a clothesline: it sways with the wind, expands in the sunshine, and contracts when it rains or grows cold. The suspension bridge engineer is being asked to design a clothesline capable of supporting six 24-ton trucks, traveling abreast.

When an engineer plans a bridge's stiffening truss, the design must first be guessed at, a process known in the trade as "cut and try." The engineer must assume the number of square inches in the structure's truss components—its lengthwise chords and joining webs—and proceed with his computations based on that assumption. When these calculations are finished, he will be able to make a better-designed guess or assumption and repeat the process. Where the load is perceived to be greater, the stiffening truss can be made deeper, heavier, stiffer. But with a suspension bridge, this is reversed. Making the stiffening truss deeper increases instead of diminishes the chord area. "This conclusion will be somewhat of a mystery" until the young engineer realizes that the limberness of the stiffening truss is increased by reducing its depth or lightening its chords. In a suspension bridge, the truss literally shirks the load, handing it over to the cable through the suspenders. The stiffer the stiffening truss, the more hesitant it is to do this. Thus the stiffening truss of a suspension bridge should actually be a "limbering" truss, as light and shallow as the deflections will permit and remain consistent with good design.

Another mystery concerns the stiffening truss used in the side spans of a suspension bridge. Assume that the designer knows his dead-load (the weight of the bridge itself) stress in the cable under normal temperature. He then fully loads the side span and computes his live-load (the weight carried by the bridge) stress in the cable for the highest temperature. Since the maximum temperature causes the greatest deflection (deviation of the truss from its radius), it also causes the greatest moment (tendency to move) in the truss. When the engineer has his total dead-load, live-load, and temperature stress in the cable computed, "he will find to his astonishment that the stress in the cable is *less* with a full live load than when there was no live load at all."

The suspension bridge, behaving like a clothesline, lengthens under higher temperature and, therefore, decreases the stress.

In every aspect of its design, a suspension bridge challenges the traditional assumptions of bridge design, but in ways that require an understanding and appreciation, an informed view, of the principles being challenged. Some things considered essential become negligible. Other things, considered negligible, become crucial.

As to the wind stresses, according to Ellis, there had been no theory published anywhere that was even approximately correct for the analysis of them. At least there hadn't been until Moisseiff's breakthrough idea of balancing the horizontal deflections in the cable of a suspension bridge with those in the stiffening truss. "Now it takes a genius to discover a new idea," Ellis conceded, "but it doesn't take very many brains to make slight improvements in it."

Basing his calculations on Moisseiff's theory of balancing stress loads between truss and cable, Ellis segmented the stiffening truss into parts. He let A, B, C, etc., equal the load that the stiffening truss carried at each point. The load that the cable carried at the corresponding points would be 1,300 pounds per square foot (the known dead load) minus A, B, C, etc. "This process," said Ellis, "involved the solution of as many independent simultaneous equations as there were points chosen; but once the equations were solved, it was unnecessary to make a second trial or guess."

Once again Ellis, in the interests of clear explanation, was oversimplifying. What he was doing was devising new algebraic formulas (he came up with thirty-three in all), working with as few as six and as many as thirty unknown quantities. Ellis's calculations had to take into account not only the shape and structure of the bridge and the forces of heat, cold, and winds, but had to calculate, in advance, the amount of stress to be borne by each of the bridge's hundreds of suspender ropes. He was not merely entering unexplored territory; he was mapping it.

In his computations for the bridge towers, Ellis was also able to incorporate recent findings that further improved the bridge's lightness, flexibility, and economy. "In considering any type of bridge," Ellis explained, "the real test is not whether the structure can be designed, nor indeed whether it can be fabricated in the bridge fabricating plants. The real test is whether the bridge, after it is fabricated, can be shipped piecemeal to the site and there be put together."

At this time, and for this type of bridge, the state of the art was the Philadelphia-Camden Bridge towers, which were made of steel cells. Ellis proposed for the Golden Gate Bridge towers that were similar in structural form, although very much larger. These could be made from either carbon or silicon steel, depending on the anticipated dead and live load.

Ellis based his calculations on a live load of 4,000 pounds per linear foot, a figure reckoned on an extreme estimate of the bridge being packed full of the heaviest automobiles imaginable—sixteen-cylinder Cadillacs, seven-passenger Packards and Buicks—bumper to bumper, and jamming all six lanes. Under these conditions, Ellis calculated the total live load would be slightly more than 3,000 pounds per linear foot. Of course, in practice, there would probably be four or five lighter Fords, Chevies, or Plymouths to every seven-passenger stretch sedan on the bridge, and the distance between cars, even in heavy traffic, would probably be ten or fifteen feet, so that in reality it would be difficult to imagine a load greater than 1,000 or 1,500 pounds per linear foot. Nevertheless, the towers were designed for a load over all spans of 4,000 pounds per linear foot.

In addition to the live and dead loads on the towers, allowances had to be made for bending, both lengthwise and sideways. When the center span of the bridge and the far side span were fully loaded and the temperature was lowest—in this climate, say, 30 degrees—the top of the tower adjoining the unloaded side span would bend 23⅓ inches toward the channel. When the *near* side span was loaded, with no load on the center and far side spans and the temperature was the highest, at, say, 100 degrees, the top of the tower would bend 6⅔ inches toward land. The total deflection, then, would be 30 inches. Ellis suggested that this be equalized when the tower was completed. "One way of doing this will be to attach lines to the top of the tower and literally pull it shoreward nine inches and hold it there while the cables are being placed on the tower." There would then be a uniform lengthwise deflection at the top of the tower of 15 inches.

The towers would also have to bend sideways, because of wind load. The bridge had already been designed for a wind load of thirty pounds per square foot of exposed surface on the floor system, including trusses and cables. And for fifty pounds per square foot on the exposed surface of the towers. The maximum actual velocity—as opposed to indicated velocity—of wind ever recorded by the weather bureau at San Francisco was, according to Ellis, fifty miles per hour, equivalent to a wind pressure of ten pounds per square foot. A wind pressure of thirty pounds per square foot is officially classified as a hurricane, strong enough to overturn empty boxcars. "We have used this wind load," Ellis commented dryly, "in designing the towers and trusses. This wind load causes transverse moments in the towers and also stresses due to the eccentric loading." Under a hurricane wind pressure of thirty pounds per square foot, the towers would deflect transversely, move sideways, about five inches. One can almost hear Ellis adding: big deal.

Ellis then considered what the stresses would be if all four of these conditions were combined at the same time. If, under some bizarre combination of circumstances, the whole bridge could be loaded with traffic to the weight of 4,000 pounds per linear foot but with only the center span and one of the side spans so loaded; and if, in addition, the temperature was either down to 32 or up to 110; and if this whole dead and live load were subjected to a wind strong enough to tip over a line of boxcars a mile long —if these four unlikely and even contradictory conditions should occur simultaneously, what would happen then?

"In an ordinary design," explained Ellis,

where dead and live loads are combined with wind loads, it is customary to increase the allowed unit stresses by at least twenty-five percent. This is done in all standard specifications. In this design, however, we have assumed that all these four extreme conditions may happen simultaneously. The structure has been designed so that in no part of the towers will the allowed unit stresses be equalled or exceeded, even should all four of these conditions happen at the same instant.

2

On March 1, Ellis had sent copies of his preliminary design specifications to the members of the Engineering Board for their comments. The engineers had approved of these early designs and figures, with certain specific changes. Strauss now directed Ellis to devote all his time to the detail computations. "Before I return West," the chief engineer wrote the three consultants, "sufficient progress should have been made on the general design to enable us to jointly determine a fairly definite schedule."

While Ellis, in Chicago, labored with the thousands of calculations for the suspension ropes, highway deck, floor beams, highway track and sidewalk "stringers," or longitudinal members, stiffening trusses, cables, and towers, Moisseiff, in New York, worked on calculations of his own. The two men checked figures with each other frequently. It was an enormous and complicated job; occasionally, when he was stuck, Ellis wired Moisseiff for help: "WHAT DO YOU CONSIDER MAXIMUM ALLOWABLE DEFLECTION SIDE SPAN?" one Ellis telegram began, followed by a description of the dimensions of the problem. To which Moisseiff wired back, "DEFLECTION SIDE SPAN SIX THOUSAND LOAD HIGH TEMPERATURE SHOULD NOT EXCEED SEVEN AND A HALF FEET." In the complexity of these communications, and their transmission of them electronically, both men were pushing at the limits of the technology of the time. It was work that, in a later era, would seem far too complex and immense for men to do without computers.

Ellis was being forced to accomplish an enormous amount of complicated and important thought in a very short time. "The pressure was constantly being brought to bear to rush the work," he confided later to Ammann and Moisseiff, "and give a definite date of completion. I always explained"—presumably to Strauss, his boss and the logical applier of the pressure—"as clearly as I could that it was impossible to make any estimate which could be relied upon; that this was not just another structure to be designed similar to many others in which the computations were more or less a mechanical process, but one that required considerable original thought."

Strauss was growing impatient with these delays. Undoubtedly he was feeling considerable heat from the Bridge District directors, who were now promoting an invisible bridge, one for which they had no agreed-upon, presentable design. He was also under considerable financial pressure. "Although I have made many sacrifices," he wrote to Moisseiff, who had complained about slow pay, "in order to insure success at the bond election, and the cost has become a considerable burden, I have not asked and do not now ask anyone else to make any sacrifices." As an employer, Strauss was sensitive to being imposed upon or taken advantage of. Impatient with theory, he had begun to feel that Ellis was magnifying his role by fussing unnecessarily over details. Hadn't Strauss put his design staff at Ellis's disposal? Couldn't Ellis delegate some of these detailed computations?

Ellis had indeed assigned some of the computations to a Strauss staff engineer, Charles Clarahan, who had been a student of Ellis's at the University of Illinois. He was, in Ellis's view, "an able chap, but this was, to the best of my knowledge, his first introduction to a problem of this nature." Clarahan, working on the computations for the bridge towers, had run into difficulty. He came to Ellis for advice. Ellis, under time pressure and busy with the bridge trusses, had intended to tackle the towers later. He told Clarahan that it would be best if, for the present, they continued to work independently, as Strauss had instructed. Clarahan spent some three months on his tower computations. When they were finished, he presented his results to Ellis. "You or I," Ellis wrote to Moisseiff, "would not have to look at them ten seconds to know absolutely that they are very much in error." Ellis did not mention this to Clarahan.

The meetings of the Engineering Board continued through the spring and summer of 1930. The engineers would gather wherever the press of business permitted them to meet, sometimes at Strauss's offices in San

Francisco, others times in the Strauss Engineering offices in Chicago, and most often in Strauss's suite at the Biltmore Hotel in New York. At these meetings, Strauss presided, with Ellis serving as secretary. Other men associated with the project, like Captain F. A. Savage, the army field engineer, and John Eberson, the consulting architect, also occasionally appeared. Strauss would report on the general progress and on the progress of the design to date, there would be a discussion and expression of opinions, and certain conclusions would be agreed upon, most of which Ellis was expected to translate into engineering design. The bridge was gradually assuming its desired proportions; but it was not happening quickly enough for Strauss.

There continued to be irritating interruptions and delays. Professor Lawson, the geologist, was recommending that the pier base go down twenty-five feet below the rock bottom surface; Strauss and the Engineering Board were trying to persuade him to settle for ten. There was a question of earthquake insurance and the desirability of an additional check of all computations on the bridge, to be conducted by Moisseiff and his staff. Nevertheless, Strauss assured Ammann, "the work at Chicago is progressing at full speed," and Strauss was diligently at work on his engineer's report, which was due, along with the completed plans and estimates, "on or about the 20th of June."

At this point, the time constraints worked both for and against the engineers designing the bridge, particularly Ellis and Moisseiff. While pressured with deadlines, they were also able to employ, to their advantage, the blank paper syndrome. Most people, confronted with a creative or design problem, are physically incapable of putting down, on paper, a solution to it. The very emptiness of the space becomes intimidating. Yet many of these same individuals, once an idea has been conceived and committed to paper by someone else, suddenly find release. Examining a preliminary drawing, often with pen or pencil already in hand, the consciousness that was blocked before now finds itself flooded with minor improvements that can be made, ways the bare-bones idea can be adorned, corrected, embellished, redirected, saved. They are suddenly eloquent.

The more time there is to do this, the more the formerly blocked consciousness may be stroked. But because of the pressures—financial, political, procedural—surrounding the design of the Golden Gate Bridge, the time and trouble required for this customary ego stroking were denied. Individuals who would have found the bridge design regrettably flawed or incomplete and in need of their personal improvement and embellishment were forced, because of the constraints of time, to let the design pass relatively unscathed. In short, they were desperate enough to buy something good.

There is another aspect to this, and it was definitely operating in the design of the bridge. Excellence stimulates excellence. The fact that an associate has done—in some cases, gotten away with—something exceptional can encourage others around him to try the same. The standard of the acceptable, the normal, is raised a notch, and the dimensions of the possible expand. The wall had been pushed back. Moisseiff's theory had done this for Ellis; now Ellis had the opportunity and the obligation to meet the standards of dedication and imagination that had been applied to the new theory of long-span suspension bridge design. The side must not be let down. The leap, once made, must be sustained. There is no turning back or reining in without plummeting to failure.

3

The most striking feature of the new bridge would undoubtedly be its towers, the tallest bridge towers ever built, rising in one of the world's most dramatic settings. Here was an opportunity and a challenge to move beyond traditional engineering design into something that could become a fulfillment of, rather than an intrusion on, its environment.

Bridge towers, even those used in suspension bridges, had traditionally been massive structures, great bulking works of steel or stone or combinations of both. In a typical design, two posts, rising from a pier, would be joined by a great arch over the roadway, or by a webwork of giant crossbeams, or again, by a combination of the two. In order to preserve the feeling of massiveness and solidity, tower details were often designed to be larger at the top than at the base to preserve the look of uniformity against the eye's adjustments for distance and perspective. Although Moisseiff's and Lienhard's theories had already changed the nature of suspension bridge cables, suspenders, and trusses, tower design tended to lag behind. The towers of the George Washington Bridge, for example, were still the traditional Erector-Set crosshatch of exposed steel beams.

At the same time, changes in metallurgy, extending the allowable narrowness and flexibility of suspension bridges, were suggesting new, lighter, less cluttered possibilities for bridge towers. In the Philadelphia-Camden Bridge, completed in 1926 and later renamed the Benjamin Franklin Bridge, the towers had been built of steel cells, compartmented honeycomblike members rather than solid steel beams. These reduced weight, while giving the bridge overall a smoother, less exposed and mechanistic appearance. Instead of a ponderous arch or a crosshatch of beams, the Philadelphia-Camden towers were characterized by three great steel-cell cross-members, each set within a rectangle, and bordered by crossbars of smaller cross-members above the roadbed and at the top of the tower. Much less massive than the traditional suspension bridge tower, these slender, simplified structures gave the bridge an aesthetically appealing, light look. The engineer of design on the Philadelphia-Camden Bridge had been Leon Moisseiff.

On the Ambassador Bridge at Detroit, built between 1927 and 1929, with Moisseiff serving as consulting engineer, this overall slimming and lightening of the towers was taken one step further. At Detroit, the rectangles were removed from around the giant steel-cell cross-members connecting the tower posts, so that the giant crosses joined each other in a way that formed great diamond-shaped windows between the towers. At the same time, the cross-members were extended below the roadbed, leaving only two smaller cross-beam borders at the top of each tower and at the top of each pier. Already a new suspension bridge aesthetic had emerged, one that emphasized light, air, and simplicity as opposed to solidity, complexity, and mass.

Ellis, the author of a textbook on the theory of frame structures, was keenly aware of these developments in his field. He was working with the man most responsible for them. Now was his chance to advance the state of the art of tower design. In the Ambassador Bridge, the giant cross-members had been extended below the roadbed; now, why not confine them there entirely? The Golden Gate had to have the highest over-water clearance of any bridge ever built: there would be room for two huge cross-members, with cross-beams, between the pier and the roadbed. With such a strong base, the towers above the roadbed, already the tallest of any bridge, could be made to seem even lighter, longer, and more graceful. If the giant cross-members were dropped from the towers above the roadbed, they could be replaced by the smaller rows of cross-beams at intervals between the towers and covered by aluminum alloy housings, which would greatly reduce corrosion and remove the last trace of Erector-Set mechanics from the super-structure of the bridge. Where the giant cross-members between the towers had been, at Philadelphia and at Detroit, there would now be great open rectangles, huge windows or portals framing the intense blue sky, the scudding clouds, and the frequent swirling fogs of the Golden Gate.

The idea of the portals and the use of cellular construction inspired another innovation, this one the suggestion of Joseph Strauss. "Referring to the towers," Strauss wrote to Ammann in March, "I have in mind the stepped-off type of architecture."

The practice of stepping off, or indenting structures as they rose in order to accentuate the feeling of soaring height, was a practice increasingly in vogue in other kinds of architecture, particularly the Chrysler Building and the proposed Empire State Building in New York. It was an idea incorporated into the towers' architectural treatment by the first of the bridge's two consulting architects, John Eberson.

Born in Cernanti, Austria, in 1875, Eberson was an electrical contractor who became a self-taught architect, designer, and construction superintendent in St. Louis between 1901 and 1904. Here he became interested in creating effects and illusions for the theater, a specialty he parlayed into the founding of his own architectural firm, specializing in designing the lavish theaters being built for the combined showings of movies and vaudeville. In this world of calculated illusion, with its elevating orchestra pits, cavernous lobbies, and elaborate Moorish or Italianate interiors, Eberson thrived, spreading the idea of the atmospheric theater across America, as well as to Britain, France, and Australia. In the course of designing hundreds of theaters in America and abroad, Eberson's firm, eventually centered in New York, introduced and popularized the elements of a cleaner, more modern theater style: floating screens, the use of color in place of ornamental plaster, cloth sidewalls, fluorescent murals utilizing black light, plastic changeable marquee lettering, and structural glass in exterior design. Eberson, who had never worked on a bridge before, was an influential voice in developing an overall design for the bridge with simple and pleasing lines. He was particularly concerned with the architecture on the towers and anchorages. From his experience in the theatrical world, Eberson understood how architectural touches can be used to heighten subtly a structure's dramatic effect.

At the top of each portal, where the housed cross-beam connected the two posts of the tower, Eberson indented the tower so that, in a complete reversal of traditional bridge design, the tower actually grows smaller as it rises into the sky, thus intensifying its soaring, ladderlike feeling of ascension. In this initial design, however, unlike the final designs, all the portals were not rectangular: the bottom portal of each tower—the one above the roadway—and the top portal, at the tip of the tower, were designed as parabolic arches, perhaps as a gesture to the earlier Eberson world of re-created mosques and minarets. Eberson also added Art Deco embellishments, like the suggestion of a proscenium, at the corners of the portals.

The bridge as now designed was, in its cleanliness of line and soaring grace, a step beyond any bridge ever built before; yet there was something still lacking: an overall consistency of vision, a firm thematic expression of its identity. This the bridge was to receive from a source that seems to have been largely unanticipated.

4

At the Strauss offices in Chicago, Ellis, in addition to calculating the dimensions of the materials for the bridge, was also computing estimates of their costs. Using figures from the George Washington Bridge job, which Ellis felt were too high, he had to make estimates for everything from the per-yard cost of excavation to the wire wrapping used on the cables. These figures also had to be checked by Moisseiff, and the added time this took further delayed the completion of the final plans and estimates, which in turn further irritated Strauss. The June 20 due date came and went; so did July 20. "I have not yet received from Mr. Ellis the final total cost," Strauss wrote Ammann on July 28, "nor can this be done until the Consultants check estimate is completed." He was determined to submit his engineers' report at the next meeting of the board in August, even if it meant that the amount of the total cost had to be left out. The possibility of completing the bridge plans in time to hold a special election on the bond issue had vanished; now they would have to struggle to make the statewide election in November.

It was at about this time that Strauss received another surprise, this time from the federal authorities, who had realigned the road leading south from the bridge in such a way that it would require a completely redesigned architectural treatment for the San Francisco bridge plaza. To make this change, Strauss, in a rush to get the bridge design done, decided to hire Irving F. Morrow as the second of the bridge's consulting architects.

Morrow was a local architect who practiced in San Francisco in partnership with his wife, the architect Gertrude Comfort Morrow. Like Eberson, he had never worked on a bridge before; unlike the New Yorker, Morrow did not possess a national reputation. There were, however, other considerations: Eberson had submitted an estimate for a complete architectural treatment that Strauss considered too high; he had been planning to replace the New York architect with someone less expensive. Also, Morrow was prominent in local architectural circles, chairman of the Commonwealth Club of California's section on architecture, president of a local Committee on the Arts, and a contributor to regional and national architectural publications. "His opinion carries great weight here," Strauss shrewdly advised Ammann and Moisseiff. "Likewise the influence of a local architect in the bond election is important."

Morrow's design work to this point had been confined mostly to houses, schools, banks, hotels, and theaters, with the exception of assisting in the design of the Court of Ages at the 1915 Panama-Pacific Exposition. What Morrow did possess, and what may not have been apparent from his work up to this time, were a talent and sensibility uniquely suited to this particular job.

Born in Oakland, California, in 1884, Irving Foster Morrow was the son of the proprietor of a metalworks. The younger Morrow displayed an early and active interest in the graphic arts, producing original drawings and paintings for an Oakland literary magazine, *The Muse,* as early as 1900. As an architectural student at the University of California at Berkeley and later at the Ecole des Beaux Arts in Paris, Morrow continued to cultivate his early, painterly sense of color and perspective, combining this with an architectural fluency that was to allow him to work on an unusually broad range of projects in the course of his career.

By the late 1920s, Morrow was working in his San Francisco office and commuting back and forth to his home in Oakland, across the bay, by ferry. On these trips, gliding out of an East Bay slip in the morning and churning away from the west-setting sun at night, Morrow had more opportunity than anyone else who worked on the design of the bridge to observe the changing interplay of light and shadow at the only break in the hills around the bay, the Golden Gate. It was a site that Morrow had been aware of since boyhood, had pondered as an architectural student in Berkeley, and had been moved sufficiently to write about, in 1919, with a quality containing elements of both poetry and prophecy:

> *On the further shore, the long peninsula of the great city and its counterpart to the north all but touch save for the narrow passage which leads between the hills to the shimmering ocean. Distance is possessed of magic powers which integrate the most disparate elements into a symbol of perfect peace.*

It was in the summer of 1930, while the rush to design the bridge was already under way, and while Ellis and Moisseiff were wrestling with the computations and engineering design, that Strauss, seeking, for whatever his reasons, an acceptable West Coast architect, asked Charles Duncan, his public relations counsel, to recommend someone. Duncan said he would consult with his brother-in-law, the respected California painter Maynard Dixon. It was Dixon, who was at this time working on an oil painting of the general perspective of the bridge, who suggested Morrow. Once again, under the pressures of time and necessity, an informed sensibility had been applied to the building of the bridge. The one man who had probably observed this site under a wider variety of conditions than any other architect alive had, almost by a fluke, been given the opportunity to work on its architectural treatment.

Morrow's contribution to the architectural plans included in the chief engineer's report, now being rushed to completion, was confined principally to the entryways and plazas at either approach to the bridge. Yet even in this preliminary proposal there are sketches with a dramatic improvement in the towers: the portals had now been made rectangular, of similar shape but varying size. This was to give the finished bridge its fascinating, painterly sense of perspective so that, seen head on, each tower seems instead a row of towers, of ascending height, extending away in depth behind the first one.

Morrow's original plans called for two elaborate plazas, one at each terminal of the bridge, with not only the usual buildings for bridge personnel and public restrooms but halls for memorial purposes and exhibits. The entry portals to the bridge itself would be faced with terra cotta and decorated with murals and sculpture. The color of these entryways—broad, angled walls narrowing toward the entrance of the bridge—would be "in fully polychrome, increasing in richness by successive steps as it focusses upon the bridge." The pylons—the last solid land structures the driver, passenger, or pedestrian passed before entering onto the bridge itself— would be in "pure metallic gold—veritably, as well as symbolically, the Golden Gate."

These proposals, conceived in a very short time and later scrapped because of cost, indicate the amount of imaginative energy the idea of the bridge released in Morrow, and suggest the striking, if simpler, visual treatment he was to add in the future.

5

While the engineering and design plans were in progress, an eager watch was kept on them by the Bridge District Board of Directors. There was first of all Derleth, who, through MacDonald, was keeping the directors informed of developments within the Engineering Board. In addition, there was the situation, common among directorships, in which a few members lead the way, with the rest of the board generally falling into line. Richard J. Welch in particular, with his moral authority as the earliest elected advocate of the project and his political clout as a congressman, seems to have been the official liaison between the directors and their engineers at this point. There was a certain grass-roots appropriateness to this: Welch had originally entered San Francisco politics from a job in an ironworks.

On July 9, Welch assured the directors at their monthly meeting that design work was proceeding rapidly and that prospects were bright for wrapping up final details "preliminary to actual construction work on the bridge structure." The directors were understandably anxious: it was now only four months to the next general election, for which they expected to propose a bond issue for a bridge whose design they did not yet have in hand. Welch may not have relieved these anxieties by reminding the directors that the project had now assumed a nationwide importance, or by predicting that the electors of the district would vote the bonds by an overwhelming majority. There still had to be a bridge.

Thus it was with understandable relief that on August 21 a call was issued for a special meeting of the board to be held on August 27 in the chambers of the Board of Supervisors at the San Francisco City Hall. Its purpose was (1) to receive, consider, file, and adopt the report of the chief engineer, and (2) to call a special election, submitting to the voters of the district a proposition to issue $35 million in bonds.

On the appointed day, it was Joseph Strauss, accompanied by Derleth, who presented the engineer's report to the board. They were also joined by MacDonald, the general manager, and Harlan, the Bridge District's legal counsel. Neither Ellis nor Moisseiff was present.

The report, together with architectural studies and the fact-finding investigation conducted by the Board of Directors, runs to some 285 pages and includes everything from a prefatory eulogy by the San Francisco poet George Sterling, who had committed suicide in 1926, to engineering drawings and specifications, to cost breakdowns, to Lawson's geologist's report. Capitalizing on the nine years of promotion, frustration, negotiation, and imagination that had preceded his report, Strauss had approached the question of the bridge from every conceivable angle. Each of the elements or factions anticipated as being in opposition to the bridge is addressed with a full armory of factual weaponry: the antidevelopment interests, the shipping interests, the ferry boat interests, the taxpayer, military, political, aesthetic, earthquake, finance, travel, traffic, commerce, and future concerns of an entire metropolitan area are touched upon in this document. Clear, direct, long on information, mercifully short on grandiloquence, yet with a firm sense of the emotions involved and a sure grasp of the momentousness of what is being proposed, the report is an example of Strauss's generalship at its most effective. Here he had marshaled his promoter's positive vigor, his salesman's sense of the territory, his Napoleonic ego, and his ability to recognize and attract to his cause people of genuine ability, all in behalf of the great work he had yearned to be part of since his boyhood days in Cincinnati.

There is, however, another aspect to this report, and it sheds a certain light on why engineers of solid credentials and national reputations had such misgivings about working with Strauss.

In volume 1, section 1, article 8, which describes the engineering personnel proposed for the bridge, Strauss is listed as possessing the "degree of C.E., University of Cincinnati, 1892." While it might be claimed that this referred only to Strauss's liberal arts B.A. in commerce and engineering, it is obviously intended to show comparable academic status to the "degree in C.E." listing for O. H. Ammann, Charles Derleth, Jr., and Leon Moisseiff, all three civil engineers of national distinction.

Additionally, Strauss, in the same article, is described as consultant to the New York Port Authority on the Bayonne Arch, which was to be completed in 1931, and the Hudson River (George Washington) bridge, which was to open in 1933. According to John Hughes, publicist for the NYPA, Strauss was not involved with the George Washington Bridge at all, had no bearing on it, and is not listed among its engineers. And while there are no records of consulting engineers on the Bayonne Arch, the NYPA's staff expert on the bridges, whose experience goes back more than 30 years, "has never heard of Strauss being connected with that or any of our other projects."

Since both bridges were directed by Ammann and, in part, designed by Moisseiff, perhaps Strauss felt he could claim consulting engineer credit for having offered a few opinions or suggestions. Or he may have assigned Ellis, his employee, to work with the other two civil engineers and, through the Strauss firm, assumed credit for Ellis's work. What is clear is that Strauss, especially sensitive on the grounds of professional credentials and attributed work, was more than capable of ethical misdemeanors and sharp practice.

There is, in this report, which was written by Strauss, no indication of Ellis's role in conceiving the overall design for the Golden Gate Bridge. Instead Ellis is described, along with Clifford Paine, another Strauss engineer, as one of Strauss's two chief assistants on the project. There is no special thanks, no individual credit, no division of labor as to who was responsible for what. The bridge design is represented as being the work of the Strauss Engineering Corporation: in other words, Joseph Strauss.

What had been achieved, up until now, in advancing the bridge project had been due almost entirely to the promoter in Strauss. The speeches, the newspaper articles and editorials, the crafting of a government framework with which to manage and finance the project, the crucial approvals by county, state, and federal governments—all this had been accomplished with precious little engineering and critical analysis. It had come largely on the authority of Strauss's reputation as an engineer and on the basis of hastily drawn estimates, erroneous ground surveys, and a paper sketch of an ugly structure of mixed parentage. Now Strauss's dream of a great bridge was close to being funded. The desired state of assured financing would require a jaundiced appraisal of the technical foundations of a project that took it beyond the limits of Strauss's ability as an engineer. What was required now was the exacting and inspired thought of an engineer who comprehended both the theory and technology necessary to make the dream true, who could make possible a bridge within the feasible price range Strauss had recklessly promised. In Charles Ellis, Strauss had found the man who had delivered an engineering design that had surpassed Strauss's most fervent hopes. Yet just as it was not in Strauss himself to conceive such a design, it was not in him to delegate or even share recognition for it. In assuming full credit for what was actually Ellis's work, Strauss the promoter, now fully ascendant, had overruled Strauss the engineer.

Moisseiff, who understood the significance of Ellis's work more than anyone else, fully appreciated the irony and potential unfairness of this situation. On August 19 he wrote two versions of a letter expressing his verification of the fundamental data on which the design of the main structure was based.

The first version was addressed to William Filmer, president of the Bridge District. It begins:

Dear Sir:

In accordance with the request of your Board of Directors, I beg to submit to you herewith, the results of the independent check made by me of the design and preliminary plans and specifications of the Golden Gate Bridge.

I and the engineering staff under my direction have carefully examined and checked the design of the Golden Gate Bridge as prepared by Joseph B. Strauss,

Chief Engineer. I have also checked the estimated quantities of this design as well as the estimated cost.

For this purpose, Mr. Strauss has submitted to me forty-one drawings together with photostats of his computations and estimates.

Moisseiff goes on to summarize his verification process, and then concludes:

I take this opportunity to express my gratification that the work done by your Chief Engineer and the staff under his direction shows careful consideration of the problems involved and has produced a good, workable design of the bridge.
                              Respectfully submitted,
                              Leon S. Moisseiff
                              Consulting Engineer

The second version of the letter was written to Strauss, Ammann, and Derleth, his fellow engineers. It began like this:

Gentlemen:
In accordance with our understanding at the meeting of the Board of Consulting Engineers, in Chicago, on June 11th, and 12th, 1930, I and the engineering staff under my direction have carefully examined and checked the design of the Golden Gate Bridge as prepared by the Strauss Engineering Corporation under the direction of Mr. Charles A. Ellis. I have also checked the estimated quantities of this design as well as the estimated cost.

For this purpose, Mr. Ellis has submitted to me forty-one drawings together with photostats of his computations and estimates.

This version concludes:

I take this opportunity to express my gratification that the work done by Mr. Ellis and under his direction with the staff of the Strauss Engineering Corporation has shown careful consideration of the problems involved and has produced a good, workable design for the bridge.

The first version is what was required according to the rules of accepted business practice. The second version was Moisseiff's attempt to set the record straight. As a consultant, in private practice, there was a practical set of rules Moisseiff had to live by; as a civil engineer, a professional dedicated to the primacy of fact, there was another set of ethics he had to live with; by writing two versions of this letter, Moisseiff had attempted to honor both.

The directors, thrilled and relieved at being given such a tangible, promotable, detailed presentation of their proposed bridge, resolved to produce 2,500 copies of the engineer's report for distribution. They also resolved that, it being necessary for the Golden Gate Bridge and Highway District to incur a bonded indebtedness, a special election would be called, for the purpose of submitting the proposition of issuing bonds to the electors, to be held the same day as the statewide election, on November 4, 1930.

"IT IS NOT MR. MOISSEIFF AND MR. ELLIS . . ."

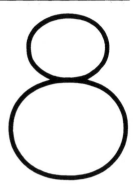

I

Now, at last, Strauss's army had its flag, the image of a bridge unlike any other in the world. From above, the perspective adopted in the Maynard Dixon painting, it was a narrow ribbon, elongated as though stretched in a tug-of-war between two mountainous and muscular points of land. Viewed head on from the roadway, the towers, with their ladderlike rise, carried the eye and the spirit upward; in profile, the towers shrank to slender ellipses, and the great cables and long suspenders turned the bridge into a giant harp, hung in the western sky. There was, in this bridge, no bad side, no unflattering perspective, no camouflaged ugly hodgepodge of mechanistic works. The most powerful emotional argument against the bridge—its looks—now worked in its behalf.

The proponents of the bridge—Strauss, MacDonald, the directors, Derleth, and others—seem to have seized immediately upon its metaphoric power: this bridge, as designed, was a statement of faith in the future, radiating confidence amid the bleakest of American economic times. The country, this bridge declared, was not at the end of things economically any more than it was geographically, but instead at the beginning of something new, unobstructed by the past, with renewed aspirations to excellence.

The metaphor was immediately adapted to the interests of promotion. Much of the design was used to promote the bridge and the bond issue before civic gatherings and among the press: lantern slides showing the overall perspective and the towers; a profile comparison to the George Washington Bridge, establishing the Golden Gate as the longest bridge of its kind in the world; the reproductions of the engineer's report and more.

These graphically powerful presentations were inaugurated none too soon, as opposition to the bridge had now fallen into more politically sophisticated hands. In May 1930, the powerful Pacific American Steamship Association had issued a statement declaring that the bridge was bound to be a hazard to navigation and a handicap to the shipping industry. The San Francisco Chamber of Commerce reacted by declaring that the chamber

11:0
Swlk
Sidewolk
11:0
60:0
Roodwiy

Top of Pier
El.+35.7

CROSS·SECTION
AT TOWER
PROPOSED
GOLDEN GATE BRIDGE
AT
SAN FRANCISCO, CALA
APPLICATION BY
GOLDEN GATE BRIDGE
HIGHWAY DISTRICT.

Sheet 4 of 4.
April 9 1930.
Scale:
0   30   60   90   120   150

Note:
Elevation are referred
to Mean Lower Low
Water which is El. 0.0.

would support the bridge only if the navigational clearances were increased and if the bridge's cost could be determined accurately "before a shovelful of earth is turned."

These objections proved strong enough to force another hearing by the Army Corps of Engineers in San Francisco in June. After reviewing testimony by representatives of the Steamship Association and by officials of the district, the army engineers announced that the federal government would raise no objections to the bridge, providing that it would not interfere with navigation. Also, in August, when Strauss submitted the revised plans for the bridge to General Lytle Brown, chief of the Corps of Engineers, General Brown gave only grudging approval. A construction permit, signed by Secretary of War Patrick J. Hurley, was issued on August 11, 1930; but the navigational issue had by no means been laid to rest.

Meanwhile, as the promotional effort in behalf of the bridge was intensifying, work continued on the details of the bridge design itself. Proposals must now be turned into specifications, overall formulas broken down into detailed computations, plans expanded into blueprints. This work was now being done in the Strauss offices in Chicago, under the direction of Ellis.

Derleth, the seasoned veteran of years of classroom discussion and debate, became the engineering point-man for the probridge forces, dispatched to address groups where questions might be raised requiring specific, quotable answers. His specialty was ridiculing the "so-called navigation interests" by pointing out that, were the bridge not built, the increase in water traffic would eventually require so many ferry boats that a *real* menace to navigation would arise, and that if the Bay Area's number-one priority was to be the elimination of every conceivable obstacle to water navigation, then the Corps of Engineers might well get busy on dynamiting Mile Rock Lighthouse, outside the Golden Gate, and the Farallon Islands, some twenty-eight miles out to sea. Then, in September, Derleth was presented with an opportunity to raise both appreciation of the bridge's design and the discussion concerning its practicality.

John Eberson's original architectural treatment of the bridge towers included parabolic arches above the roadway and near the tower tops—probably a legacy of his years of designing movie theaters.

On September 18, the first conference of the National Academy of Sciences ever held on the Pacific Coast was to be convened on the University of California campus at Berkeley. Almost one hundred of the most prominent scientists in America would be attending the conference, which would include discussions covering scientific subjects ranging from anthropology to molecular physics. It occurred to Dr. Robert G. Sproul, president of the university, host for the occasion, and an enthusiastic supporter of the bridge, that the conference represented a unique opportunity to present, describe, and explicate the "insoluble" engineering problems that had been overcome in the design of a bridge at the Golden Gate, a site dramatically visible from the California campus. Sproul proposed the idea to Derleth, who responded enthusiastically: not only would such a gathering provide an ideal forum for presenting the engineering and design story behind the bridge, Derleth could arrange for the address to be made by the man most qualified to tell it, the design engineer, Professor Charles Alton Ellis.

On the opening morning of the conference. the scientists gathered at the Greek Revival Hearst Memorial Mining Building on the university campus. After a welcoming address by Sproul and a response by Dr. Thomas Hunt Morgan, president of the academy, Derleth spoke briefly, introducing Ellis.

Ellis's speech ran the better part of an hour. Its tone was modest and understated; Ellis's delivery was droll, wry, anecdotal. Like a practiced university lecturer, he led his class step by step through the thickets of theory, pausing to point out reassuringly familiar landmarks along the way. The design of a suspension bridge was no great mystery—once you knew what you were doing, Ellis assured the scientists in his own dry way. Then he proceeded to lead them through a ten-minute explanation of Moisseiff's theory of wind stresses and his application of it to this bridge.

This was Ellis's subject; the idea he had been incubating all these months now had been hatched and was seen to be not only perfectly formed but with the possibility of becoming aesthetically beautiful. His pride in what he had done, his absorption and application, in an era before computers, of a phenomenal amount of theory and mathematics in an incredibly short time, his role in applying all he had learned about frame structures to the design of a work of engineering that also promised to be a work of art, could not be entirely concealed. Although in his talk Ellis deferred repeatedly to Strauss, referring to him, with respect, as "Mr. Strauss," treating him always as the initiator and overseeing eye of the project, when it came time to talk of the design itself, the author in Charles Ellis stepped forward:

"at this point, Mr. Strauss gave me some pencils and a pad of paper and told me to go to work."

To Joseph Strauss, a man with an ego as touchy as an inflamed corn, the news that a man who worked for him, an underling, was to address a gathering of the most prominent scientists in America on the subject of *his* bridge must have been felt with the excruciating pain of an elephant's tread. All the old rejections—of his shortness, his Jewishness, his lack of peer or academic recognition as an engineer—must have welled up in him. Under such circumstances, before such an audience, to such a personality as Strauss's, Ellis's statement about taking a pad of paper and going to work could have appeared to be an attempt to claim sole credit for conceiving a bridge already a step beyond any ever designed before. It was also uncomfortably close to the truth. The design of this bridge had entered realms of theory and mathematics in which Strauss was entirely unfamiliar; it had replaced the earlier, cherished, and patented design of his own; in order to be part of it he had accepted the shared authority of a consulting board and a reduced rate of pay. Now he was faced with the possibility of being denied conceptual credit for the bridge at all. It was something that Strauss, a dedicated scrapper, would have to do something about.

Feelings of this nature, rooted so deeply in the personalities involved, tend to put every aspect of a relationship on edge. Every word, each gesture becomes part of a pattern, an attitude, a scheme of some sort that reduces the usual complex mix of emotions that people inspire in each other to a simple, final, and usually unfavorable judgment.

On October 16, before another important audience, there was another blurted admission of who, exactly, had done what, and this time Strauss was present. The Commonwealth Club of California, which conducted the most influential speakers' forum in San Francisco, had commissioned an independent check of the plans for the bridge. The conclusions, viewed from the point of traditional engineering practice, were critical of the bridge's design in several important respects, and the chief engineer was invited to respond. A luncheon crowd including prominent civic and business figures heard Strauss open with a few digs at the motives of the outside experts responsible for the report, then settled back to listen to Derleth respond to the specifics.

Speaking impromptu, his words taken down by a stenographer, Derleth addressed himself to three issues raised by the report: the thorny matter of navigational clearance, the questions of overall cost, and the threat of earthquakes. "You know," said Derleth, with an edged amiability, *all projects of this kind seem to arouse groups of antagonists. I call them my friends, the objectors. Oh, I met so many of them and they made good friends for me when we built, against terrific obstructions, the Carquinez Bridge. And we expect the same obstruction here. And we must do missionary work, and I, as a professor, must help to instruct these men that their fears should be alleviated.*

The shipping interests' experts had insisted that the height of the bridge roadway was insufficient to allow clearance of certain ships under it. The navigational clearance beneath the bridge, as designed, was 220 feet. According to Dr. Wilhelm Fuerster of Hamburg, designer of the superliners *Europa* and *Bremen*, who had lectured at Berkeley in March, the era of tall ships with "ostrich plumes" of elaborate funnels and masts was over. "We must have ships shaped like dolphins," Derleth quoted the designer as saying, "and get rid of all this rigamarole of millinery."

There were some men now in the room, Derleth recalled, who were present at the Engineers Club when Dr. Fuerster had appeared there in March. One of them, an engineer for the shipping interests, had tried to draw Fuerster into the argument over the bridge. Was it sufficient, the engineer had asked Fuerster, to have 220 feet of clearance for the Golden Gate Bridge? "And before the man knew what was happening—he was very much disappointed—Dr. Fuerster said, 'That is much too much, 200 is enough.'" Improvements in ship design, the increasing sophistication and compactness of radio equipment, the development of hinged masts— all these made enormous clearances for navigation irrelevant. Yet the "shipping-interest gentlemen" still insisted on a clearance, at the Gate, of 250 feet. There must be other motives, Derleth suggested, at least for some of them, in trying to cloud the issue.

Derleth turned to the matter of earthquakes. The panel investigating the plans had cited Professor Lawson's geology report on the lack of tensile strength in the rock at the San Francisco anchorage and suggested that the bridge, oscillating in an earthquake, might rip that anchorage out. The gentlemen did not know, said Derleth, "that we have made two designs for the anchorage, one of which, in accordance with Professor Lawson's recommendations, is a gravity anchorage, which makes it entirely unnecessary to consider the sealing strengths in these terrifically unfavorable rocks."

If, because of an earthquake, the bridge were to fall, Derleth advised his audience, "you won't care. There will be nothing else left."

On the matter of costs, the engineering critics had pointed to the lightweight stiffening trusses proposed for the bridge and suggested that, in comparison to those used in other bridges, they had been designed to be too light, in the interests of lowering the estimate. They were approaching the "mysterious things" associated with suspension bridges that were so perplexing in terms of conventional bridge engineering. Here, Derleth, not all that comfortable with the details of the bridge's design himself, felt obliged to present outside credentials.

The original design for the bridge, he explained, "was made in Chicago, under Mr. Strauss's direction and ably assisted by a man named Charles A. Ellis, who stands high as a structural bridge engineer." Ellis, he explained, had been a professor of engineering first at Michigan and then at the University of Illinois, "where they have some of the best-known experts in the world. From that he went to Mr. Strauss's Engineering Corporation."

"Mr. Ellis," Derleth declared, as Strauss and the audience listened,

was in charge of that first design. And then those results were taken to New York. There are engineers here who will corroborate what I am about to tell you, that Mr. Leon Moisseiff is recognized in Europe and America as the greatest living analyst of stress and computation in long span bridges. And he is the one who has made the advances possible in long span work, so that now we can build suspension bridges of long spans at costs which are far below what anyone dreamed of when they used the old methods.

The new lighter trusses specified for the bridge by Ellis were in accordance with this new theory and could only be considered insufficient by people measuring this bridge by materials used in other bridges built according to antiquated thinking, Derleth said. Viewed this way, it was only natural for "some of my friends here, who have collaborated in this report," to look at the design for the Golden Gate Bridge and conclude, "Here is a mistake, here is an error. Somebody made a blunder."

"Well, I want to assure you," declared Derleth, "that it is not Mr. Leon Moisseiff and Mr. Ellis, even if Mr. Strauss and Mr. Ammann and I do not know anything about the subject."

At last, responsibility for the bridge's design had been revealed. Strauss, seated at the speakers' table, was publicly endorsing it. The public admission of his own ignorance of the theory and computation behind the design of his most famous work must have made him cringe. He would devote a considerable part of his energies for the rest of his life to the cosmetic task of papering it over.

2

As recently as August 1929, proponents of the bridge had approached the Hoover administration proposing that the federal government assume part of the financial burden of construction. With the onset of the Depression and the Hoover Commission still studying the possibility of a state- and federally-financed bridge across the bay between San Francisco and Oakland, the prospect of governmental financial assistance of any kind had evaporated. So it was that in a time of grave and deepening depression, with the value of individual investments and property shrinking almost daily, the Bridge District directors were obliged to propose to what was now a universally threatened electorate a $35 million bond issue.

The opposition understandably pounced upon this fear. "Taxpayers!" shouted a half-page newspaper ad, paid for by the "Taxpayers' Committee Against Golden Gate Bridge Bonds." "Before you vote to borrow $35,000,000 to go into the toll bridge business, study these questions and answers."

The questions were all the old ones, plus a number of newly pertinent gut financial issues. The bonds, if voted, would represent a first mortgage on all taxable property in the Bridge District, 85 percent of which must be borne by the citizens of San Francisco. The construction and operating costs of the bridge were expected to be paid for by bridge tolls, yet there were now three toll bridges on distant parts of San Francisco Bay, each of which operated at a deficit. Moreover, J. B. Strauss, head of "a Chicago concern," who had apparently originated the idea of a taxpayer-financed bridge district, had also engineered the Longview-Columbia River Bridge recently opened between Oregon and Washington. Net income for that bridge in its first six months of operation had been less than one-fifth of its preconstruction estimate and was not enough to pay bond interest. Not only had thirteen leading engineers of the West opposed, in a public statement, the Golden Gate Bridge project as financed, the Commonwealth Club had issued a report comparing the Golden Gate and Longview bridges that concluded: "There the stockholders and bondholders are holding the sack. Here the taxpayers of the District would hold it."

The ad, which was supported by a flight of radio commercials, was signed by a taxpayers' "Central Committee." Among its members were William P. Roth, president of the Matson Navigation Company, Roger D. Lapham, president of the Pacific Steamship Association and a future mayor of San Francisco, and perhaps most significantly, San Francisco City Engineer M. M. O'Shaughnessy.

The times had not been kind to Michael O'Shaughnessy. In an era of dwindling civic income, with city officials struggling to come up with money enough to provide adequate unemployment relief, O'Shaughnessy had been forced to go to the electorate again and again with appeals for more bond issue money to complete the great water project at Hetch Hetchy. The *Examiner,* riding him hard now, had tagged him "More Millions," and repeatedly accused him of selling out the city's private power resources to the Pacific Gas and Electric Company. Mayor Rolph, O'Shaughnessy's patron, was about to embark on a campaign for the California governorship, leaving his city engineer exposed. Now, for the first time, serious investigations of the "Czar of City Hall" were either threatened or pending. O'Shaughnessy responded to these imputations about his integrity with outbursts of professional pique. In the spring of 1931, when an independent investigation by an outside engineer, retired Army Captain John A. Little, was critical of O'Shaughnessy's recent management of the Hetch Hetchy project, the touchy city engineer responded by filing a formal complaint, charging "gross incompetency" against Little with the State Department of Professional and Vocational Standards. The charge backfired, instead producing among the San Francisco Board of Supervisors, O'Shaughnessy's old pursuing pack of hounds, demands for the city engineer's ouster and public questioning of the legality of his job. O'Shaughnessy, it seemed, wearing his detachable wing collar and inevitable boutonniere, was still living in the era of "Sunny Jim" Rolph when that sun, in fact, had set.

As election day approached, the frequency of the appeals against the bridge bond issue increased. The deepening Depression turned even the navigational clearance into a bread-and-butter issue: the bridge, claimed the Steamship Association, was a huge artificial barrier that would throttle the economic life of a great seaport. Most poignant perhaps was the plea of a group of artists and patrons of the arts, including the novelist Gertrude Atherton and the clubman-sculptor Haig Patigian, reminding the citizens of San Francisco that "The Golden Gate is one of Nature's Perfect Pictures" and urging them not to "Disfigure it."

Against this well-organized and substantially financed opposition, the bridge proponents wielded their most powerful weapon: the bridge itself. Even in the simplified caricatures of editorial cartoons, the bridge looked slender, soaring, graceful—and strikingly original. It was, even in its most primitive representation, so strongly itself, so freshly unreminiscent of any other bridge anywhere, that it had a realized quality beyond that usually associated with any proposed structure. A vote for such a bridge was a vote of confidence in the future, a declaration of faith in progress that would itself help bring that progress about.

The opposition—the Senior Council of the San Francisco Chamber of Commerce, the senior executives of the steamship and ferry boat lines, O'Shaughnessy—were shrewdly summarized as the "Old Guard" and depicted as selfish, grasping, and reactionary. The San Francisco *Call-Bulletin* rendered the Old Guard as a giant, potbellied ogre, sweeping aside a graceful bridge tower with one beefy arm, to allow a fleet of ferry boats, towed with its other hand, to pass in its stead. To the San Francisco *News*, the Old Guard was a military force, outfitted in shako hats, epaulets, and muskets, ensconced in a castle, behind a moat and drawbridge, grumblingly watching the March of Progress parade by. On another day, the *News* rendered the Old Guard literally, as a bearded, ancient football lineman, propped up on crutches and about to be steamrollered by a flying wedge of hulking backs, one of whom is toting a football labeled "Golden Gate Bridge Bonds." It was the local newspaper of Sausalito, in Marin County, that successfully wedded the bridge's economic and aesthetic appeals. A giant social-realist muscleman, wearing a block cap tagged "Labor," stands waist deep at the Golden Gate, gently cradling in his outstretched hands the superstructure of the bridge, the Marin tower of which has been replaced by a giant lunch bucket, labeled "Work for Thousands." "Labor, Unleashed," the cartoon headline advises, "Will Build the Bridge."

In the end, when the arguments had all been aired and the endorsements had been fought for and decided, it was the bridge that sold itself. Against the pervasive fear of such a time, the overall feeling of futility, the apathy of spreading idleness, here was an opportunity to act, to make a positive move in behalf of local investment and local jobs now. The fact that it would leave the community and its heirs not with some hulking monstrosity but with a work that might match and even enhance its setting was enough to offset any distant, long-term financial fears. It was, ultimately, in the privacy of the polling booth, an emotional decision, and the emotion in this case proved overwhelming. On November 4 the bridge bonds were carried by a three-to-one margin, passing in all six counties of the district. The majority in San Francisco was the largest for any bond issue in the city's history. The people of the district, in an act of either mass foolhardiness or considerable courage, had put themselves at risk, in the service of a metaphor.

9

"THE STRUCTURE WAS NOTHING UNUSUAL . . ."

I

In the sobering morning light following the passage of the bond issue—there had been fireworks, parades, and a bonfire burning of Old Man Apathy—the bridge proponents were faced with a cold reality. It had been one thing to win approval to offer $35 million worth of bonds in the teeth of what had become a worldwide economic depression. Getting people to buy them was another matter entirely. In a time of unprecedented and growing investment losses, bank failures, business collapses, and property foreclosures, how many people were there left who possessed the combination of capital and confidence required to support so ambitious and optimistic an undertaking?

There was also an enormous amount of planning still to be done on the bridge. The approaches on both sides had yet to be designed, the structural details of piers and towers and framework and cables had to be worked out, and everything had to be broken down into the contract drawings required for the letting of bids. Eventually, there would have to be 130 different plans drawn, using the entire staff of the Strauss office in Chicago and the just opened Strauss office in San Francisco, before the bids could be let.

In February 1931 Ellis, still performing at an intense level of inspiration, went to work on one of the crucial underpinnings of the bridge, that supporting the San Francisco entrance to the roadway. Here, above the site of the old Fort Point, which had at one time been scheduled for demolition to make way for the bridge, a structure would have to be built that buttressed the bridge while allowing room for the fort, a situation fraught with possibilities for compromise and an aesthetic hodgepodge. Instead, Ellis responded with another gem: a graceful arch made of three steel truss spans, curving over the fort and under the bridge in such a way that detracts from neither and serves as a transition between the lightness and strength of the steel bridge and the Romanesque red brick arches and vaults of the fort. Ellis's design was converted into detailed plans by the Strauss Chicago drafting staff under the direction of the firm's managing engineer, Clifford Paine. The bridge had been moved a graceful step closer to realization.

To allay voters' fears of an unlimited power to tax, the Bridge District directors had been forced to sign a pledge that none of the bridge bond money would be spent until it was certain that the bridge would cost no more than $35 million. Now, following the all-out effort of the election campaign, the district was all but without funds. To get the money to begin work, the directors would have to present a final bill in advance.

While plans were barely under way to issue a call for construction bids, the directors received a tempting offer of what amounted to a buy-out. A group of investors, including the firm that managed the Market Street Railway, the successor to Patrick Calhoun's gouging United Railways, offered to take over the entire project and build the bridge within the $35 million figure. This single-stroke solution was immensely appealing, and the Marin County Board of Supervisors went on record in favor of it. But the directors, led by Francis V. Keesling, a San Francisco attorney and chairman of the Finance Committee, saw things differently. The harsh economic times, in addition to shrinking the bond market and reducing the money supply, had also reduced the costs of materials and labor. There was a good possibility that the bridge, which people had once feared would cost over $100 million, might now actually come in at substantially less than $35 million—a difference that the Bridge District could keep for itself. It was agreed to keep the bids for, and control of, the project in the hands of the board, which began advertising for bids in April 1931. Construction, on the Marin pier, was scheduled to begin on on March 1, 1932, and the estimated completion date of the entire bridge was now July 1, 1935.

The directors, who were asking for bids on ten separate contracts, had correctly gauged the buyer's market now prevailing in the construction business. The Bridge District received more than two hundred requests for plans and specifications, and when the bids were opened on July 17, it was found that twenty-seven different firms had submitted more than forty bids. Adding up the low figures, the total estimate for the bridge's main structure, excluding approaches, was just under $25 million. The directors, immensely relieved, issued provisional contracts, to be confirmed upon the successful sale of the bridge bonds.

It had now been more than a dozen years since the initial resolution in behalf of a bridge at the Golden Gate had been passed by the San Francisco Board of Supervisors. The dance of proposal and approval had been dragged out for so long that the repeated interruptions and delays seemed to suggest that someone must be fundamentally out of step. In these long, disruptive pauses, any doubt might be turned over in one's mind again and again without being laid to rest. In the terrifyingly vacant, idle months and years of the early Depression, when people often wondered whether circumstances had been altered only temporarily or if instead society itself might be undergoing some profound and lasting transformation, there was both the time and the inclination to dwell upon any decision appearing to influence the future. The activity of an entire work-oriented society had been slowed to the point of hibernation, and the resulting inactivity made bridge baiting an absorbing occupation for any number of restless minds.

All the old unanswered questions were revived, most of them because they were unanswerable. No one could ever know with absolute, guaranteed certainty what effect the geology of the floor of the strait, or the earthquake fault, or the traffic revenues would actually have on the bridge until it was built. And as long as the bridge remained a design and not a fact, there was room for the often mercifully distracting activity of dialogue and debate.

The most persistent of these unanswerable questions concerned the mysterious, unseen floor beneath the Gate. A series of rumors, encouraged if not actively spread by a still sullenly active opposition to the bridge, maintained that the floor of the Gate could not possibly support the weight of the bridge's San Francisco pier and tower. There had been a cover-up, seemingly confirmed by the whispers of Professor Lawson's resignation and rehiring on the eve of submitting his geology report, and the report, it was said, had been doctored to produce a favorable conclusion on the strength of the bottom rock.

In fact, Lawson had been under considerable pressure to alter certain portions of his report. His insistence on the depth to which foundations should be carried had been resisted by Strauss, who turned his considerable powers of persuasion on the elderly geologist. "I have been unable," Strauss confessed to Derleth, "to get him to modify his position, but have suggested to Messers. Ammann and Moisseiff that they get in touch with him and see what they can do." Derleth himself had trouble with his obdurate colleague over Lawson's statement concerning rust and replacement of the bridge, which he thought "a gratuitous contribution to a geological report." Derleth recommended that these statements of the professor's be deleted. Lawson remained firm, however, and his comments appeared in the report as written.

According to the rumors now circulating, the engineers and bridge directors, protecting their own interests, had gone along with a falsified report and were now seeking to compound this crime by foisting off on the public bonds for a bridge that was sure to fail. The burden of this failure would fall entirely on the taxpayers, who had already been paying taxes to finance the project for two years, without a lick of actual construction being done.

To keep these rumors from undermining the sale of the bonds, the board initiated a completely new survey of the foundations for the bridge's piers and towers. The panel overseeing this study included Strauss and the Engineering Board—Ammann, Derleth, and Moisseiff—plus Professor Lawson and a consulting geologist, Allan E. Sedgwick. At the insistence of George T. Cameron, a Bridge District director and publisher of the San Francisco *Chronicle,* the panel also included Robert A. Kinzie, a mining engineer who had been formerly employed by a cement company in which Cameron was a stockholder and an executive.

While the test borings were still in progress and only half the pier area had been explored, Kinzie made it known that he would render an unfavorable report, which he submitted to the board on May 20. The San Francisco pier and tower as now planned would, Kinzie charged, rest on a "rotten rock foundation" that resembled "plum pudding." More specifically, Kinzie alleged that the existing pier and tower plans had been based on inadequate data; that there was a deep cavern at one corner of the San Francisco pier site that made it unsafe as a foundation; that there were continuous fissures in the rock, extending to the sea; and that the pier site was underlain with "blue mud" that would squeeze out over a cliff into the sea, causing the pier to settle.

Strauss submitted his own report, in rebuttal, on June 10. The report is Strauss at his feistiest, his most combative, his most lawyerly. Marshaling the evidence of his experts—the consulting engineers, the two consulting geologists, the bidders, the drillers—he addresses and refutes Kinzie's charges one by one. There is no cavern and no blue mud—no one else has found any evidence of either. "Mr. Kinzie himself," Strauss points out, "directed the location of the first hole at the point where he predicted the existence of a cavern, and he carried this hole down 107 feet into the rock before he conceded that no cavern existed."

There had been water leakage during the drilling. "It is the opinion of the drillers and all except Mr. Kinzie," comments Strauss, "that the water loss (in the casings) is due to loose connections of the casing to the rock surface. Mr. Kinzie, however, overlooks this practical detail and so conceives a spectacular theory of his own: namely a foundation honeycombed by fissures leading to the sea. Professor Sedgwick explicitly states that there are no such connecting fissures."

Kinzie, he charges, has also contended that hydration—the water formation of the rock—is still going on, when it is established geologic fact that serpentine is an igneous rock, whose hydration here would have ceased ages ago. Indeed, Strauss charges that the almost visibly retreating figure of Mr. Kinzie "seems terror-stricken by the very word serpentine." This is, Strauss suggests, because Kinzie is neither a bridge builder nor a geologist, but a mining engineer, and his world is full of caverns and threats of cave-ins.

As to the possible motive Kinzie had in making such charges, Strauss points to Kinzie's recommendation that the Bridge District retain the engineering consulting firm of Moran and Proctor to make final decisions instead of the existing engineering staff. It is most remarkable, observes Strauss, that Mr. Kinzie says that he will abide by whatever Moran and Proctor, if employed, say. "This means that he is willing to surrender the opinions he had so steadfastly maintained throughout, upon the say-so of certain men whom he does not know, and who are not geologists."

The dissenting engineer's report, which may or may not have been part of a plot to unseat Strauss, had obviously failed. "The Kinzie report," observed the San Francisco *News,* "has already taken its place in the limbo of disregarded things, along with the 1930 estimate of City Engineer O'Shaughnessy that the bridge would cost $100,000,000."

2

Like the seasoned pug who is obliged to take on all comers at the county fair, Strauss had no sooner seen the ring cleared of Mr. Kinzie and his rotten rock when fresh opposition stepped through the ropes. This time it was the bond buyers of the California banks and investment houses who, on the day before the bridge bonds were to be offered for block sale, announced that they would boycott the offering.

The threat summoned up here was another unanswerable question: the actual income from tolls on the bridge. Suppose bridge traffic proved light? What if significant numbers of drivers preferred to use the ferries? What would happen if the income from tolls was not enough to pay interest and principal on the bonds? Any bond buyer would insist on knowing—for sure—that the bonds would be made good from tax revenues. The only way to determine the district's power to tax, in advance, was through a court ruling—which was what the bond brokers were now demanding.

MacDonald, the bridge general manager, was infuriated at the threat of a boycott. Months before, at the request of the local bond brokers, the bridge board had paid for an attorney's review of the legality of the bonds.

The lawyer had vetted the bonds; the district had already levied half a million dollars in taxes. If the bond buyers were going to insist on a Supreme Court decision, why hadn't they made this known earlier?

It was too late to cancel the opening of bids. The boycott held: only one offer was received, from the brokerage house of Dean Witter, and that was so unfavorable to the district that it was rejected.

As they had with the rumors of unsafe geologic foundations, the bridge directors chose to give the suspicions of unsound financial ones a deliberately public and methodical response. They would arrange a test case, by having the bridge district secretary, W. W. Felt, refuse to sign the bonds. By declining to sign, Felt would make the bonds unsalable. The district would then take its own secretary to court, thereby guaranteeing a swift judicial review. This news alone was enough to force an opening wedge into the boycott. On July 16, the district received an offer on its opening block of $6 million worth of bonds from a syndicate composed of two major banks and an investment house. The syndicate would pay a premium for the bonds, on the condition that the district received a favorable ruling from the Supreme Court before November 16. The condition, and the offer, were accepted.

On the same day, at the same directors meeting, Felt, the district secretary, announced that he was declining to sign the bonds. The directors then authorized their attorney, George Harlan, to take Felt to court in order to force him to sign. The district, in attempting to respond to its public opponents, had thus opened itself to legal action from all its hidden foes, including the most powerful and most immediately threatened private interest in the state.

From the completion of the transcontinental railroad until the regulatory reforms of the Progressive era after 1910, the Southern Pacific Company had dominated California politics at every level. Assemblymen, state senators, and governors were on the Southern Pacific payroll. So were scores of mayors and county supervisors. So had been Abe Ruef. The gratuities ranged from cash bribes openly handed out in the state capitol, to passes for influential officials granting them unlimited free travel on the railroad. The struggle to break the political hold of the Southern Pacific on California had been long, bitter, and of Oedipal intensity: Hiram Johnson, leader of the antirailroad Progressives in the state, had been inspired partly by hatred for his father, a longtime legislative minion for the SP.

By the 1930s, the Southern Pacific's overt political hold on the state was a thing of the past, but the company, the state's largest employer and its largest private landholder, was still capable of throwing around enormous economic weight. The afterimage of the Southern Pacific as a sinister, manipulative force had lingered on into the Depression era, and not always without justification. The railroad's term for itself, "The Friendly SP," was used by the public it was required to serve, more often than not, with an edge of bitter irony.

The Southern Pacific had originally entered the ferry business on San Francisco Bay to connect the city with its transcontinental terminus in Oakland. Through ruthlessly competitive tactics ranging from price cutting to price gouging to outright violence, the SP had all but cornered the market on public transportation between San Francisco and points east. To the less-settled north, the Southern Pacific's interest had been much slower in developing. The railroad, working through a subsidiary, Golden Gate Ferries, Ltd., had provided passenger service between San Francisco and Marin reluctantly, operated it unsatisfactorily, and, when it was discovered that the run was profitable, met attempts to break its monopoly truculently. There was little doubt that the Southern Pacific had been a major contributor to the "shipping interests" that had opposed the bridge bond issue on navigational grounds. There were also suspicions among Strauss and the directors that the SP had encouraged the "rotten rock" and shaky tax-power rumors that had brought the bridge again into question. Occurrences like the crucially timed bond buyers' boycott reinforced these fears of a ruthless and conspiratorial opposition. "Such opposition did develop," Strauss was to recall years later, "with a vehemence and persistence unique in the annals of bridge construction."

On August 15, 1931, Warren Olney, Jr., an attorney and former California Supreme Court justice, entered the Bridge District's case before the court. Representing clients whom he declined to name, Olney contended that the action to mandate sale of the bonds should be dismissed because the issue was not a real controversy but one conceived for obtaining a decision "on other legal phases of the authority assumed by the district directors." These "other legal phases"—the Bridge District's power to tax, the legality of the district's boundaries—Olney challenged, as well as the validity of the original petition in behalf of joining the district submitted to the San Francisco Board of Supervisors, which, through a technicality, had been found to be 129 names short of the required 10 percent of the city's registered voters.

The obstructionist spirit of this action, with its legalistic hairsplitting and refusal to name clients, summoned up images of the Old Guard at its most reactionary and defensive. All the old election campaign bogeymen had seemingly leaped out of the editorial pages of the San Francisco papers, which were now demanding that Olney identify his unnamed clients.

The lawsuit was now expanded by the entry, in behalf of the Bridge District, of the San Francisco Board of Supervisors, whose spokesman, E. Jack Spaulding, accused Olney's retainers of acting "for the sole purpose of delaying the start of work, for purely selfish purposes," and by the entry of the San Francisco Chamber of Commerce, whose own Old Guard had, on this issue, apparently been dismissed.

On September 15 Olney, in arguing his motion before the court, disclosed the identity of his clients. He listed ninety-two "taxpayers," most prominent among them the Southern Pacific-Golden Gate Ferries Company. The other ninety-one names ran heavily to property development firms and their executives, interests particularly subject to pressure from the railroad.

In presenting his argument, Theodore J. Roche, attorney for the district, took a jab at Olney, who, when a member of the court, had concurred in decisions of a nature identical to that of the current petition. Roche also reminded the court that any great delay would prove costly to the district in the form of terminated contracts and increased materials costs.

The district, in fact, had just about gone broke. In order to pay salaries and meet office and legal expenses without imposing more taxes upon the public—a political impossibility in the present climate—the district was forced to borrow money from the syndicate that had entered the one valid bid on the bridge bonds. That syndicate, consisting of Bankamerica Company, R. H. Moulton & Co., and the American Trust Company, had set a deadline of November 16 for the expiration date for its offer. As the Supreme Court conducted hearings on the questions involved—the amending of the original bridge act to permit the collection of taxes before the construction of the bridge, the legality of a tax district including noncontiguous areas like Del Norte County, the validity of the San Francisco citizens' petition—and then pondered the evidence, time ran out.

On November 25, 1931, the California Supreme Court, in a twenty-one-page decision, ruled against Olney's petition to dismiss the suit filed by the district against its own secretary. The Bridge District had won on every count. Its right to levy taxes was established. The validity of its bonds was guaranteed. Only now, once again, there were no purchasers for them.

The representative of the syndicate, on November 16, citing the expiration of the offer, had asked for the return of the bond group's performance guarantee. The bridge directors were forced to hand back a desperately needed $120,000 and found themselves without a firm offer, facing a bond market that, in the last six months, had drastically declined.

The opponents of the bridge, well financed and organized, realized by now that they could not beat the bridge down, but they might still wear it down. On November 28, representatives of a San Francisco-based property development firm, the Garland Company, and the Del Norte Company, a north-state lumber organization, requested an injunction from the federal district court prohibiting the Bridge District from selling any of its bonds. The lawyers for the companies named contended that the imposition of taxes for the building of the bridge would be confiscatory and in violation of the Constitution.

That the two complainants were nothing but stalking-horses for the Southern Pacific was apparent to all but the most naive. "The Southern Pacific-Golden Gate Ferries," began the San Francisco *Examiner's* article reporting the suit, "once more attacked the validity of the Golden Gate Bridge and Highway District's $35,000,000 bond issue."

In attempting to wear down the proponents of the bridge, the Southern Pacific and ferry interests had worn out the public's patience. An outcry against railroad interference now arose that recalled the most insistent reform demands of the Progressive era. The San Francisco Board of Supervisors, following a speech by Supervisor Dan Gallagher that accused the Southern Pacific of first opposing the establishment of regular ferry service to Marin, then buying control of Golden Gate Ferries when it proved to be profitable, unanimously passed a resolution calling upon corporations and other interests to drop their opposition to the bridge. Local business, social, and promotional groups ranging from the Motor Car Dealers Association to the American Legion responded by urging a business boycott of the Southern Pacific and contacted manufacturers and shippers in the East and Midwest, urging them to divert their freight to other carriers.

Stung by the breadth and intensity of this public reaction, the Southern Pacific was forced to respond. On December 4, Southern Pacific President Paul Shoup issued a baldly disingenuous statement in which he claimed that the Southern Pacific, as only one of many stockholders in Golden Gate Ferries, would not interfere with the legal rights of other stockholders in the ferry company who wished to oppose the bridge. The idea of the Southern Pacific, which owned 51 percent of the ferry company's stock, as merely a consenting minority in this matter was a palliative that no one would swallow.

Still, the bridge's opponents had reason to believe that time might be on their side. They were doing battle with an admittedly underfinanced Bridge District, and in hard economic times. And the public's patience, especially where its pocketbook was involved, could not, even in behalf of the bridge, be infinite.

3

While the means to begin the bridge remained in question, the plans for its completion continued to progress. A signed contract had been received for the steel superstructure from the McLintic-Marshall Corporation, a Bethlehem Steel subsidiary, and fabricators of both the George Washington and Ambassador bridges. In San Francisco, Irving Morrow was incorporating his own insights into the architectural details, while at the Strauss offices in Chicago, Charles Ellis had plunged into the detailed computations required for the final design of the bridge towers.

Relations between Ellis and his boss, Strauss, never more than businesslike to begin with, had been brought to a certain pressure point by the delays and frustrations in obtaining legal and financial approval to start construction. The two men were so unalike temperamentally that any differences that developed over the only thing that bound them, the bridge, had the potential of widening into a permanent breach.

For Ellis, comfortable with abstract thought, the towers presented problems, not apparent on the surface, that at first seemed insoluble in terms of the present theory of structures. The procedural delays in setting a start-work date for the bridge presented a welcome opportunity for the intensive study that might yield a solution.

To Strauss, the practical man of affairs, out in the world doing battle with people like Kinzie, trying to anticipate and parry the powerful and often camouflaged moves of the antibridge interests, attempting to maintain an office staff in the face of a slow-paying Bridge District and a seemingly evaporating market for bridge construction, Ellis's painstaking reasoning had begun to seem like an intellectual exercise, indulged in at the expense of the Strauss Engineering Corporation.

Beginning in October 1931, Strauss, from San Francisco where he was contending with the bridge's opposition, and from various other points of the compass where he was trying to scare up new projects, began sending Ellis letters urging him to propose a definite program, giving dates, when each part of the work would be finished. Ellis responded, at great length, in a letter to Strauss explaining the difficulties, stating once again that the bridge was not just another structure to be designed by merely adapting existing theory, pointing out why the solution proposed by the assistant, Clarahan, could not be correct, and asking for Strauss's cooperation in arriving at a proper design.

By November, Strauss, increasingly impatient with legal delays, had become extremely critical, complaining that the manner in which Ellis had handled the work of designing the towers had resulted in a great waste of time and money. The extensive research that Ellis had suggested, Strauss now contended, was unnecessary, and the assistant, Clarahan, should have been brought into the work at an earlier date.

Each man now spoke from a position so representative of his personal philosophy of engineering and of life that any sort of resolution, short of a sudden announcement of a legal and financial go-ahead for the bridge, seemed impossible. Strauss, impatient with abstractions, wanted to get on with the work; Ellis, devoted to engineering ethics, would not be rushed into consenting to a design that he was convinced was in error.

In late November, Strauss wrote Ellis urging him to start on his two-week vacation and suggesting that he turn his work over to Clarahan. Ellis responded, again in detail, explaining again where he thought Clarahan's work was wrong and insisting that he could have no thought of a vacation until the design of the towers was completed. Strauss's reply to this letter was a telegram instructing Ellis to start his vacation at once.

On December 5, 1931, Charles Ellis, design engineer of the Golden Gate Bridge, left the offices of the Strauss Engineering Corporation for what turned out to be the last time. "When I left the office," Ellis later recalled, "the staff in the draughting room had been reduced to a very few men, and no one except Mr. Clarahan was working on the Golden Gate job." Strauss had instructed Clarahan to proceed independently on the towers, and he was busily at work on computations of his own.

Three days before his vacation was up, Ellis, at home in the Chicago suburb of Evanston, received another letter from Strauss, critical of his work, and declaring that "the structure was nothing unusual and did not require all the time, study and expense which [Ellis] thought necessary for it." Strauss closed by instructing Ellis to turn over all papers concerning the Golden Gate job to Clarahan and to take an "indefinite vacation without pay." He had been fired.

4

In San Francisco, the surrogate struggle between the Bridge District and the Southern Pacific had intensified into direct conflict. A volunteer organization calling itself the Golden Gate Bridge Association sponsored nightly radio broadcasts, urging listeners throughout Northern California to join the boycott and offering, through a speakers bureau, the services of some two hundred lecturers, available to carry the boycott message personally throughout the Bridge District's six counties. In these counties, at meetings of local chambers of commerce, Lions clubs, and merchants' associations, resolutions were passed castigating the opponents of the bridge as selfish and unfair obstructionists. A discussion forum, sponsored by the California Automobile Association at San Francisco's Fairmont Hotel, turned into a probridge rally, with the chair ceded to the president of the Golden Gate Association.

On December 9, 1931, the president of the ferry company, a man bearing the unfortunate initials of S. P. Eastman, attempted to assume sole responsibility for the most recent attack on the bridge, claiming that it was the ferry company and only the ferry company that had encouraged the Garland and Del Norte firms to seek an injunction against the bridge. This attempt to divert the boycott was ridiculed in the San Francisco press. "Not all Mr. Eastman's words," said the San Francisco *News,* "will conceal the fact that responsibility rests in the final analysis with the Southern Pacific." Less than a week later, Eastman proposed a compromise. If the bridge could be built and operated by a public body, the newly created California Toll Bridge Authority, now involved in planning and financing the San Francisco–Oakland Bay Bridge, and if the project were backed by state revenues rather than property taxes imposed by the district, the ferry company would withdraw its opposition. MacDonald, general manager of the district, emphatically rejected this "compromise," pointing out that the abandonment of the district and its replacement by a public body would begin the proposal-and-approval dance all over again, ensuring a delay in actual construction of between ten and fifteen years.

The San Francisco supervisors now called attention to the fact that the franchise of Golden Gate Ferries, as well as those of the Southern Pacific ferries running between San Francisco and Oakland, had originally been granted by the citizens of San Francisco. The board recommended, unanimously, that the city attorney immediately investigate these franchises and determine whether they might be revoked. On January 1, 1932, the most direct threat yet was issued against the railroad-backed ferry system with the announcement of the formation of Marin Municipal Ferries, Inc., a publicly owned ferry system to run in competition with the Southern Pacific's boats until the completion of the bridge, whereupon it would be sold to the Bridge District.

Despite this pressure from business interests, the press, the Bridge District, the Golden Gate Association, local political bodies, and an aroused citizenry, the officials of the Southern Pacific and its ferry subsidiary continued to press their suit. It was, in a strange way, almost a matter of principle now, as if all the continent-mastering, public-be-damned, ruthless energy and stubborn drive of the era of the railroad robber barons had been gathered for one final stand at the last limit of the West, as though the bridge would have to be built over the dead body of an earlier industrial giant.

5

While the federal court hearings were getting under way in February, in the heat and glare of the most intense outpouring of public support in the history of bridge building, the man most responsible for the design of the bridge was languishing in idleness and obscurity.

For all his personal differences with Joseph Strauss, Charles Ellis's firing had come as a brutal and unanticipated shock. For more than three years he had thrown himself, body and soul, into the task of engineering the design of a bridge unlike any the world had ever seen. It had been the challenge of his life, calling upon everything that Ellis was and everything that he had ever learned, under the most severe pressures of time and money and careers —other people's and his own—and he had met it, only to find himself dismissed from participating in its realization. Perhaps if he had been less absorbed in the intellectual problems posed by the bridge, or if he had spent more time diplomatically cultivating Strauss, this would not have happened, but in order for those things to be different, Ellis would have had to be a different man.

Now, at home in Evanston, with plenty of time at last for thinking, he pondered what had happened to him and considered also the future of the great bridge that had once existed in only his mind.

Ellis's firing had come as a shock not only to himself but also to Derleth and Moisseiff, the two members of the Engineering Board with whom he had been most closely associated. Since Ellis was Strauss's employee and not a consulting engineer or board member, there was little either consultant could do to influence Strauss's personal hiring and firing policies. Both men did, however, attempt to help Ellis get placed elsewhere.

Ellis was now fifty-five years old, a senior engineer without seniority, with a wife and daughter to support, job hunting in the bleakest economic climate of his lifetime. He hoped to find work on another bridge-building project, or to return to the engineering faculty of a university, but opportunities in these fields were, if anything, worse than in the economy generally. On the San Francisco-Oakland Bay Bridge, where Derleth had made a strong recommendation for him, at Columbia University, where Moisseiff had taught, and at Harvard and Yale, where Ellis's work was known and his textbook on framed structures was in use, the story was essentially the same. Were economic conditions only different, a place would be made for him, but at present the policy was to make no additions even to fill vacancies left by resignations, retirements, or death.

Ever the gentleman, with a deeply ingrained reserve, Ellis remained reticent about his reasons for seeking work. "In all my interviews," he wrote Derleth, "I have simply expressed a desire to return to university work and have in that way avoided any reference to unpleasantness in my present position."

This personal downeaster's refusal to make a fuss also prevented Ellis from contacting MacDonald, manager of the Bridge District, whose contracting firm had successfully bid for work on Boulder Dam. MacDonald knew and respected Ellis, but the thought of approaching him threw Ellis into an ethical quandary: "I should of course say nothing," he confided to Derleth, "except that I was looking for employment, but even then it might not be wise to write him."

With prospects of immediate employment dim, and in order to maintain his mind's intellectual rigor and avoid brooding over his personal misfortune, Ellis turned his trained faculties to the absorbing problem of the bridge towers and their design.

"In my previous computations," he confided to Derleth, "I had suspected that something was wrong with the bracing, especially with the diagonals below the floor, for the results did not check with sufficient accuracy. The reason was soon apparent."

Ellis had assumed that the force of horizontal winds on the bridge towers would be borne entirely by the diagonal braces between the tower posts. "This assumption," he now concluded, "is far from true, because the tower legs, being so very stiff, carry a considerable portion of the shear." The problem, as Ellis now saw it, was to discover some method whereby the correct participation of the tower legs and diagonals in carrying the shear could be introduced into the solution. "As I proceeded with the problem," Ellis wrote Derleth, "I found that it was as elusive as it was interesting."

Ellis desperately needed this mental occupation in order to avoid succumbing to despair. It distracted him from the realization that the work that had challenged and come to matter most to him in life had been taken from him, probably forever.

"To speak very frankly, I just do not know what to do in case the project goes ahead and I find myself out of it. I have no intention of attempting to embarrass Mr. Strauss, or of entering into any unnecessary arguments."

Still, Ellis possessed a full and realistic appreciation of Strauss's ability to command the machinery of publicity, and of the tendency of his hungry ego to take credit for any achievement and to delegate all blame. "If any criticism of my work be made, or any delay be attributed to my manner of handling the work, I hope that in justice to myself, the Board will give me notice of the same, and the opportunity of being heard." Against the weight of time that now pressed in upon him, Ellis buttressed himself with the sustaining discipline of his work. At home, alone with his drafting paper and engineer's pencils, he plunged back again into his calculations for the bridge towers, devoting weeks, then months, to the sweet, absorbing agony of revision.

6

On July 16, 1932, an opinion on the legality of the organization of the Bridge District and its power to tax was issued by Federal Judge Frank H. Kerrigan. In the overheated atmosphere of the dispute, the decision was a remarkably cool and rational display of common sense. Refusing to grant the injunction requested by the Garland and Del Norte firms, Judge Kerrigan upheld the district as representing a new trend in governmental organization, dictated by necessity. The bridge would be of great benefit, not only to the entire state, but to the people of Oregon and Washington, and "a single unit, because of sentiment or prejudice peculiar to itself should not be permitted to obstruct an undertaking of great public importance." It was simply the Lincolnesque honoring of the greatest good for the greatest number.

The probridge forces exulted in the legal confirmation of their cause and threatened to extend and intensify the economic and political pressure if the Southern Pacific and the ferry line continued to pursue the issue. Both the San Francisco Grand Jury and the city attorney were threatening to require "extensive pier modifications" and other improvements if the ferry's franchise were not to be revoked.

On August 9, in separate statements united by a spirit of grudging gracelessness, both the ferry company and the Southern Pacific announced that they would abide by Judge Kerrigan's decision. The bridge project, warned Eastman, the ferry president, in a parting shot at Strauss and the directors, "is largely a promotion . . . ill founded and ill advised, and will impose a burden on taxpayers out of all proportion to benefits." Paul Shoup, the Southern Pacific president, still insisted that the SP had had no relationship to the suit, except as a stockholder in the ferry company. The SP, in his version, was yielding to the conclusion reached by the president of the Golden Gate Ferries.

Later that month, Angelo J. Rossi, who had succeeded Rolph as mayor of San Francisco and who had helped dissuade the ferry company and the Southern Pacific from appealing the case further, announced that the Garland Company, and later the Del Norte Company, had each waived its right to appeal. The last legal obstacle in the way of the bridge had been removed. "Now the bonds can be sold," said the ebullient Mayor Rossi, formerly a florist. "Now the Bridge can be built."

10 "THE RESPONSIBILITY WOULD BE PLACED AT MY DOOR . . ."

I

With the subsconscious persistence of a dream, the design of the bridge continued to develop, an idea with a powerful existence of its own, a life apart from the grinding daily tedium of law and finance. Like Charles Ellis, Irving Morrow, the consulting architect who had come aboard in a rush in 1930, looked upon the various procedural delays as an opportunity to refine and improve upon the quality of thought that had gone into the bridge. Unlike Ellis, he was not viewed as a competitor by Strauss, who encouraged Morrow and approved of his work; it was a situation whose possibilities Morrow intended to realize to the fullest.

No one had spent more time pondering the aesthetics of the site, the shifting interplay of light and shadow on the Marin hills, the waters of the strait, the changing color and texture in complex harmony with weather and sky, the contrast between the bare Marin headlands and the deep green of the Presidio on the San Francisco shore. "It is caressed by breezes from the blue bay throughout the long golden afternoon," Morrow had written of the Gate in 1919, "but perhaps it is loveliest at the cool end of day when, for a few breathless moments faint afterglows transfigure the gray line of hills." Morrow, heading up from the Oakland ferry slip each evening to his home in the hills, knew this subtly changing scene by heart, fully appreciated its epic quality, and realized that he must literally live with the outcome of any design made manifest here, where "the slow passing of evening into night evokes the stillness of innumerable lights which twinkle around the harbor shores and along the avenues leading from point to point." The bay, especially at night, was a vast amphitheater, with a hush like that of an audience's anticipation before the promised revelations of great drama. "Across and far up and down the bay faint lights reveal communities unsuspected in the daytime haze. A vast map is traced upon the night."

In 1930–31, while Ellis worked out the details of the engineering design, Morrow refined the architectural treatment, most strikingly in his own charcoal drawings, done more than two years before construction began.

Morrow had the ability to listen to the voices of others, as well as his own subconscious. In refining the details of the bridge's appearance, in particular its lighting and its color, he sought out and absorbed the opinions and feelings of people like himself—artists, teachers, the painter Maynard Dixon, a forest service pathologist—individuals who combined deep personal ties to the area with a strong aesthetic sense. These views he incubated, along with his own, until what emerged was a remarkably responsive and insightful design, boldly innovative, yet with an inherent rightness about it: the architectural extension of Moisseiff's wind stress theory and Charles Ellis's engineering design.

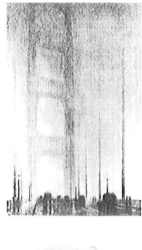

In the preliminary plans submitted with the engineer's report in August 1929, the bridge's pedestrian walkways were bordered by an elaborate wrought-metal grillwork railing and streetlamps that suggested a turn-of-the-century elevated train station. Morrow, who had hiked frequently in the Marin hills and who understood that the bridge would be not only the object but the vantage point of spectacular views, now eliminated this embroidery to allow anyone passing over the bridge, even people rushing past in cars, full access to the view to either side. By simplifying the railing to a line of uniform posts, spaced slightly farther apart than usual, he provided waist-high security for pedestrians, with a barrier that dissolved into a scrim to people passing by in motorcars, allowing an unobstructed view of the bay, its islands and cities on the one hand, and the Pacific all the way to the horizon on the other. The antique-looking streetlamps Morrow replaced with slender, angular standards that curved over the roadway and echoed the angled look of the towers. These touches, subtle in themselves, gave the bridge design a clear and uncluttered quality that made it even more at one with its surroundings.

The bridge, in its proposed setting, would run northwest-southeast, thus catching almost broadside both the rising and the setting sun. Morrow, with his acute painter's sense of light and perspective, set out to heighten, intensify, and add dimension to this effect. To embellish the steel-plate housings that covered the towers' horizontal portal bracing struts—the cross-frames for the giant windows of the towers—he designed a pattern of vertical fluting, which would catch the light in all its aspects on one row of faces, while adding the deepening dimension of shadow in the other. The more angled and tinted the light, particularly at sunrise and sunset, the more profound the effect.

This architectural motif of vertical fluting, which added to the absorbing quality of depth already present in the towers, with their stepped-off giant

Irving Morrow's use of vertical fluting gave the bridge its unifying visual identity and incorporated into it, as in a sculpture, changes in light.

rectangles varying in size as if in distance as well as height, Morrow extended to the concrete pylons at the north and south ends of the bridge, the concrete piers under the bridge towers, and the anchorages. This fluted motif, with its enhancement of the power of light and depth, gave the bridge a consistent architectural imagery, an expression of the bridge-as-idea echoed and reechoed throughout its components, designed to take maximum advantage of its setting, just as a statue does. With these touches, Morrow had transformed the bridge into an environmental sculpture, the largest ever built.

In these architectural changes, which enhanced the bridge without over-elaborating it, which carried on the angular Art Deco look of the towers without getting carried away by it, Morrow fixed the bridge in its time so strongly that it became an expression for all time, a declaration of willingness, persistence, and optimism from the depths of the most economically discouraging of American eras. "Who would do this today?" the San Francisco writer Jon Carroll observed of Morrow's architecture in 1981. "Who, today, would care this much?"

The color that the bridge was to be painted was, at this time, still in question. The engineer's report had recommended that it be painted with an aluminum paint, "which will give the bridge a silver lustre, remarkably effective by day and still more so by night." The Navy Department, according to Charles Ellis, had announced that it wanted the bridge painted in yellow and black stripes, to assure visibility. Morrow himself was considering painting the bridge in a variety of tones, a basic tone for the towers, growing slightly darker in the bracing below the deck, in the stiffening trusses and in the arch over Fort Point, slightly darker than this for the approach viaducts and cables, another tone darker, or a contrasting color for the handrails and light standards, and a neutral gray for anything that needed to be "painted out": airway beacons, strut housings, and the like. Since a decision on the bridge's final color was not necessary at this point, there seemed to be a consensus that initially it would be best to go with a simple "red lead shop coat," while testing panels of other colors and combinations for suitability and weathering.

2

While the idea of the bridge remained pristine and intact, the maneuvering, bargaining, and pressing for advantage that were required to bring it into being continued to swirl around it. The Bridge District directors, who, led by their Finance Committee chairman, had gambled on declining Depression prices in turning down the tempting offer of a construction buy-out in 1931, decided in July 1932 to try their luck again with the construction bids. Since the deepening slump had continued to drive contractors to the wall, why not let the previous bids expire and call for new ones? This hard bargaining, with its kick-'em-while-they're-down opportunism, was justified as shrewd business practice, an example of the kind of treatment the Bridge District had been receiving at the hands of the ferry

Bridge opponents claimed that such a structure in such a place would constitute a navigational hazard. Bridge supporters pointed out that without a bridge there would be even more water traffic and inevitably more shipwrecks.

151

line, the bond buyers, and the Southern Pacific. The bids, except that by McLintic-Marshall on the steel superstructure, which was from one of the only two firms capable of such large-scale steel fabrication and was considered low anyway, were allowed to lapse. The bridge may have been sold to the voters as a boon to Depression-racked construction and labor, but heaven help the contractor who had taken anything for granted.

To further strengthen their bargaining position, the bridge directors now decided to reverse their earlier procedure and arrange financing first before advertising for new bids. On July 19, William L. Filmer, president of the Bridge District, wrote to the Hoover administration's sole functioning Depression relief organization, the Reconstruction Finance Corporation, requesting that the agency allocate $35 million to purchase the bridge bonds. Soon after this the directors, still strapped for operating capital, imposed their third annual tax on the property owners of the district. It was a measure they had deliberately avoided while the court case challenging the legality of the district had been pending, and it seemed to justify the most pessimistic predictions of the Golden Gate Ferries and Southern Pacific executives and their allies.

Now it was the Bridge District's turn to be the target of public protest. Opposition to the new tax was so intense that the directors quickly reversed course, electing not to wait for approval from the Reconstruction Finance Corporation but instead to call for private bids on the bridge bonds once again from the banks and investment firms.

On August 31 the district opened three legitimate bids. One, from the R. H. Moulton investment house, offered to pay ninety-five cents on the dollar for the first $6 million block of bonds, with an option to buy any or all of the remaining $29 million in bonds at ninety-seven cents. The two other bids were received from a syndicate consisting of the Bank of America and four investment houses. This group offered, in one bid, to pay slightly less than the Moulton figure for the first block, with an option on the remainder of ninety-five cents on the dollar. The second bid was for a slightly lower figure, with no option, on the first $6 million in bonds alone.

At a closed meeting with the bridge directors, representatives of Moulton and Company and the Bank of America group were each invited to make a cash advance to the district on their offers, pending the expiration of the legal deadline for appeal of Judge Kerrigan's decision in favor of the legality of the district and its power to tax. The Moulton representatives, stating that such a cash advance, to ensure the awarding of a bid, would be unethical, refused to give the directors the money. The Bank of America representatives, exhibiting no such qualms, agreed to advance the district $200,000, providing the directors accept the second of their offers, on the first $6 million worth of bonds, with no options.

On September 2, with this advance money in the till and the purchase contract for the first block of bonds in the hands of the Bank of America syndicate, the bridge directors voted to withdraw the application now before the Reconstruction Finance Corporation and to rescind the announced tax on the district's property owners. Once again the bridge had been given a go-ahead—this time under questionable circumstances—and both the legal clearances and the financial means to begin work were in hand. The district vice-president, R. H. Trumbull, announced that construction would get under way within ninety days.

3

At his home in Evanston, Charles Ellis struggled to maintain an even keel after veering suddenly from the most active and productive period of his life to its most idle. His career possibilities seemed to be receding from him in a way that reduced both his accomplishments and his prospects to insignificance. Like so many managerial and professional people cast out in the Depression, he was forced to adapt his life to conditions he had never anticipated. Like other families, the Ellises cut back, used savings, made do as best they could—and hoped for a change in the times.

Ellis tried hard to remain in contact with his work, to continue the sustaining exercise of his profession. Earlier in the year he had written his discourse on Moisseiff's theory, which was to be published, along with Moisseiff's paper, in the *Proceedings* of the American Society of Civil Engineers. It was oddly prestigious company for an engineer who at the time could not find a job. After this, he had found himself "with nothing in particular to do and with the thought of the condition of the towers constantly on my mind."

As the man who had made all the computations and design and who had written the specifications for the bridge job, Ellis was certain that when the condition of the towers became an issue, "as it surely must, the responsibility for any errors or delay on account of them would be placed at my door." Concerned for the welfare of the project and for his own reputation as an engineer, Ellis had plunged back into the computations for the towers.

Ellis had been reasonably familiar with the bridge tower question to begin with, and based on his previous studies, he had anticipated that a month would be time enough for a complete review. Instead, he found himself working on the problem continuously, ten hours a day, seven days a week. It was nearly five months before this work was completed. Ellis now considered himself in full possession of the facts and "in a position to speak with some assurance and authority concerning the design."

Now, in the fall of 1932, Ellis read in the San Francisco newspapers, copies of which reached him regularly, that the work on the Golden Gate Bridge was at last likely to get under way. New bids were being let; the consulting board was meeting; the project was going forward without him, without the benefit of his new calculations. Errors were about to be incorporated into the permanence of steel.

Ellis began writing letters to the other consulting engineers, to Derleth and Ammann and Moisseiff, explaining in detail what he considered the urgency of the situation. In a letter to Derleth, he included a rough sketch of his own for a proper bracing of the towers. In another letter, Ellis asked Derleth whether it would be possible for him to have an interview with the Bridge District Board of Directors. He would, in effect, be going over Strauss's head.

"Mr. Ellis realizes that he may be placing me in an embarrassing position," Derleth confided to Moisseiff. "However I would never hesitate to do what is right to all parties concerned, and . . . to defend the integrity of our design."

Right or not, Ellis was an outsider now, without the protective mantle of a job. He was asking to address laymen, who would have to accept on faith the seriousness of what he was telling them, against the word of the chief engineer, who, as Ellis's boss, had known him longer than any of the directors had and had fired him. The directors would undoubtedly look for guidance to the consulting board, who would be forced to take sides in what could be viewed as a personal dispute. They would risk alienating Strauss, who, as chief engineer, was the man with whom they still had to get along. "My relations with him are very friendly," Derleth advised Moisseiff, "and I desire that such a relation continue."

Perhaps, Derleth suggested, an independent check, at least of the highlights of the design, might be made by Moisseiff and his staff, to establish or dismiss the legitimacy of Ellis's objections. "It would be my hope that Mr. Strauss would not object or be offended were this decision made public. It ought to rather relieve his mind and please him." This would, Derleth hoped, effectively depersonalize the issue.

The consulting engineers had begun to fall away from Ellis, the man upon whom they, like Strauss, had depended most. Whatever Ellis's protests, he was an outcast now, unemployed, a man intellectually active all his life who now had too much time on his hands and who was giving perhaps too much thought to the "interesting problems" concerning the towers. Was it not possible that a man in Ellis's present frame of mind might find problems where, in fact, there weren't any? "These grave differences," Derleth suggested to Moisseiff, "may not really exist."

Ammann and Moisseiff agreed. "Mr. Ammann and myself," Moisseiff wrote to Derleth on September 19, "are of the opinion that we cannot exert any further pressure on Mr. Strauss. We have done so before and without avail. We do not see what can be done in this matter in behalf of Mr. Ellis."

There was an air of defeat about Ellis's letters now, a seeping desperation that could prove inflammable. "I have worked three years and more on this job," Ellis wrote to Moisseiff and Ammann on October 3, "and I cannot face the possibility of having the work start off on a basis which my five months' intensive study has convinced me needs many revisions." Should Ammann and Moisseiff be passing through Chicago on their way west, Ellis suggested, perhaps they could set aside two or three hours for a meeting, "for I wish you to have all the facts clearly before you so that you can in your own mind arrive at a conclusion whether the situation is of sufficient importance to warrant your doing something about it."

"I do not wish to put the consultants in an embarrassing position by asking their advice," Ellis insisted, "but I do feel it a duty to the project and myself to present my findings to someone who not only will adequately understand the situation but will see to it that the tower design in its final form is what it should be."

The consulting engineers, Derleth particularly, sensitive to the fact that what had happened to a man of Ellis's stature and ability could conceivably befall any of them, had tried to find him a job in an era when there were no jobs. They had done what they could; any personal, moral, or professional obligations had been fulfilled: now it was time to cut loose, before Ellis jeopardized the entire project.

"I am much vexed by the actions of Mr. Ellis," Moisseiff wrote Derleth on December 7, "though not surprised. The ancients said that one sin generates another." Moisseiff, along with his staff, had been working on the calculations for the towers and had found "nothing wrong in the theory and methods used by the Strauss Engineering Corporation."

"Mr. Ellis," Moisseiff concluded of his former colleague's insistence on

155

the unsafe design of the towers, "started on an inclined plane, and accelerates himself accordingly."

"I sympathize deeply with Mr. Ellis," Derleth wrote his fellow consultants, "but there are times when we are helpless to aid."

In firing Ellis, Strauss had taken a calculated risk. Ellis was known and respected by the Bridge District directors and general manager, and he had been welcomed and accepted as an equal by the consulting engineers. There was no one else with Ellis's qualifications in the Strauss organization; yet Ellis's work on the bridge, the engineering design, was essentially done. All his abstractions, his formulas and computations and unknown quantities, had been translated into contract specifications. In many ways, there would never be a better time to turn the work over to someone else.

Strauss even had a man to replace the indispensable Ellis, within his own organization. Clifford Paine, already a Strauss associate, was an engineering graduate of the University of Michigan and a civil engineer. A tall, lanky man like Ellis, Paine, who had spent some twenty years in railway and bridge construction, was primarily a detail man, an organizer and administrator rather than an originator. He was not a theoretician. He was, importantly to Strauss, not a professor. He was also more adroit politically than Ellis had been. Paine was a man whom Strauss could rely on to carry out his directions faithfully, without representing a threat to his touchy boss. Yet Paine was no patsy; he had once quit Strauss in protest over his boss's interference with his work. Strauss had hired him back, assuring Paine that he would, as much as possible, leave him alone.

Paine understood that handling Strauss was at least as important as handling the details of his job, and he devoted considerable attention and skill to the task of cultivating his employer. Strauss, who had always been "*Mister* Strauss" to Ellis and to the consulting engineers, became "Joe" to Paine, whom he addressed as "Clifford." Within less than four years, Paine was to become full partner in the renamed Strauss firm, and he was in many ways Strauss's successor. Paine, who survived into his eighties, served as a consultant to the Bridge District for more than three decades after the bridge's completion. In his later years, having outlasted the other engineers involved, he was frequently credited with being the man who had "really designed the Golden Gate Bridge," a conclusion that Paine did nothing to discourage: his obituary, in the Chicago *Tribune* of July 15, 1983, identifies him as such. In more than one sense, Clifford Paine was the real beneficiary of the work of Charles Ellis.

"WE'LL TAKE THE BONDS."

I

In accepting a crucial cash advance of $200,000 from the Bank of America, the Bridge District Board of Directors had also accepted a certain degree of interest, on the bank's part, in the district's operations. It was quickly made known that the bank's president, Will F. Morrish, was opposed to the continued presence of MacDonald in the post of general manager. MacDonald, it was felt, had attempted to swing the sale of the first block of bridge bonds to the R. H. Moulton investment firm instead of the bank. Moreover, as a longtime Northern California construction executive, MacDonald was seen as subject to the influence, if not the orders, of San Francisco's behind-the-scenes political figures, Tom Finn and Murphy Hirschberg.

Finn was a survivor of what now seemed the Pleistocene age of local politics, the pre-earthquake era of labor organization and its attempted suppression, the rise of union influence, the election of the nation's first big city labor administration, and the excesses of the Schmitz-Ruef regime. A stableboy who had organized the stablemen's union, Finn, "the boy who made friends," was carried by the closing ranks of local labor into a seat in the California State Assembly in 1901. Representing a workingman's district, Finn consolidated a labor-based support he never lost and established a growing network of influence among local unions and within city politics. When the Union Labor party assumed control of San Francisco in 1902, Finn was among its candidates elected to the Board of Supervisors. Appointed by Eugene Schmitz to the Fire Commission, Finn escaped the scandals of the graft trials, went on to serve as sheriff and deputy tax collector of San Francisco, surviving administration after administration, expanding his private influence and acquiring the reputation of being the city's political boss.

Through his influence on the powerful Building Trades Council and, through Hirschberg, his bagman, on the Board of Supervisors, Finn was now rumored to have acquired a monopoly on the insurance that all bridge contractors would be required to carry. This monopoly was allegedly guaranteed by the Finn-Hirschberg choices on the Bridge District board: San Francisco supervisors Warren Shannon and William P. Stanton, and MacDonald, the construction company executive who had become general manager.

Indeed, some of the early construction bids on the bridge had given off a peculiar odor. When in 1931 the directors had accepted contractors' bids for cement, which they had intended to buy in bulk and then furnish to the individual contractors as required for use in the bridge piers, roadways, and anchorages, the bids, when opened, turned out to be identical. To avoid this sort of obvious collusion, the directors, on seeking new bids in the fall of 1932, announced that those contractors desiring to work on the bridge would be required to provide their own cement.

The bridge was now the prize in a struggle between two kinds of financial influence. One was the old-time blue-collar squeeze, focused on the physical construction of the bridge, applied by Finn on the contractors. The other was a more modern, white-collar kind of pressure, centering on the bridge's financing, and applied by the Bank of America on the Bridge District's management. Although the means, and the style, of the boss and the bank were very different, their ends were more alike than not. Both sources felt, by right, a proprietary interest in the building of the bridge, and it remained to be seen which would be able to dominate, and then eliminate, the other.

The immediate priority, in the fall of 1932, was for working capital, and the only way to get it was to sell the bonds. In the election of 1930, voters had approved the issue of bonds that would pay an interest rate limited to no more than 5 percent. Under the terms for the first block of bonds, to be sold by the Bank of America, the bonds would be bought so far below par that the purchasers would, in fact, earn in excess of 5 percent. There was a question whether, in terms of the ballot measure, bonds with such a yield would be legal, and the matter had been submitted to a firm of New York securities attorneys for an opinion. Meanwhile, the directors had already accepted the advance on the basis of bonds yielding in excess of 5 percent. Bids had been called for, and received, for construction contracts. A court ruling would invite yet another legal challenge, which would postpone the start of construction indefinitely.

Emboldened by desperation, Strauss and the directors decided to appeal directly to the Bank of America's legendary chairman, A. P. Giannini. Forming themselves into a delegation, they arranged to call on Giannini at the bank's main office at One Powell Street in San Francisco.

More than any other individual, Amadeo P. Giannini had originated modern banking in America. Founding a bank based on the small depositors that other banks had largely ignored, Giannini had already acquired more customers than any other private financial institution in history and was in the process of building Bank of America into the nation's largest bank.

A maverick among bankers, Giannini had a strong populist streak that manifested itself repeatedly in his business policies. He introduced, defended, and established the idea of branch banking in America, reduced the standard mortgage rate to small borrowers, was the first American banker to advertise loans actively, and invested in and encouraged businesses he considered important to the future of the West, among them such high-risk categories as California agriculture and motion pictures. Convinced from experience that when his bank's constituents did well the bank did well, Giannini, as a matter of policy, had the Bank of America bid on every school bond issue offered in the state of California.

Opposed on principle to what he referred to with contempt as "idle money," Giannini had devoted considerable energy over the past few months to urging Depression-ridden Californians to get the economy going by "keeping money moving." In his own bank's advertising space, and in radio addresses, Giannini urged people to put money in the bank—"any bank"—in order to get it circulating, building homes, farms, schools, creating jobs.

Exasperated at the Republican lack of initiative in coming to grips with the Depression, Giannini had refused a personal request for support by Herbert Hoover in the 1932 election campaign and quietly supported Franklin Roosevelt instead. Now, with the country in the worst doldrums of the Depression, with the rejected Hoover still in the White House and Roosevelt not due to take office until March, Giannini pondered the future of his country, his community, and his bank.

When they met, Strauss and Giannini were both sixty-two years old. Each man had struggled for years, against powerful and entrenched opposition, to sell an officially resisted, yet broadly popular idea. Giannini had spent most of the past year fighting off a takeover of Transamerica, the holding company for his bank; he knew what Strauss had been put through in terms of the frustration and expense of legal maneuvering and delays. And he believed in the power of investment to stimulate the economy. "Dollars at work create credit," Giannini had told Californians over the radio, "and credit creates business and business creates jobs."

Giannini listened to Strauss's summary of what was now a fourteen-year struggle to begin the bridge. Now the Bridge District was all but out of money; it couldn't sustain its existence through another court test. Any funds that might have been available from the Reconstruction Finance Corporation had been committed to the San Francisco-Oakland Bay Bridge; there was no possibility of the government committing a similar sum to another project in the same area. It was a case of sell the bonds now or never.

"We'll take the bonds," Giannini said simply at the end of Strauss's summary. "We need the bridge."

What Giannini had agreed to do was extraordinary. The bank would itself buy the first $3 million of the bonds at slightly more than 96 cents on the dollar, giving the bonds an even and legal yield of 5 percent. The California Supreme Court would meanwhile rule on the question of the legality of a higher yield. Regardless of the court's decision, the bank committed itself to buy another $3 million in bonds at 5 percent on March 1.

Before Strauss left Giannini's office, the banker asked him a familiar question, one that Strauss had been asked by individuals and before committees and press conferences and groups both in favor of, and hostile to, the bridge, ever since his first presentation of the original proposal more than eleven years before. How long would this bridge last?

"Forever," was Strauss's practiced reply.

2

The man to whom Strauss had first made this pledge, and who had first suggested to him the possibility of the bridge that had absorbed his life, was now in the penumbra of his political eclipse. Michael O'Shaughnessy, once the avatar of Progress in San Francisco, had become the symbol of die-hard resistance to change. His public opposition to the bridge had linked him with the most resented of entrenched local interests, and his outspoken fears of extravagant costs, made groundless by the cold reality of construction bids, had brought into question not only his political but also his technical and managerial competence. With Rolph's departure from the mayor's office, the politically untouchable city engineer had become fair game, and he was now under fire, it seemed, from every quarter. There were charges in the Board of Supervisors of cost overruns on the Hetch Hetchy project, conflict of interest, and the use of O'Shaughnessy's staff for outside work. There were also personal incidents that added fuel to the various flickering feuds O'Shaughnessy had sparked over the years, combined with an overall public desire, intensified by the Depression, to share and dramatize bad luck by seeing the once-mighty brought low.

In December 1931, the investment property that O'Shaughnessy had bought in Marin County, a Mill Valley cafe that he had leased to a pair of

Italian proprietors, was raided by federal agents as a speakeasy; there was a possibility that O'Shaughnessy might be held responsible for the fine. His dispute with Captain Little, the consulting engineer hired by the supervisors to check on the Hetch Hetchy pipeline figures, had been aggravated into an ugly row by O'Shaughnessy's contesting the engineer's competence and, in a letter to the state engineering board, his reference to Supervisor James R. McSheehy as a "political buccaneer and agitator." The letter, and the remark, were made public, and the Board of Supervisors passed a resolution censuring O'Shaughnessy.

Beneath these surface ripples was a deeper tide of public change. People were tired of O'Shaughnessy and his ceaseless demands for money to complete his distant, unseen Hetch Hetchy project. They had grown weary of this combination engineer-official who always seemed so sure he alone knew what was best for San Francisco. The times had passed him by. The city had rebuilt itself thoroughly from the earthquake. The new bridges, one to the East Bay and one to the north, promised the dawning of a new era of predominantly regional and national, as opposed to local, interest. O'Shaughnessy, even though he had prophesied and helped bring out this new age, was too strongly associated with the old. By giving himself totally to his city, he had engineered himself out of a job.

In 1932 San Francisco adopted a new city charter, eliminating the post of city engineer as occupied by O'Shaughnessy. In its place, a chief administrative officer system was introduced, with Alfred J. Cleary, formerly an O'Shaughnessy assistant, as the nonelective CAO. O'Shaughnessy was retained as a consulting engineer, with his duties confined to reduced responsibilities on the Hetch Hetchy job, which was finally winding to its conclusion. Thus it was Cleary, in his new capacity as the city's chief adminstrator, who began meeting with the Bridge District officers and the consulting engineers, ironing out the details for beginning construction of the bridge.

Within the chambers of the Board of Supervisors, there was unconcealed relief, mixed with what had to be a certain apprehension, at O'Shaughnessy's departure from the center of local politics. If the age of Michael O'Shaughnessy's influence in San Francisco had come to an end, could the waning of the power of Tom Finn and Murphy Hirschberg be far behind?

Some elements of the press and within the Bank of America were convinced that time had already come. The beginning of the bridge, it was argued, should represent a fresh start, a clean break with the past, a stop to the continuation of old, discredited personalities, interests, and habits.

In the fall of 1932, shortly after the bank had agreed to buy the first block of bonds, Will Morrish, Bank of America president, wrote a letter to the bridge board, reemphasizing the bank's substantial financial and moral interest in the bridge and demanding the dismissal, on grounds of misman-agement, of MacDonald, the general manager.

According to the San Francisco *Examiner,* MacDonald, along with Shan-non and Stanton, the two San Francisco supervisors who had been originally appointed to the board and who had refused to resign when accused of being Hirschberg's picks, represented the Finn interest within the board. It had been Stanton, the *Examiner* pointed out, who had originally nominated MacDonald for the general manager's job, and the two supervisors were now supposed to be the last pro-MacDonald holdouts on the board.

There was, it must be admitted, a certain Laurel-and-Hardy quality about Shannon and Stanton, a teamlike propensity for political scrapes that at times edged over into physical comedy. In January 1933, following a Lions Club dinner at the Alta Mira Hotel in Sausalito honoring the bridge directors and officials, Shannon attempted to board a return ferry and, missing his step, fell into the bay. Held afloat by his coat collar, which had been grabbed by Strauss and A. R. O'Brien, another director—and a Shan-non rival—on the board, Shannon had to endure the indignity of being fished out by the employees of Golden Gate Ferries, Ltd., the same ferry line that he and the other directors had been attacking for years as being selfish and incompetent.

It was just a few weeks after this that Shannon, working in tandem with Stanton, attempted to use the formal groundbreaking ceremony celebrating the beginning of actual construction on the bridge to settle old political scores. While Shannon, chairman of the day, and Stanton, general chairman of arrangements, packed the front rows of the grandstand with their friends, relatives, and supporters, the other directors, including District President Filmer and O'Brien, one of the men who had helped hold Shannon afloat, along with most of the northern county directors, were unceremoniously ushered to the rear of the stands, or were told that they had no tickets. Reduced to watching from the periphery as Shannon and Stanton hogged the spotlight, the slighted directors, most of whom had traveled to San Francisco's Crissy Field from considerable distances, were consoled some-what by the ceremony itself degenerating into a fiasco. "The bands didn't play as scheduled," reported the *Examiner,* "the twenty-one-gun salute was never fired, and arrangements for handling the crowd were such that it was necessary to halt the exercises when they were half finished—but Shannon was chairman of the day, and he was in the front row." With support of this nature, MacDonald hardly needed a determined opposition. Yet that was exactly what he had. "The time has come," O'Brien announced on December 28, the day of the monthly bridge board meeting, "to break loose from 'Murphy' Hirschberg's snake charming. Hirschberg got MacDonald his job and demands that he be kept in it at all costs."

In advance of the meeting, MacDonald had been presented by the directors with a redefined set of requirements for his job, including a 50 percent pay cut and an insistence that he devote full time to the district. The directors were still in session when a messenger arrived with the general manager's letter of resignation. It was accepted immediately, with Stanton the only director dissenting.

"We have gotten along after a fashion up to now," O'Brien remarked, "but the time has arrived for a definite working organization."

With what seemed like the final gasp of the old politics, construction of the new bridge was at last ready to get under way. A week after the directors' meeting, steam shovel excavations began on the Marin shore for the pit that would hold the bridge's massive concrete north anchorage. The long years of proposals and debates and plans, the legal and financial stratagems, the time of intellectual bridge building, was giving way at last to the simple, exhilarating, physical challenge of the greatest bridge man had yet attempted.

On the evening before the groundbreaking, a formal banquet, in dedication of the bridge, was held at the Fairmont Hotel in San Francisco. It was to be Joseph Strauss's night, and his friends and associates had done their best to see to it that the man they so admired would not be denied the recognition he deserved. Strauss was to make the keynote speech, and, as with the groundbreaking ceremonies the next day, the audience had been strongly stacked with Strauss supporters. One of the northern county directors, told he would have to pay for his banquet tickets, was so irate he threatened to pack his bags and leave and had to be persuaded to remain for the formalities.

For Strauss, this evening and his speech, in many ways his valedictory, took him back more than forty years, to the stage of the opera house in Cincinnati, where as a young graduate he had addressed his class's commencement audience. Once again he was presenting a bold proposal for a great bridge, only this time, the project was real.

This return to his earliest ideals and hopes, the transcendence of all the years of uninspiring work between, with their endless promoting and haggling, negotiating and political wheel-greasing, had released the boy poet that had always abided in Joseph Strauss. He began his address with a quote from Tennyson:

BREAK, BREAK, BREAK,
ON THY COLD GREY STONES, OH, SEA!
AND I WOULD THAT MY TONGUE COULD UTTER
THE THOUGHTS THAT ARISE IN ME.

"Tonight," said Strauss, "I am face to face with such a moment. Tonight ears long accustomed to harsher sounds have been filled with the harmony of friendly voices and the music of testimonial greetings." The depth of his feelings was such, Strauss maintained, that language itself was inadequate to express them; and yet he was expressing them, for all his self-involvement with the bridge had turned the stammering, inhibited Strauss who had haltingly addressed the Sausalito City Council eleven years before into a compelling after-dinner speaker. The bridge had rendered him eloquent.

It had also, it was apparent now, exacted a terrible price. Strauss, a short man, had become physically frail. His hair had gone completely gray. Although accompanied constantly by a man who was rumored to be a physician, H. H. "Doc" Meyers, Strauss did not seem to show any increase in stamina or resistance to disease. Indeed, Meyers, a man who seemed to have no medical practice or other visible means of support, appeared to devote most of his skills to healing over the feelings of people injured by Strauss's rude, abrupt manner. Not so obvious, but just as precarious as the state of Strauss's health, was that of his financial condition: his commitment to the bridge had cost him, Strauss would later estimate, a million dollars in rejected jobs.

And yet he had survived, and there was honor in that. He had won the long war with the powerful ferry boat interests, "a ferryland," Strauss quipped, "within the limits of your magnificent Bay. Tonight that ferryland is back in your possession. The tocsin of a 13 years' war fought to recover it, has been forever silenced. Martyrdom"—he was apparently referring to his own physical state now—"has run its course."

Strauss had outlasted not only his opposition but also his associates and one-time friends. There was O'Shaughnessy, once the most powerful man in San Francisco, who had first interested Strauss in a bridge at the Golden Gate, had collaborated with him on the design for such a bridge, only to turn against the idea; he was now merely a marginal local figure, another officeholder out of a job. There had been Ellis, the academic, whom Strauss had enabled to enter the real working world, build real bridges, make decent money; a gifted man who had done outstanding work but who had been disloyal and had egotistically insisted on the importance of his own contribution. Most recently there was MacDonald, the general manager, locally well connected everywhere, from the Bohemian Grove to the backroom bars presided over by Tom Finn. He had attempted to dictate policy to Strauss and the directors, and he too had fallen. Nevertheless these and other men of ability and power were already being forgotten—their contributions to the bridge went unmentioned in Strauss's speech. But Strauss had persevered. Survived. Prevailed. As with the building of his great bridge, San Francisco, so cruelly ravaged by nature, would now prevail.

"In these giant arms of steel," Strauss addressed his audience with fullbore grandiloquence now, "clasping hands above the tumbling waters of the Golden Gate, San Francisco returns to her bosom, her own flesh and blood, torn from her by the Gods of Fire and Water in the shadowy long ago."

At last, the city would be physically tied to the vineyards and redwoods to the north; the constricted community would expand into the residential hills and valleys of Marin. With the finishing touch of the bridge, California would present her answer to the call of her people, "and her rebuke to the irreconcilable iconoclasts and the gospel of inertia."

Strauss had even outlasted his own now-orphaned original design for the bridge, a proposal now so completely forgotten, so thoroughly replaced by the slender, soaring imagery of the new bridge that it seemed to be the work of another age, another man entirely. It was as though destiny, not haste and economy, and Joseph Strauss, not Charles Ellis, had been responsible for conceiving the design about to grace the Golden Gate.

Once more, as in the Golden Age of Greece, "the ideal and the altruistic march side by side with the material and the commercial." Once again, a work had been conceived that would stand beside "the marble pillars of the Parthenon, the spires of the Cathedral of Milan, the ruined temple of Karnak, and the inlaid walls of the Taj Mahal." Here at the Gateway of the Pacific, at the Crossroads of the Redwood Empire, "will stand the Bridge of Ages," the embodiment of silent power and ethereal beauty, capable of stirring ordinary men to attempt the language of poets:

> OH, THE GLORY OF ALL TOGETHER,
> IS COMMONPLACE TO ME,
> WHEN I STAND AT THY RAIL, OH MIGHTY SPAN,
> AND DREAM THY DREAMS WITH THEE.

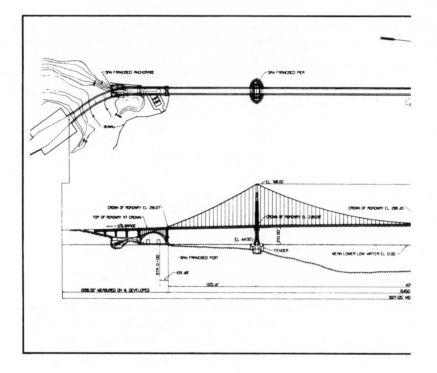

TWO

BUILDING

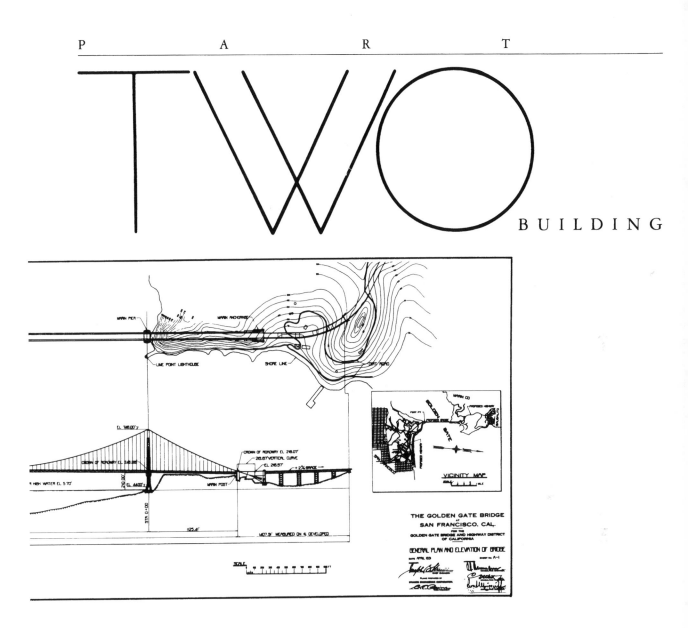

You couldn't help
know it.
You were building
the greatest
structure in
the world.
You knew one
thing for sure—
there wasn't
going to be
anything else like it.

Harold McClain,
Bridgeman,
June 9, 1985

"BEGINNING TO FEEL LIKE MY OLD SELF . . ."

I

Joseph Strauss, who had never before proposed the building of a suspension bridge, had also never supervised the construction of one. Now, for his firm's entry into the category, the Strauss Engineering Corporation was charged with overseeing the building of the largest bridge of its kind ever attempted, with the tallest steel towers ever erected, the longest, largest cables ever spun, and the most enormous concrete anchorages ever poured. There were, in addition, other problems unique to the colossal nature of the bridge and the site, the Pacific Ocean entrance to San Francisco Bay, among them the construction of a concrete fender, the dimensions of a football stadium, in the open sea. Once again Strauss, who had got the chief engineer's job on the strength of being able to assure Ammann's and Moisseiff's participation as consultants, looked outside his own organization for the credentials he needed.

In the winter of 1932–33, while the final financial and beginning construction preparations for the bridge were getting under way, Strauss concentrated his considerable powers of persuasion on getting the man he most wanted to be the resident engineer to coordinate and supervise the work of ten different contractors in the day-by-day building of the bridge.

Born in Ottumwa, Iowa, in 1897, Russell G. Cone had studied bridge engineering at the University of Illinois, where he had been a pupil, and an admirer, of Charles Alton Ellis. Within a week of his graduation, with honors, as a civil engineer in 1922, Cone had gone to work for the bridge engineer Ralph Modjeski, whose firm, Modjeski & Chase, was then building the Philadelphia-Camden Bridge. Starting as a junior engineer, Cone, a remarkably commanding and charismatic young man, soon became resident engineer, responsible for piecing together the side and central spans of what was then the longest suspension bridge in the world. From that job, Cone advanced to resident engineer, in charge of all construction on the Ambassador Bridge at Detroit, a span that surpassed the Delaware bridge.

Cone possessed the ability to direct men without having to make a lot of noise. A veteran of the World War I Rainbow Division, where he became a

Russell Cone in leather hard hat. The Golden Gate Bridge was the first major construction job on which the wearing of hard hats was required.

lifelong drinking buddy of Charles McArthur, coauthor of *The Front Page,* Cone, burly and vigorous, was fearless out on the job, climbing catwalks, descending into caissons, unwilling to send anyone to a place he wouldn't go himself. A member of the American Society of Civil Engineers, he contributed thoughtful, lucid papers to its journal; he also sang with a fine tenor voice. In a time when most people were grateful to have any job at all, men who could meet the physical and intellectual rigors of what were then just about the only expressions of national optimism of the day, the great construction jobs, had acquired an almost Renaissance-figure glamour.

In 1936, the film director Frank Borzage met with Cone and drew, from his personality, traits incorporated into a film character later played by Spencer Tracy. At about the same time Cone, along with a number of other engineering and construction figures, among them the resident engineer of the San Francisco-Oakland Bay Bridge, appeared in a newspaper ad for Camel cigarettes, a pre-Marlboro attempt to merchandise tobacco by being butch.

In 1932, Cone was managing the Tacony-Palmyra Bridge between Pennsylvania and New Jersey, a steel arch bridge which Modjeski & Chase had built during Cone's tenure in Detroit. It was a holding position as the Modjeski firm was between major bridge construction jobs. Starting as early as 1931, Cone began receiving first inquiries, then offers, from Strauss. The first offers were flatly refused. As a former pupil of Ellis's, Cone undoubtedly knew of Ellis's problems in working for Strauss. Ellis was something of a hero to Cone, and he would surely have sympathized with his professor's position in Ellis's firing. Also, Cone had always worked under the auspices of Ralph Modjeski, a suspension bridge engineer of the stature of Ammann and Moisseiff. To go to work for Strauss, essentially a builder of cantilever bascule bridges, would, to him, be a serious loss of prestige. Cone's refusals only encouraged Strauss, like a spurned suitor, to press his case more vigorously. By the fall of 1932, he was offering Cone $10,000 a year— more than double what Cone had been paid as resident engineer on the Ambassador Bridge, but Cone, still skeptical of Strauss's reputation, didn't take it.

The Modjeski firm was, at this time, entering a transition. Ralph Modjeski, the founding engineer, was an old man. Frank Masters, one of his partners, was essentially a promoter, like Strauss. Clement Chase, the other partner, who had become Cone's mentor, saw an advantage to Cone's going West, even if it was to work for Strauss. Simply by working on a project of the stature of the Golden Gate Bridge, Cone would enhance his reputation.

He could take his own team of engineers with him, exercise a certain autonomy in the vacuum of the Strauss firm's inexperience in suspension bridge building. When the job was done, he could return to Philadelphia and become a full partner in the Modjeski firm. He would have sacrificed nothing in the East by accomplishing whatever he could in the West. It was only after these discussions with Chase and with other of his friends, possibly including Ellis, that Cone suggested he might be willing to come to San Francisco.

Meanwhile, as the questions of financing the bridge were resolved and the start-work construction dates approached, the Bridge District directors, the bankers, and the insurance people increased the pressure on Strauss to have experienced suspension bridge engineers overseeing the job. Finally, to close the deal, Strauss at the end of 1932 sent Clifford Paine to New Jersey with $2,000 in cash to cover the moving costs of Cone and his family. With the prospect of nothing to lose from the association, Cone took the job and in February 1933 headed the family car west.

Cone arrived in San Francisco on Valentine's Day: February 14, 1933. The following morning he reported for work at the just-completed District Field Office built on pilings near Fort Point on the San Francisco shore and got his first look at the site that was to dominate his thoughts for at least the next few years. Here he, and the complex of engineers, contractors, inspectors, foremen, and gangs he was to work with, were charged with erecting the largest bridge of its kind ever built, in a location where that bridge and the men who worked on it would catch the worst of an ocean outlet that combined cold weather, constant yet shifting winds, blinding fogs, deep water, powerful tidal currents, and frequent violent storms.

Already, elements of construction crews were at work on both the Marin and San Francisco sides of the strait, demolishing buildings, putting in access roads, runways, and power lines, and excavating earth for the cable anchorages, while on the water and beneath it, contractors had started building the loading and unloading wharves and piers that this unusual job required. Between the two shores ran a speedboat, a fifty-five-foot exrum-runner with a shallow draft and an uncomprehending Swedish pilot that made her tilt and slam against the swells of the Gate until passengers were drenched and seasick.

The water of the strait was so consistently rough that the use of floating equipment for building the crucial San Francisco pier, which was to support the south tower of the bridge, was out of the question for anything but dredging. Instead, an access trestle would have to be built from the San

Francisco shore, extending some 1,100 feet out to the pier site, where materials could be carried and set in place and concrete poured from a strong, stable base. Construction up to this point had been with wooden piling, and the contractors, Pacific Bridge Company, were now preparing to drill and blast holes for the trestle's supporting steel framework.

On the Marin side, the same firm, Pacific Bridge, was doing preliminary work on the cofferdam necessary to shut out the sea so that work on the north pier could begin.

Section by section, players and instruments were assuming their proper places. Strauss, who on occasion would affect a Sol Hurok slouch hat and over-the-shoulder topcoat, was very much the impresario; Clifford Paine, exacting, thorough, even-tempered, was his orchestrator and arranger, while the thirty-six-year-old Russell Cone was the whiz-kid conductor, charged with directing the performance of the work of the now all-but-forgotten composer, Charles Ellis.

To run the hall, the Bridge District directors had appointed a new general manager, James Reed. A former navy commander and construction engineer with an advanced degree from MIT, Reed, who had most recently been general manager of the Schlage Lock Company, was determined to be nobody's fool, and he wasted no time in taking on Strauss. When the feisty chief engineer proposed that inspection of the bridge metalwork materials, which were to be fabricated at the McLintic-Marshall shops in Pottstown and Steelton, Pennsylvania, and at the John Roebling plant in New Jersey, be centered in Chicago, headquarters of the Strauss Engineering Corporation, Reed firmly objected. "Based on my own experience in charge of material inspection in the Eastern District of the U.S. for the Navy Department," Reed notified the directors and engineers, "I do not concur with the opinion of the Chief Engineer." The headquarters for inspection were established at Philadelphia.

Reed, sensitive to the maneuvering that had undermined MacDonald, directed Strauss and the consulting engineers to supply him with copies of all their correspondence, including telegrams. He sent Ammann a copy of Strauss's contract with the district, asking the New York engineer to review its terms in comparison to those considered standard with the Port Authority in New York. He also reestablished the relationship between the general manager's office and Derleth, the directors' man within the Engineering Board. "I am pleased to note what you say," Ammann wrote Derleth on February 7, 1933, "relative to the manner in which Commander Reed takes hold of matters." Reed, described by people who knew him as "100 percent Navy," was definitely not going to be elbowed off his own bridge.

2

The man chosen to inspect the manufacture of the bridge's most expensive component, its steelwork, was Herbert J. Baker, chief inspecting engineer of the New York Port Authority. Working under Ammann's direction, Baker and his staff would devote full time to overseeing the rolling and fabrication of the important parts of the bridge's metallic construction, preparation for which was now getting under way at plants in Roebling near Trenton, New Jersey; Bethlehem, Pottstown, and Steelton, Pennsylvania; and Sparrows Point, Maryland. The question of who would pay for this additional staff of inspectors, their travels, and the maintenance of an inspection headquarters office was left undecided and would ultimately be the subject of a lawsuit brought by Strauss against the district.

The enormous scale of the bridge made the manufacture and fabrication of steel for the towers—the largest and highest in the world—and for the floor's five miles of twenty-five-foot-deep stiffening trusses—a job that challenged the capacities of even the enormous mills and plants of Pennsylvania. According to E. J. Harrington, designer of the Dumbarton lift bridge over the southern arm of San Francisco Bay and an early opponent of the Golden Gate, the bridge could not be built because there was no plant large enough to fabricate the steel needed for its construction.

The steel for the towers and for the suspended structure had to be rolled in four different forms, in four different rolling plants. This rolled steel was also of several different types: silicon steel, used only in the towers and the suspended structure; carbon steel, which would be used only in the architectural treatment of the tower bracing above the roadway; and copper-bearing carbon steel for the expansion joints of the floor structure and in the railing. Cast steel would be used in the cable saddles that went on the tower tops, the strand shoes that would help anchor the cables, and the cable bands that would hold the strands of the cables in compression and support the suspender ropes. Forged steel was to be used for rollers, pins, bolts, rope sockets, and collars.

To assure the consistency of the grades of steel, the rolled slabs were to be stamped, while white hot, with a number indicating the melt and a letter pinpointing the slab's location in the ingot. The structural steel was rolled in a variety of shapes and thicknesses, and at least two tension tests and one bend test had to be made from each melt and variety: plate, shape, flat, rivet rod. To help identify them at the fabricating shops, the ends of silicon-steel products were painted green, and copper-bearing steel yellow.

Fabrication of the steel into sections for the tower columns took place in two Pottstown and Steelton shops, long shedlike buildings with the interior space of vacant manufacturing plants. Even these were not large enough to accommodate the complete assembly of a tower leg or of a floor truss. Also, such work would have involved great difficulty because of its complexity. Instead, assembly operations were conducted on the parts of each unit.

The tower legs were made up of plates and angles, which were combined into a cellular design, with each cell forty-two inches square. These were shop-riveted together in column sections varying from $22\frac{1}{2}$ feet to 45 feet long. Photographs from this time of these recumbent tower sections, side by side, crowding the gloomy sheds, the slanting factory-window light falling across their flat angled surfaces dotted with rivet heads, suggest the basement laboratory of Doctor Frankenstein, where the enormous creature sleeps, ready to be awakened, lurch to life, and stalk towering across the continent.

The various plates and angles were fitted together by jigs—positioning forms with adjustable cranks and planed beams—to make sure they were straight and true. Held by the jig, the pieces could be angled or planed to ensure a perfect fit.

The parts of a column section were then "boxed"—the cells were assembled into a section, kept square with diaphragms and jacks—and the entire cell group was then riveted together.

The accuracy of the results of this procedure had to be checked by erecting at least 60 percent of the cell groups of the base tier of one tower leg. Because of its size, this base had to be erected outdoors. It towered over the shed, a hulking giant child, so out of fit with its surroundings that it remained exposed until it eventually had to be cleaned of rust.

As a final step, tower and suspended structure members were painted by spray or brush with up to three layers of shop coat, originally red lead but later cut with a mixture of synthetics. The fabricated steel would eventually be loaded, in sections, onto railroad flatcars and hauled to Philadelphia. From there it was sent by ship through the Panama Canal to the Pacific Coast, where its arrival would be timed to coincide with the construction schedule on the bridge.

3

On both the Marin and San Francisco sides of the strait, excavation involving blasting and the use of steam shovels was under way for the great pits that would accommodate the enormous pylons and anchorages to guide and tie down the great cables. Each of these anchorages and pylons was a construction job on the scale of building a stadium or an amphitheater, requiring earth removal, bracing, the webwork installation of steel reinforcing rods, and, later, eyebars for the cables, the building of wooden forms, and the pouring of concrete in such quantities that separate batching plants were erected on either shore. For the first time, trucks would be used that mixed, as well as carried, concrete. On the San Francisco side particularly, where the slope was mainly serpentine, a massive gravity anchorage was planned as a kind of counterweight to the earthquake fears of the still-active bridge opposition.

In early June, an earth slide at this site would reveal an old tunnel, "What the Opposition," Derleth remarked, "would have doubtless been pleased to call a crevass." Entering the tunnel himself, Derleth found it to be an old water tunnel, used as part of San Francisco's water supply system before the Civil War. "We found the tunnel," he reported, "with two feet of water standing in the bottom. Toward the shore of the Golden Gate, the tunnel was disintegrated. But at the site of the anchorage, though exposed all these years, the serpentine in the walls and roof was clean and intact and still showed the pick marks. There is no evidence of softening."

On the Marin side, where the land was mostly sandstone and shale, the excavation, though just as massive, proceeded with less delay. Day by day, the pits were deepened into great rock caverns, to be filled again by crews of carpenters, building what resembled a matchwork maze of wooden forms. It was labor-intensive work whose costs the contractor, the San Francisco firm of Barrett & Hilp, controlled by taking maximum advantage of an employer's market and hiring men as pieceworkers.

"There were no steady work crews on construction," recalls Albert "Frenchy" Gales, who worked on the anchorages on both sides of the strait. "They got men as they needed them. The foremen for Barrett & Hilp would pick guys—there'd be a breakdown, and they'd let men go. You'd only get paid for the time you put in—an hour, two hours—no matter how much time you'd spent waiting for work."

The jobs, at least at this point on the bridge, were mostly unskilled and nonunion. "There were other guys standing by the office waiting to go to work. You could see them, waiting to take your place. The pushers wouldn't let you go to take a leak, have a smoke. If you did, you were gone."

Men would be hired, used, let go, then rehired often by the same contractor. A carpenter who got two or two and a half days' work a week would consider himself lucky.

The same sandstone and shale that permitted the rapid excavation for the Marin anchorages also provided a solid base for the foundation of the bridge's north tower. On the solid shelf of rock that the drillers had discovered at Lime Point on the Marin shore, preparations now began for erecting the first element of the bridge itself. A three-sided frame cofferdam was barged in from a Sausalito shipyard, filled with crushed rock, and sunk on steel pilings at the pier site, forming a kind of artificial peninsula, jutting into the strait. Additional rock was used to shore up the space between the cofferdam and the Marin bluff, and the sides of the cofferdam were covered with overlapping sheet-metal panels. To seal out the water, the cofferdam was caulked—with wheat.

"I had the caulker's job," recalls Frenchy Gales. "They'd lower you on a rope, and you'd pour half a sack of wheat. The waves would come in. Hit you. You'd get soaking wet. We'd pour wheat down—it would swell up and caulk it up."

When the cofferdam was sealed, the interior was pumped dry with enormous centrifugal pumps "so powerful they could suck up a rock you couldn't lift." The day the full peninsula emerged, the cofferdam was full of fish, "bass, cod, rays, sharks, everything. Everybody dropped everything and went to grab some fish." Then, within the now-dry rock base, the work of excavation could begin, quarrying down into the rock for the firm, level foundation for the tower's supporting pier.

Across the strait, on the San Francisco side, work was under way on the access trestle that would have to be completed before construction of the south pier could begin. The original plan had been for a steel trestle, or wharf, extending from the San Francisco shore to the pier site, with its deck fifteen feet above the waters of the strait. To seat the framework for the trestle, the initial plan had been to drill holes in the bottom rock slightly larger than the steel support columns and to fill in the extra space with concrete. This strategy was abandoned when no drills could be found that would clear the underwater holes as they were drilled. Instead, the Pacific Bridge engineers developed an underwater blasting system resembling the dropping of a depth charge. Through a large pipe, a black powder bomb would be dropped on the bottom at the pile location. This charge would be driven into the rock by a metal cylinder, or follower. Then it would be fired, shattering the bottom material with successive blasts until it was possible to drive a steel tubular-pile footing into the rock.

The crew on the San Francisco trestle was a mix of ironworkers and piledrivers and included the Pacific Bridge job superintendent Jack Graham and his brother Jim. This shirtsleeve participation was partly dictated by necessity. To get the job, the Grahams had cut Pacific Bridge's bid to the bone: when the bids were opened, it was discovered that Pacific Bridge's bid was nearly a million dollars lower than the next bidder. When Jack, the elder of the Graham brothers, found out, according to a woman who was there, "he nearly jumped out the window."

Like the Grahams at the San Francisco pier, each of the contractors had his own engineers on the job. Russell Cone and his staff, which at first consisted of a secretary and two other engineers, were there to oversee and coordinate the work, sort out conflicts, schedule the arrival of men and materials, enforce safety and production standards, and file a detailed daily report on each aspect of each contracted job. Despite the fact that Cone was able to attract other young engineers of outstanding promise and ability, like Ted Kuss from MIT and John Blondin and Ed Davenport from the University of Illinois, there were never really enough supervisory people for a job that was to grow progressively more complex, as Moisseiff, Ammann, and Reed all protested. For the present, Cone was stuck with long hours, the prospect of no vacation, and, in effect, no days off.

While Cone oversaw the day-to-day progress on the job, Clifford Paine was in charge of the preparation and quality of materials, as well as the operation of the Strauss Engineering office in San Francisco. Because of the peculiar nature of the job's financing, where the costs had to be guaranteed before the bonds could be sold, specifications for the individual contracts had to be revised and rewritten in the course of construction, an enormous amount of additional work for the Strauss staff.

Also, in response to Ellis's criticism about the adequacy of the tower design, the consulting engineers agreed to have a model made and tested by Professor George E. Beggs of Princeton University. The model, made of stainless steel, and the tests would cost more than $12,000, and although the engineers did not consider the model test "an essential part of the design," they nevertheless felt it was "desirable" that it be done. Charles Ellis might be wrong about the safety of the towers, but the other engineers weren't willing to gamble that he was all *that* wrong.

With the on-site work moving along on schedule under Cone, the office and inspection work proceeding efficiently under Paine, and the steel fabrication under way in the East under the experienced eye of Baker, there didn't seem to be any pressing day-to-day need for the participation of Joseph Strauss in the project at all. This was probably why, following Strauss's address at the groundbreaking banquet on April 4 and a visit and inspection by the Engineering Board to the bridge site on April 10, several weeks went by before anyone noticed that the chief engineer had disappeared from sight entirely. It was early June before the Bridge District directors, trading notes among themselves, realized that none of them had seen their chief engineer for more than two months. When inquiries were made as to Strauss's whereabouts, his personal representatives, Charles Duncan and "Doc" Meyers, told the directors that Strauss had suffered a nervous break-down and that, after consulting Dr. Arthur Bloomfield of Stanford University Hospital, the chief engineer had taken a recuperative cruise through the Panama Canal to New York, followed by several weeks' rest in the Adirondacks. He was now in New York, the directors were informed, en route to Washington.

The directors, already annoyed that Strauss had left without informing them, were particularly irked that he had found time, while in New York, to grant an interview to several New York newspapers.

"If Mr. Strauss is well enough to be interviewed," observed Francis V. Keesling, head of the bridge board's Building Committee, "it is my opinion he is well enough to return and pay some attention to this major project. There are plenty of matters right now needing his consideration."

Dr. Bloomfield, contacted at Stanford, said that he had been unable to determine the extent of the engineer's disability, since he had made no examination. Amid accusations that Strauss had been "playing hooky," the directors adopted a motion urging the chief engineer to return at once and "get busy." All this was gleefully reported in the San Francisco newspapers, along with repeated references to Strauss's (actually the Strauss Corporation's) $1,080,000 fee.

On June 30 Strauss wired William Filmer, president of the bridge board, that he was now "beginning to feel like my old self." On his doctor's advice, Strauss reported, he would return to San Francisco "by leisurely stages," arriving about July 20. "Have kept in close touch with Chicago and San Francisco offices," he assured Filmer, "and, as you know, work is going along smoothly and satisfactorily."

On July 21 Strauss, back in San Francisco, issued the following state-ment: "Although I am still under the care of a physician I am 100 percent on the job again. I have found that work has progressed to my perfect satisfaction during my enforced absence."

The "physician" in this case was "Doc" Meyers, a man of extremely dubious medical credentials. Six years later in 1939 Meyers, under indict-ment in Tacoma, Washington, for selling fraudulent oil leases through the mails, would be investigated by inspectors of the U.S. Post Office. "Dr." H. H. Meyers, as the postal inspectors referred to him, had been represented in an oil and gas lease selling campaign "as a very wealthy financier and businessman, and a principal in the Strauss Engineering Corporation, builders of the Golden Gate Bridge." Meyers's attorney had indicated to the court and jury, said the postal inspectors, that "Dr." Meyers had been instrumental in procuring the appointment of Ammann and Moisseiff as consulting engineers in cooperation with Strauss. It was an association that Moisseiff and Ammann quickly and vehemently denied, as well as any knowledge of the details of Meyers's relationship with Strauss.

In his statement announcing his return to work on the bridge, Strauss had neglected to mention the most significant change in his personal life. He had remarried.

On June 9, 1895, in Chicago, Strauss had married May Van, a friend from Strauss's early days in Cincinnati. They had two sons, since grown, one who had attended West Point, and the other Annapolis. Now, on June 26, 1933, Strauss, who gave his age on the marriage license as sixty but was actually sixty-three, was married in Prince Georges County, Maryland, near Washington, to Ethelyn Annette Elworthy Hewitt, forty-seven and a widow. Annette Strauss, as she came to be called, was from Sacramento, California, originally, but had most recently lived in Los Angeles, where she had been trained and had performed professionally as a singer. She was a close friend of "Doc" Meyers, who may have introduced her to Strauss.

"She was a much younger woman than Strauss," remembers Izetta Lucas Cone, Russell Cone's wife at the time, "very attractive, beautifully groomed, with prematurely white hair." The Strausses moved into the Park Lane Apartments on San Francisco's Nob Hill, situated cater-corner to the Fairmont Hotel. From his apartment, Strauss commanded a view of the bay from the Golden Gate all the way east to the hills behind Berkeley. The view from here was like that from his long-demolished Aeroscope from the 1915 world's fair, only this time Strauss's project was both significant and permanent. Here he could keep a satisfied eye on his great life-work unfolding at the Gate, while stirring from home only when necessary. A car and driver were available to take him and Annette wherever they wished to go, but with the bridge so obviously in good hands, and with the justification of his mandated gradual recuperation from his breakdown, there seemed to be little reason to go anywhere. Strauss looked out at his bridge, was looked after by his younger, attractive wife, and alternately fondled and chided his dog, a cream-colored Great Dane "the size of a calf." Strauss wrote poetry, dreamed of new projects, and puttered with inventions. Annette's pet name for him was "G.G.," their code for "Golden Gate."

Although the Strausses rarely entertained, they had assumed a certain prominence in San Francisco social and cultural circles. In 1935, Annette Strauss sang in recital at the Western Women's Club in San Francisco, later the Marines' Memorial Theatre. Her performance was preceded by an introductory address by Mrs. Pierre Monteux, wife of the conductor of the San Francisco Symphony. Joseph Strauss's own 1932 poem, "The Redwoods," was published as a song and performed, along with two patriotic marches, "Flags Aloft," dedicated to the U. S. Military Academy, where one of his sons had graduated, and "America Our Own, Our All," by the San Francisco Municipal Band at a Sunday concert in Golden Gate Park. Strauss might never achieve, in engineering circles, the intellectual eminence of a Moisseiff, a Modjeski, or an Ammann, but in San Francisco, at least, he was acquiring something of the aura that had arisen around that other great bridge builder, John Roebling, who, as tradition had it, had given his life in the course of building the great Brooklyn Bridge in New York. The story of John Roebling and his son and successor, the reclusive Washington Roebling, who oversaw the construction of the Brooklyn Bridge largely from the confines of a sickroom, was a legend that Strauss was to grow closer and closer to as time wore on.

4

From the picture window of his Nob Hill apartment, Strauss could, by the summer of 1933, savor the sight of the first indications of the "great etching in steel" that was to appear on the mighty canvas of the Golden Gate. On the Marin hillside, the wooden forms for the northern anchorages and pylons were completed, and the concrete was now being mixed, trucked, and poured in a river of aggregate that sometimes flowed continuously, day and night.

Below this, at the Marin shoreline, the concrete pier for the base of the north tower was beginning to appear above the edges of the cofferdam, while on the opposite shore the most noticeable change was the trestle, being extended slowly into the strait toward the site of the San Francisco pier.

At the hillside anchorages, the scale of construction was like something out of the building of the Pyramids. Collapse in response to dynamic failure at the cable anchorages was part of the behavioral history of suspension bridges. In 1848 a man named Charles Ellet, Jr., built a wire suspension bridge over the Ohio River at Wheeling, West Virginia, the first suspension bridge with a span of more than a thousand feet. Six years later, Ellet's bridge collapsed in a hurricane that ripped ten of its twelve cables from their anchorages. At the Golden Gate, to resist a pull of more than 63 million pounds, each cable would have to be moored to a weight of more than double that of the pull. In order to keep such a mass of concrete from slipping into the Gate of its own weight, each anchorage was designed and built in three interlocking blocks. First the pit for each anchorage was excavated in steps, like those in an amphitheater, ascending toward the waters of the Gate. A base block of concrete was then poured into this hole and finished with its own stair shape. Here, deep in a pit behind a hill, working with concrete that was poured in by "tremie"—a crane-operated combination of a hopper and a tube like a giant elephant's trunk—men labored in gangs at tasks whose dimensions dwarfed them.

"I worked on a night pour at one of the anchorages," recalls Frenchy Gales:

The massive anchorages at either side required a matchwork maze of wooden forms—and so much concrete that pouring continued day and night.

S.O.Co. Photo

There were guys down in the cement. You'd walk on it to level it off. They took a count and we were short one guy. Everybody started stabbing around in the cement. We couldn't find the guy. We had to notify his family. The timekeeper asked me if I'd come with him. It was 1:30 in the morning. We went to his home. Knocked on the door. The guy answered in his pajamas. The timekeeper nearly fainted. The guy said, "I got tired. I went home and went to bed." That was the end of him on the bridge.

On top of the base block, there was poured an anchor block, at the rear of which—at the foot of the "stairs"—were set eighteen huge steel girders or "anchor chains," tilting upward and away from the Gate. Into the openings, or links, in these girders were fitted steel support members 134 feet long, like giant needles with enormous "eyes" at one end. These eyebars—sixty-one pairs at each anchorage—stretching out toward the Gate, buried almost up to their eyes in concrete, would eventually be threaded with the strands of each cable, stitching the cable to the shore. The entire massive structure—the base block, girders, eyebars, anchor block—would be topped by a poured concrete weight block that, interlocking with the two other blocks, eliminated any possibility of slippage.

Below the Marin anchorages, at Lime Point on the shoreline, men working behind the cofferdam on the foundation for the Marin tower found that getting to bedrock was a more complicated process than anticipated. The solid rock required for the foundation's southwest corner was not found until they had drilled and blasted more than thirty-three feet below the level of the water outside the cofferdam, or some thirteen feet lower than the rest of the foundation base. To fill this space with concrete, they poured continuously, trucking each batch from the Marin plant in mixing trucks that dumped concrete into the steel forms at the pier site. The pier rose in four-foot layers until it reached forty-four feet above the water's surface; built into it was a webwork of steel rods and dowels that would support the pier and connect it to its steel tower. On June 29, well within its budget and two months ahead of schedule, the Marin pier was presented to the Bridge District as the bridge's first completed element.

Out on the choppy waters of the strait, the construction trestle was stretching from the San Francisco side toward the pier site 1,100 feet off shore. Here a barge, the *Ajax,* and divers were at work excavating the bottom for the San Francisco pier's foundations and for the concrete fender that was to shelter the pier from the combination of the strait's tides and the Pacific's ocean waves.

To keep the pitching of the *Ajax* in the rough water of the strait from breaking her anchor lines, the bow anchors were attached to towing engines aboard the barge; the *Ajax* could haul in or slack off as the waves required.

The method of blasting at the pier site was an expansion of the depth-charge technique developed for the access trestle. A fourteen-inch blasting tube, 120 feet long, was held in place by a portable scaffold tower on the *Ajax*. A pair of steel legs, extending about 4 feet beyond the end of the tube, supported it on the bottom. The bombs dropped down the tube were fitted with automatic detonators, timed to explode a minute and twenty seconds after the fuse was lit. Once the bomb was in the tube and lit, a spud—a percussion-drilling mallet at the end of a cable—was dropped on it, driving the small bomb a short distance into the rock formation, where it went off.

"They'd put charges in, a lot of little ones at first," recalls George Albin, who worked for Pacific Bridge on the barge, "then there'd be a big one, black powder." This was a bomb made of eight-inch pipe and loaded with 100–350 pounds of 60 percent dynamite. Each bomb had a steel point six inches long and was about twenty-two inches long overall. The bomb would be driven into the blast hole made by as many as a dozen of the smaller bombs, below the adjoining rock surface, with a cable tool follower nine inches in diameter and weighing 2,500 pounds. There was a blasting cap attached to the bomb, with wire leads that were recovered by a diver in the classic deep-sea outfit of domed metal helmet and rubberized canvas suit. The leads were attached to an electric blasting machine on the surface and the plunger would be pushed.

"You'd see a lot of bubbles," George Albin recalls, "then there'd be a swell, the deck of that barge would lift sometimes six feet, and you'd see some guy turn green. There was a lot of seasickness on that barge."

At times, several large bombs were detonated together. The blasting tube would be hoisted up completely and the *Ajax* shifted several hundred feet to the east. At such times, seagulls, anticipating what was coming, flocked around the area, where they knew they would soon feast on stunned fish.

The rock loosened by the blast was excavated with a clamshell bucket with specially reinforced manganese-steel teeth and lips. Still, the rock— the much maligned serpentine bottom—was so hard that the work went very slowly and the bucket showed enormous amounts of wear.

The operator of the bucket could not see what he was digging. Moreover, the ebb tide could carry the bucket thirty feet out of position as it descended through the water. Tidal currents, rolling boulders and broken rock along the bottom, swept these into the excavations, so that much more material was dug up than could be accounted for by the swell of the rock and the slope of the underwater diggings.

The muck from the excavation was deposited in a 500-cubic-yard-capacity barge and towed out to deep water, where it was dumped. In rough weather, the bucket could swing so violently that it was impossible to position it over the dump barge. Instead, the loaded bucket had to be dipped below the water to calm its vibrations. Then it would be dumped.

The hardness of the rock, which was delaying excavation at the pier site, was "very gratifying" to Derleth in one respect: it would put the lie to the continuing rumors of shifting foundation stone and underwater crevasses. He suggested that additional soundings and contours be taken, to have on hand for an official reply to anticipated "propaganda."

Despite the resistance of the rock, the pitching of the barge, and the treacherous tugging of the tides and currents, work on the San Francisco pier progressed steadily through most of the summer. By early August, the trestle reached nearly all the way out to the pier site, and the bottom excavation had quarried out 85 feet of the required 100-foot foundation depth. Work was actually proceeding ahead of schedule when Cone and the Graham brothers fell victim to one of the peculiarities of the Golden Gate as a construction site.

Middle and late summer, in most places a builder's golden time of abundant light and warm or hot weather, often find the Golden Gate plunged in cold, wet gloom. As the largest and lowest of the gaps in the coast range, the Gate becomes, each summer, a caldron of wildly fluctuating winds, moistures, and temperatures. As the pressure area known as the Pacific High moves north with the summer sun, the upwelling of cold ocean waters off San Francisco increases. This combines with stronger ocean winds into giant moving walls of fog.

Meanwhile, in California's Central Valley, shielded by the Coast Range, the north-moving sun is cooking the farmlands in temperatures frequently above 100 degrees. This warm air rises, sucking the cool heavy wall of ocean air into the valley through the gap in the mountains that forms the entrance to San Francisco Bay. The fog, tumbling through the Gate as swift as smoke, can bring a drop in air temperature of 30 degrees in a matter of hours and reduce visibility from a matter of miles to one of yards.

It was in one of these dense, drizzle-laden summer fogs, on August 14, 1933, at 2:25 in the morning, that the McCormick Line freighter *Sidney*

M. Hauptman, outbound for Portland and some 1,000 yards off course, crashed into the trestle with such force that a 120-foot section was carried away entirely, and the remainder of the trestle deck was driven a full six feet inshore. The *Hauptman,* whose bow was crushed by the impact, was able to back off and return to a San Francisco pier, while Cone and the Grahams estimated damage to the trestle would take a month to repair.

The crash ended a certain honeymoon construction period on the bridge. It reminded everyone concerned of the precariousness of the conditions and the size and complexity of the job. Construction had been proceeding with a misleading smoothness. Weather, tides, current, navigational hazards, financial problems, and political infighting had all remained dormant for more than six months. Now the fog and the *Hauptman* had delivered a message: at the Gate, where approval for a bridge had been dragged out for more than thirteen years, nothing good would come easily.

"I'LL NEVER FORGET THAT DAY . . ."

I

Looking back on it all afterward, Russell Cone ranked the building of the San Francisco pier as the most difficult of the problems encountered in the construction of the bridge. It was certainly the most frustrating. All the nerve ends associated with the project—political, financial, geological, climatological, navigational—seemed to be concentrated at this one sensitive spot. And there was no way to avoid striking any or all of them at one time or another.

On August 23, while engineers, foremen, and workmen scrambled to repair the collision damage to the trestle so construction on the fender and pier could begin, the Engineering Board revised the specifications for the cement to be used in both. Because of the greater exposure of the concrete in the San Francisco pier and fender, it was now required that only high-silica or "pink" cement, a type that supposedly increased the resistance of concrete to seawater, be used in this part of construction. There was only one company in the area capable of supplying cement manufactured according to this formula: the Santa Cruz Portland Cement Company, whose president, George T. Cameron, was publisher of the San Francisco *Chronicle* and a director of the Bridge District.

When notified of this change in specifications, the officers of the Pacific Portland Cement Company, suppliers of more than 75 percent of the cement to be used in the bridge anchorages, pylons, approaches, and roadbed, decided that they smelled a rat. On September 13, the Pacific Portland officers and their attorney, Charles Ruggles, appeared before the bridge board and charged Cameron with conflict of interest. The directors, after being informed by their lawyer that since the cement was being supplied on a subcontract it could legally be furnished by Cameron's company, voted eleven to one in favor of the change in specifications.

In its account of the directors' vote the following day, the San Francisco *Examiner* took advantage of this opportunity to turn the heat up under the publisher of its rival newspaper. "Eleven directors of the Golden Gate Bridge," the *Examiner* editorialist scolded, "voted yesterday to give their fellow director George T. Cameron a $250,000 cement contract. . . . Is any bridge director so innocent of the ways of this world that he thinks the public wishes such a state of affairs to continue?" The resignation of Cameron from the bridge board was not demanded, the *Examiner* concluded, "It is expected."

The dispute stirred Strauss, always warmly responsive to combat, out of his benedict's seclusion. Once again, he wrote the directors on September 15, the bridge was being subjected to "such wild, exaggerated, and absurd criticisms as beggar description." Once again, "the less the deprecator knew about it, the more violent [was] his defamation." Again, a "selfish interest . . . heedless of what is best for the Bridge District" was deliberately broadcasting "alarmist and threatening assertions in order to intimidate the Board of Directors."

The specifications for the bridge, Strauss contended, were purposely made flexible to allow the engineers to take advantage of improvements in materials and construction. The high-silica cement was brought to the attention of the Engineering Board "in a proper manner" and was investigated over a period of many months. The board made its decision in the light of exhaustive tests by the materials and research engineer of the state of California and "by my own staff." Pacific Portland's lawyer's charges of inferiority and danger were either misstatements or founded on ignorance. "I am astonished," Strauss fumed, "that reputable concerns would be so unethical." The lawyer's charges, Strauss contended, "border on libel" and sprang from nothing more than selfish interest and a desire to promote trouble.

Ammann, Moisseiff, and Derleth, as well as Reed, the general manager, backed Strauss in the dispute, while Cone commented that the idea of lawyers telling engineers how to do their jobs "is a big belly laugh to me."

Strauss's vehemence did nothing to lower the temperature of the debate. On October 11 Ruggles, Pacific Portland's lawyer, filed a thirty-three-page petition for a court order rejecting the change in specifications, restraining the Bridge District from paying for any work involving pink cement, and holding the district officials liable for any money paid to Cameron or his cement company.

Pressured by the threat of a lawsuit and wounded by the continuing sniping of the *Examiner,* which hired a panel of engineers to buy high-silica cement and test it against ordinary cement in a competition that established, according to the *Examiner,* that the pink cement "would be in danger of not only probable but practically certain cracking," Cameron submitted his resignation as a director on November 20. The "pink cement" controversy, however, would outlast his term of service on the board.

As political turbulence around the bridge increased, the climate at the Gate itself began showing signs of the approaching Pacific winter storm season. Unanticipated swells, generated by storms far out at sea, coming out of nowhere in otherwise calm weather, would sweep over the trestle, lift and tip the barge crazily, and disrupt the work of the divers. Nevertheless, repairs on the trestle were completed and the stretching of the dockwork out to the barge and pier site resumed.

By October 19, the trestle had been extended out to the location for the concrete ring, or fender, that was to enclose and protect the pier site. To guide the forms for the fender into place underwater, the final frame section of the trestle was built in the form of a steel tower, weighing some fifty tons and jutting up at an angle from the straitside end of the trestle. This

❙The end of the trestle, with its 50-ton steel
❙form. Not long after this photo was taken, both were destroyed by storms.

guide tower, which was 115 feet long, was built to the "giants' causeway" scale of the construction that was scheduled to begin underwater. Rails on the outside of the tower would be used to lower forms for the first section of the fender. The tower was fitted into place in the same manner as the foundation framework for the trestle. First four pipes were driven into holes blasted into the bottom rock by small pilot bombs. Then, using a crane mounted on the trestle, the tower was lowered and the legs were guided into the pipes by divers. The size and weight of the tower, and the depth at which the divers worked—as much as ninety feet below the surface of the strait—made this work extremely difficult and hazardous.

The combination of tides and currents in the strait restricted the divers' working time below the surface to four twenty-minute periods of slack water each day. It was, suggested Pacific Bridge's Jack Graham, like trying to build a bridge in the middle of a river. The schedule was so tight that the divers would be brought to the surface all at once, without allowing for the usual gradual decompression in the water, increasing the risk of caisson disease, the painful nitrogen deficiency known as "the bends."

Finally the tower was set in position, on the sloping south side of the underwater excavation for the pier. The lower or north legs were founded at ninety feet below sea level, or about ten feet above the bottom of the pier excavation, and the upper or south legs at minus seventy feet, on the adjacent sloping surface. After the legs of the tower had been entered into the pipes, they were concreted firmly in place.

By October 30, the steel form for the first fender unit, weighing about forty tons, had been successfully placed in the tower's guide rails and lowered until its top was awash in the swells of the strait. Two more sections were added, and these three sections were then riveted together, while supported by the guide tower. Then, on October 31, while the tower was still supporting this eccentric load, an unexpected storm began rolling huge waves into the Gate.

"I'll never forget that day," Cone recalled later. "A fierce storm at sea brought in immense waves, which struck the steel forms with tremendous force." The waves, breaking through and over the deck of the access trestle, buffeted the trestle and tower until the fifty-ton tower began to shudder, six feet forward and back. This oscillation worked the foundation pipes for the tower loose. "The trestle groaned and creaked and then suddenly a mountainous wave, higher than the others, hit the forms like a cyclone, and swept it and the end of the trestle into the Golden Gate."

The entire tower and the forms attached to it had been torn away.

Cone and the Graham brothers, their Pacific Bridge divers and ironworkers immediately set to work recovering the guide towers and forms. By November 22, Strauss was able to report to the directors that a portion of the tower and the fender sections had been returned to the Moore Dry Dock Company for repairs and that "the Pacific Bridge Company has returned the fender sections and restored the trestle bay that was destroyed, and is getting ready to re-erect the structures."

Then, on December 13, another storm brought another succession of tremendous rollers sweeping through the Gate. "It tore the remainder of the trestle from its foundation," Cone recalled, "and dropped it, a tangled mass of wreckage, between the shore and the pier site." All but 600 feet of the access trestle had been destroyed. It had also just about taken the heart out of Cone and the Grahams.

"I can remember how Jack Graham and his brother Jim and the foreman, Harry Ericsson, stood there dejectedly. It was a sick-looking little group who watched the trestle disappear beneath the waves. There, in a matter of minutes, went ten months of the hardest kind of work."

2

In the wake of the wreck of the San Francisco trestle, Strauss and the directors were forced to look to the Marin side for consoling signs of progress. There were, at the same time, certain feelings of resentment at the ease with which things had proceeded on the more remote and less populated end of the project. As often occurs when disaster visits certain aspects of an enterprise, leaving others unscathed, there was an increasing managerial readiness to delegate blame.

On November 20, General Manager Reed accused the management of Barrett & Hilp, prime contractors on the anchorages on both sides of the strait, of transferring "your principal activities from the San Francisco side to the Marin side, in spite of the repeated requests of the Chief Engineer and the Resident Engineer that you concentrate certain of your activities on this side . . . and prosecute work with all possible dispatch." It was bad enough to suffer construction delays and traumas, but to have them concentrated on the side of the bridge most visible to the press and people of San Francisco was too much undiluted bad news to bear. Partly to distract the directors, Strauss urged them to visit the Marin anchorages, where they could observe the setting and aligning of the anchor bars before the bars were covered by concrete.

In fact, with the completion of the Marin pier, the contrast in progress between the two sides was about to escalate dramatically. Instead of a reassuringly symmetrical rise of towers on both sides of the strait, it now appeared that the Marin tower would be completed while work on the San Francisco pier was still going on, giving the overall project a lopsided look for months. In October, Clifford Paine had gone east to inspect the steel-work for the towers, being fabricated at the McLintic-Marshall works in Pennsylvania. He was met by Reed, and the two men then visited the Princeton offices of Professor Beggs, who was conducting stress tests on the model of the bridge tower.

There had, in the past, been models made of bridge towers to study the effects of stress and the need for bracing. On the George Washington Bridge a celluloid model had been constructed, and the Bayonne Arch had been reproduced in a test model made of brass. The stainless steel tower model of the Golden Gate, however, 12½ feet high and built on a scale of fifty-six to one, was a step beyond these, using material with elastic properties similar to actual bridge steel. On a testing rig at Princeton resembling a drop forge, Professor Beggs was able to reproduce conditions caused by both longitudinal and transverse loads and to measure the effect on the tower of both wind and earthquakes.

On January 20, 1934, a group of 200 engineers was invited to a program at Princeton featuring the model of the Golden Gate Bridge towers, whose building and testing were summarized by Professor Beggs. O. H. Ammann discussed the place of model testing in current engineering practice, then Leon Moisseiff described the evolution of the tower design.

"Mr. Moisseiff," reported *Engineering News-Record,* "pointed out how the stiff frame, without diagonals, was adopted for the sake of appearance, and how this complicated the analytical design." An ingenious method, Mois-seiff went on, had been worked out by C. A. Ellis of the Strauss Engineering office. What Ellis had done, explained Moisseiff, was to combine Williot diagrams, a trigonometric means of measuring structural strain, along with more than thirty algebraic equations, plus the moment equations, and to utilize all this information simultaneously. Ellis's analysis allowed the prob-lems of the towers to be worked out as a unit rather than as an assemblage of connected parts. The soundness of the analysis was demonstrated by computing the stresses in the model according to the theory, then checking the results by testing the model.

The engineer whose analytical abilities were being celebrated at the gath-ering in New Jersey was, at the time, at home and unemployed in Illinois. Charles Alton Ellis had now been out of a job going on three years, and his dominant role in the engineering design of the Golden Gate Bridge was fast fading from public and professional memory. In fact, Moisseiff's salute to him before their engineering colleagues at Princeton was to be the last public mention of Ellis's contribution by anyone officially associated with the Golden Gate Bridge. An article, by Clifford Paine, in *Engineering News-Record* in 1936, on the design of the bridge towers, would make no mention of Ellis and would attribute the mathematical computations to Ellis's assistant and replacement, Clarahan.

It wasn't until the autumn of 1934 that Ellis found another position, this time on the engineering faculty of Purdue University at Lafayette, Indiana.

Through his own reticence, Ellis may have conspired in denying himself recognition for the great work of his life. Perhaps he realized that his understated style had little chance of prevailing against first Strauss's command of the machinery of publicity and then the vested interest of others in the Strauss hagiography. Or perhaps, gauging the cost to his own values of a determined public recognition campaign, Ellis simply decided it wasn't worth it. There was compensation of a kind in the fact that he had returned to teaching and could begin inspiring another generation of student engineers.

"When he was at Purdue," recalls Professor Marion Scott, "Professor Ellis told his staff and students that he was the designer of the Golden Gate Bridge. But he never communicated any bitterness. He agreed that he and Strauss didn't get along, but he never made a big thing about it. He was one of my favorite professors. I was a real fan of his."

Charles Kring, who as a young engineer had worked on the Golden Gate job, met Ellis at Purdue in 1938. Asked about his involvement in the famous work they had in common, Ellis told Kring, "I designed every stick of steel on that bridge."

Nevertheless, Charles Ellis, design engineer of the great work now unfolding at the Golden Gate, would not be given the opportunity to design another bridge.

By December of 1933, on the northern side of the Golden Gate, the first of the giant towers that Ellis had envisioned was rising above the twin concrete tops of the Marin pier. The steelwork was being lifted and connected into place; for the first time, the enormous dimensions of the bridge were apparent to anyone who looked toward the Gate.

The steel sections had been fabricated at McLintic-Marshall's plants in Pennsylvania, where about 60 percent of the sections of one tower leg had been erected at the shop and the assembly checked for accuracy. Then this was dismantled and sent by rail along with the rest of the fabricated steel to Philadelphia, where it was loaded aboard ship and sent, via the Panama Canal, to San Francisco. After passing through the Gate, above which it would soon stand, the steel was unloaded and stored at Alameda, on the eastern side of San Francisco Bay. From here, the steel was barged out as needed to the Marin pier in 400–500-ton loads.

To prepare the twin tops of the pier to receive the tower base slabs, the concrete surface was ground to a level plane with a carborundum wheel. Then a layer of red lead paste was applied to the finished surface immediately before the base slabs of the tower were set. "I helped set the base plates for the Marin tower," recalls ironworker Walter "Peanuts" Coble. "The plates were heavy, five inches thick, set in red lead paste. What a mess!" There were nineteen of these base slabs to each tower shaft, and they were fitted to the steel dowels imbedded in the concrete.

At the base of the Marin pier, an Erector-Set-like derrick 85 feet high unloaded the barges with a 100-foot boom that lifted the steel-cell sections onto the pier and stacked them at the foot of the tower's concrete base. From here the steel sections would be hoisted into place with a giant, spidery, movable erection crane, or "traveler." Positioned between the tower legs, designed to climb the tower as it rose, the traveler supported two stiff-leg derricks, each with a capacity of eighty-five tons, and equipped with 90-foot booms. "Each crane had a carrier that boomed out with a load block on it," George Albin remembers. "Each chunk that came up weighed fifteen to thirty tons. You'd grind the surface of the piece. When it was about eighteen inches from where it was to set, they'd drop it. It would shake your eye teeth."

The steel was erected by gangs of men working in teams. First a raising gang, communicating in a code of bells and hand signals, would use the traveler to lift and guide each steel section. The fifteen-to-thirty-ton hunk of steel would be lifted, swung, caught, grinded, pushed, nudged, cursed, and dropped into place. Then it would be inspected. "We had an inspector," says Albin, "with a feeler gauge. If he could find space anywhere, he would say, 'Take it up and grind it.' It might have fit fine in Pottstown, but it had to fit perfect here."

A connecting gang, usually four men, each equipped with a spud wrench and a beater—a maul, or small sledge hammer—would work the joints. When the steel section was in place, the bolter-ups would slip a temporary bolt into the holes drilled into the steel pieces at Pottstown, stick a wrench in another hole, and run a nut on it. Then about half the holes would be fitted with drift pins or bolts, and these would be tightened and adjusted until the opening between the milled ends of adjoining steel sections was .006 inch or less.

As the tower rose, in sections, the traveler was "jumped up" to predetermined elevations about thirty-five feet apart. Two beams would be stretched across the uppermost tower sections, one at each end of the traveler. Each beam was fitted at the corners to correspond to the lower four corners of the traveler, and a one-inch wire rope, leading to the hoisting engines on the pier, was run through both. Each of the traveler's two derricks had its own motor-driven hoist, raising gang, and electric light signal system. "There is a lot of coming and going when you are jumping hammerhead cranes," recalls Albin. "Sometimes it took a day and a half to jump 'em."

After the steel sections had been lifted into place and connected, steel to steel with pins and bolts, riveting gangs, men working in teams both inside the tower cells and outside the tower entirely, bonded the sections together with lengths of hot steel. On coal forges set on scaffolds outside the tower, rivets were warmed white by the "heater," who sent each rivet into the cell through a pneumatic tube. Inside the maze of cells, perhaps a hundred feet away from the heater, the rest of the riveting team worked in poorly ventilated darkness in a space cramped to 3½ square feet per cell.

"You'd have a pipe in one end," says Albin, "a hot rivet would come up to you inside it, you'd catch it with a cone, and grab it with tongs." This man, the "buckerup," would fit the hot rivet into a hole and back it with an iron bar. When the end of the rivet appeared on the other side of the cell wall, the riveter would use his gun or airjam to forge the rivet's ideally rounded head.

A proper rivet would be just big enough to fill the hole and leave a head. "I had a hammer to see if a rivet was loose," recalls Alfred Finnila, who inspected riveters' work on the bridge. "I'd tap the rivet with a hammer, and if it was loose, they had to cut it out and put in another." The riveters were expected to drive a certain quota of rivets each day. By the time a loose rivet was cut out and replaced, the riveter could lose fifty rivets. "A good guy," says Finnila, "would only have four or five loose in a whole day. A poor guy, ten or twenty."

Inside the towers, just getting to your work could be like negotiating a gigantic maze. The steel cells, which numbered nearly a hundred at the base of each of the tower's legs, were reduced to twenty-one at each tower top by the stepped-off design. Inside the tower, in the shafts or tunnels formed by the stacked cells, it was easy to wander into a dead end.

"In a raising gang," says Peanuts Coble, "everybody likes to get to the top first. We used to shinny up inside the cells. Once you get it—there is a sort of pattern—you could go right up. The riveting gang got so they could go through there just like mice."

For those who'd rather not shinny up to work, there was a "skip," or elevator, which could take ten or twelve men up at a time. At quitting time, there could be a twenty-minute wait for the elevator down from the tower.

"It was my job to straddle the cables up there and tighten the holds [at the end of the day] so the large rigs wouldn't topple out," says Coble. "It was a safety measure. I'd slide down the cable and stick my legs through it. So one night, I just gripped it with my hands and feet. Dropped say fifty feet, and stopped. I went down another fifty feet, and continued like that to the pier. It didn't take five seconds. After that, the whole damn raising gang went down that way."

In addition to the stress of darkness, cramped space, heights, difficult access, and production quotas, the riveters and buckerups inside the tower faced serious health hazards. The confined space intensified the racketing noise of the riveting guns, increasing the odds of the occupational deafness so common it has come to be known as "riveter's disease." The closed-off honeycomb construction also exposed the inside men to a danger of an even more sinister kind: lead poisoning.

The Golden Gate Bridge was the first structure built using red lead paint in a confined space. A rivet would come out of the tube inside the cell white hot. As the buckerup fit it into the hole with his tongs, the rivet would hit the red lead and give off fumes that the men working around it couldn't help but inhale. "The raising gang was sometimes five or six sections—150, 160 feet—ahead of the riveting gang," says Peanuts Coble. "The red lead would come up like out of a smokestack. It would dissipate before it got to us." The men working inside began reporting symptoms—nausea, pain, cramps—that were first diagnosed as mass appendicitis. An investigation by the California Industrial Accident Commission established that there had been at least four positive cases of lead poisoning among men working in the Marin tower, with the possibility of more.

"Lead which enters the body through the respiratory tract is the most toxic and difficult to treat, because of its great solubility in blood plasma," the commission's supervising engineer reported. The commission recommended that "all splices and connections be shipped unpainted, and no lead paints applied where field riveting is necessary until said riveting is completed."

To eliminate, or at least reduce, the red lead fumes, the unriveted steel sections already on hand were reamed free of paint around the holes, while those still on order from Pottstown were painted instead with iron oxide. To improve ventilation, compressed air was pumped inside the tower cells, and the riveters and buckerups were given filter masks to wear. "I didn't wear a mask," says George Albin. "You could suffocate with one of those things on." Nevertheless, lead is an accumulative poison, and many men were to suffer and die from the exposure they received then.

To do this work—raising, rigging, connecting, riveting—required experienced men, ironworkers, bridgemen, in such numbers that the strict residency rules, adopted by the directors to assure as much employment as possible to residents of the Bridge District counties, had to be stretched.

Drawn by the magnet of two great bridges—the silvery towers of the Bay Bridge, for which Derleth and Moisseiff also served as consulting engineers, had already risen between San Francisco and Oakland—ironworkers were drawn to the Bay Area from the East, South, and Southwest. "Guys from New York, Pennsylvania, Louisiana, Oklahoma, Texas—the best ironworkers in the country," recalls Al Zampa, who worked on both bridges. They needed the work, but there was something more involved: a desire, rooted deep in the nature of bridgemen, to have life braced by challenge, change, risk, achievement. "Building buildings is almost like doing piecework in a factory," says Harold McClain, who worked on the Empire State Building and the George Washington Bridge before coming to the Golden Gate. "But bridgework is always changing. Each bridge is different. It was never just a job."

A man would contact the ironworkers as they came into San Francisco. "He'd take care of your residence establishment. He did it for union and nonunion men. They were crying for good hands."

From rooming houses in San Francisco, rented houses or apartments, and sometimes from lodgings in Sausalito on the Marin side, the bridgemen, most of them single, some with families left in the East, rode, sometimes by water taxi, to work, where they climbed, with the steel, a little higher each day.

By December 27, 1933, the Marin tower had reached an elevation of almost 237 feet, or the approximate roadway level of the bridge. The great cross-braces between the tower legs were in place, more than 6,000 tons of steel had been erected and more than 29,000 rivets driven into the tower legs. The base had been laid; now the tapering sculpture of the tower could be gradually unveiled.

Looking down on the cells inside the north (Marin) tower. Bridgeman "Peanuts" Coble hung by his legs from an overhead derrick to take this photo.

14

"PUDDING STONE."

I

On December 13, the same day that Cone and the Graham brothers had watched helplessly as the storm demolished their trestle, Strauss met with the bridge directors, to whom he delivered a package of even worse news.

Because of "engineering contingencies," Strauss informed the directors, the plans for the San Francisco pier and fender would have to be completely revised. The construction changes would cost an additional $350,000 and would delay completion of the bridge by nearly a year.

The original plan for the fender, Strauss explained, had been to base it on bedrock at a depth of 65 feet. Then, inside the fender's protective concrete ellipse, excavation for the pier itself would be carried down to 100 feet. However, the Pacific Bridge engineers and crews, while blasting to excavate for the fender sections, had so shattered the bottom rock that now, to provide a solid footing, the entire excavation for fender and pier would have to be taken down to 100 feet. The additional blasting and digging would cost $10,000 a foot, and since Strauss and the Engineering Board had approved the blasting, Pacific Bridge had disclaimed liability. The increased costs would have to be absorbed by the district.

The directors fumed and blustered, but there was little they could do but accept. Supervisor Stanton was particularly angry at having been informed after the fact. "Three weeks ago outsiders told me what was happening," he complained, "but it wasn't dropped into our laps until the last minute. Everybody in this town seems to know what's going on but our engineers!"

Once again, soundings were taken of the bottom of the Gate at the site of the south pier's foundation. As with the tower, a model was built of the part of the project in question, and the floor of the Golden Gate was soon on display in the rotunda of San Francisco's City Hall. The New York firm of Moran and Proctor, specialists in foundation engineering, were retained as consultants to Pacific Bridge. Important decisions needed to be made quickly, in an atmosphere where people were eager to point fingers. Although he doubted any deliberate delay on Strauss's part, Moisseiff warned Derleth, "There may be what is known as subconscious ducking."

Strauss assured the directors that, despite the obstacles and changes, work on the pier would press forward vigorously. The contractor's proposal, he reported to the board, "contains a clause binding him to work a twenty-four-hour day, seven days a week throughout in order to expedite the work . . . to employ as many men and as much equipment as can reasonably be employed."

While the additional excavation at the pier site was being grudgingly accepted, construction of the trestle leading out to it was once again under way. This time it was decided to base the trestle not on steel sections but on rounded timber pilings, which would be quicker to prepare and also offer less resistance to the strong tidal currents. The deck of the trestle would be raised five feet higher to keep the stringers, or lengthwise floor members, clear of the tops of waves, and the plan for the fender was revised so that the concrete sections could be built up in blocks, with the legs of the guide tower incorporated into the first section.

The work of restoring and extending the trestle advanced steadily, despite the effects of the winter storm season, both above and below the surface of the strait. On December 30, a Pacific Bridge diver, trying to recover anchor pipe casings from the barge, fell off a steep bank some 225 feet off the east side of the San Francisco pier; on January 2, another diver attempting the same work had to abandon the job when heavy seas began rolling large boulders around on the bottom so violently that the diver narrowly escaped being pinched between them.

As the trestle resumed its extension out toward the fender and pier, the infighting about what type of concrete was to be poured from it continued within the bridge board, in the press, and before the courts. On November 29, a little more than a week after Cameron's resignation from the board, the directors had elected to rescind the change in specifications for high-silica cement voted while Cameron was a board member. They then passed another resolution that, since Cameron was no longer a director and therefore not strictly in conflict, legally incorporated the same change in concrete specifications. This amounted, suggested the *Examiner,* to a $250,000 going-away present to Cameron from his fellow board members.

The dispute next entered the courts, where attorneys for the Bridge District maintained that no understanding for the awarding of the cement contract had been reached with Cameron before his resignation from the board. Indeed, it was Pacific Portland that were the real villains here, the district's lawyers insisted, since the company was attempting to enforce a monopoly on the supply of all cement for construction of the bridge.

On May 4, 1934, Judge Maurice Dooling of San Benito County—a man chosen, no doubt, in hope of geographical impartiality—ruled that the original contract, awarded Cameron's company while he was still a bridge director, was in conflict and therefore illegal. Following Cameron's resignation, however, his Santa Cruz Portland Cement Company could legally supply high-silica cement for the bridge. Judge Dooling assumed "that the remaining directors were actuated by a sense of public duty and proceeded in good faith" in awarding the contract to Cameron's company.

At this time, the high-silica cement in question was already being mixed, trucked out, and poured into the fender blocks for the San Francisco pier. The new trestle had been completed out to the pier site on March 8, 1934, with the last three frames or "bents" of the structure made of steel, rigidly braced both above and below the water. On March 22, the first concrete had been poured in the fender, where it was used to imbed the legs of the guide tower. The fender units could then be lowered into place and concreted.

Designed to extend 115 feet above the strait's bottom, with concrete walls 30 feet thick, forming an elliptical shape enclosing a space the size of a football field, the fender would be like a stadium, built upon the floor of the open sea. Revised to give the fender a wider base, the plan now required building a considerably larger surface of framework underwater. This made it necessary to develop lighter forms. At the base of the fender, from 100 feet below the level of the strait, up to 80 feet, each form was built as a relatively light wood-and-steel frame, lowered into position using a crane and divers, and guided into place underwater by means of triggering lines. Once the frame was in place on the bottom and fastened to the adjoining units, the sides were built up with wooden panels five feet wide. These panels were weighted with concrete blocks to overcome buoyancy and were fastened with a substance adapted to quick handling underwater by the divers, who had to tie themselves off to do stationary work in the tricky current.

The bottom form for the first unit included the legs of the guide tower. Then the first steel "box" was lowered and rested on the bottom form, after which the remainder of the first unit and the box were filled with concrete, in a continuous pour, by tremie, from the end of the trestle. This combined concreting of base block and first box, or initial tier, was used on all the remaining bottom layers of the fender.

Pacific Bridge diver, ready to descend, inside San Francisco fender.

On top of this layer a second steel box, taller, tapering slightly, was concreted along with an adjoining box, so that the second course of the fender—from sixty to forty feet below the surface, was built of blocks thirty by sixty feet, and twenty feet high.

From forty feet below on up, the boxes were replaced by skeleton steel frames or form panels of weighted timber or precast concrete. These were used to form the outside of the fender, where a smooth concrete face was required to decrease resistance to tide and waves. The fender wall was bonded together with keys between the adjoining blocks. In addition, the top part of the wall, from two feet below the level of the strait to fifteen feet above, was heavily reinforced with steel bars to tie the units together and form a sturdy arch of resistance in case the fender was rammed by another off-course ship.

2

The various trials—nautical, seasonal, legal—and delays associated with the San Francisco pier encouraged the revival of all the old suspicions and rumors about the project's fundamental soundness. Even now, while the concrete blocks of the massive fender were slowly and laboriously being poured, the pier received, from a seemingly authoritative source, its most sweeping attack yet.

In early April, the assistant engineer of the state of California in charge of administering Public Works Administration funds, for which the Bridge District had applied, received an unsolicited letter and memorandum from Dr. Bailey Willis, professor emeritus of geology at Stanford University. According to Dr. Willis, the foundation rock at the site of the San Francisco pier was unstable to a degree likely to endanger the entire bridge structure. The serpentine rock, observable in the vicinity, was subject to landslide under natural conditions, a situation that had been aggravated by blasting and "would be gravely augmented by the weight of the structure it is posed to erect on the foundation of the south pier."

At first, Strauss, Reed, and the directors assumed that Professor Willis was fronting for the same interests that had previously opposed the building of the bridge: the Southern Pacific Railroad and its various associates and dependents. Willis had been one of a group of thirteen prominent scientists and engineers who had signed ads against the bridge at the time of the bond issue, and there were familial ties between the Southern Pacific and Stanford; but as the controversy simmered on, it became apparent that Professor Willis, who was seventy-eight years old, was either riding, or being ridden by, an obsession. Still, because of his academic credentials, Willis's charges could not be dismissed or ignored. Moreover, he had an agitator's gift for the adhesive epithet: the sheared serpentine, upon which the pier foundations would rest, Willis said, would be turned, by submarine landslides, into a kind of pudding. The bridge would rest on "pudding stone."

Professor Andrew Lawson, the Bridge District's consulting geologist, after studying Willis's memorandum, termed it "pure buncomb." "No reputable geologist would consider it seriously or attempt to convert the professor from the errors of his thesis. The latter excites only a smile of contempt."

Willis had argued that the engineers, in designing the pier, had been unfortunately ignorant of geology, while the geologist, Professor Lawson, was insufficiently versed in engineering. Professor Willis, on the other hand, could speak authoritatively in both areas, since he held degrees in geology and civil engineering.

Derleth, leaving the response to Willis's geological findings to Lawson, found the Stanford professor's engineering criticism "utterly incompetent." Professor Willis's recommendation that the foundations be sunk 300–400 feet below the level of the channel bottom by excavating wells and filling them with concrete columns Derleth considered "puerile . . . it cannot be taken seriously."

"The mere fact that Professor Willis received the degree of Civil Engineer at Columbia University more than fifty years ago," Amman wrote to Strauss, "does not qualify him to pass upon important engineering questions. Without subsequent experience, he can at best be qualified as an embryonic engineer."

In the face of this sort of rejection, Willis not only remained undiscouraged, he turned his attack into a broadside, sending letters and telegrams to Derleth, Reed, the director of engineering for the Public Works Administration in Washington, and even Charles Blyth, of the investment house of Blyth & Co., associated with the selling of the bridge bonds, stating his fears for the future safety of the bridge.

Professor Willis could not be refuted because he would not be confined to any facts. Invited to visit the Bridge District offices and inspect the findings of the various soundings, the geological core samplings, or the on-site investigations of the divers, Willis limited his appearance to a conversation with Reed and Director Keesling in which he said that while there was no direct danger to the pier itself from earthquake, the resulting swaying of the tower, combined with the stoppage of the massive concrete base, would cause slippage. "The fatigue of the serpentine," said Willis, "will create a slide and it is almost certain."

Strauss, extremely irked, held his fire. His previous conflict with an academic, Ellis, had made him wary. Also, the professor had avoided addressing Strauss directly, writing just about every other figure associated with the bridge, so that no individual reply was required. Strauss was convinced that Willis's intellectual arrogance, his "happy faculty of reaching his conclusions without taking the trouble to get the facts," and his apparent conviction "that he is right and that his knowledge is superior to that of anyone else," would eventually enable the Bridge District to "sew him up."

Unembarrassed by his colleagues' disapproval, calmly superior to facts, Willis now became the Cassandra of the Gate, prophesying disaster for the project even as it progressed, always available for quotes when anything seemed to go wrong. To the public and the press, the continuing dispute with a geologist from Stanford on one side and one from Berkeley on the other took on some of the entertainment aspects of a football game, with someone big and important certain to go down to defeat.

For the present, the best antidote to Professor Willis's grim forecasts about the bridge was the reality of the rising Marin tower, the largest bridge tower ever built. Here the true proportions of the bridge were first displayed and its thrilling combination of mass and grace first revealed. Below the roadbed the tower legs had been squat and thick, the space between them blocked by the giant cross-braces common to any ordinary bridge; but as the towers gradually moved above this pedestal, the massive structure subtly changed, tapering upward, in a way now revealed to have been suggested all along, into a soaring enframement of water, earth, and sky. The fluting along the sides of the great tower shafts, the gradual stepping off of the towers as they rose, the four enormous portals that opened in the wake of the traveling crane's rise—all these gave the bridge, even as a work-in-progress, a tremendously *realized* quality, a reassuring sense of great complexity being contained and managed by the force of a single strong idea. To residents and visitors on the San Francisco side, it was already a great new landmark, taller than any building in the city, while to commuters on the Marin ferry, who passed it twice each day and who would be among the bridge's regular customers, the tower was indeed a sculpture that changed not only every day but each hour in the way its notches, openings, and indentations caught and refracted shadow and light and in the way entire parts of it could simply be amputated by fog.

For the men working on the bridge, the tower, as it rose, exposed them to the full force of the Gate's wind and cold. "The first day I worked out there, I was on signals," recalls Harold McClain. "I was warned how cold it would be, but I didn't believe it." McClain, after all, had worked on the George Washington Bridge and on the Empire State Building. "I wore all the gear I could carry. I was giving signals for the raising gang by hand. I'd never been so cold in my life. I don't think I've ever been as cold since. A guy brought me an overcoat, and I put it on. They had to see my hands and shoulders. Eventually, I worked standing in half a barrel."

There was, at the Gate, no such thing as a completely calm, still day. There was almost always wind, and the men could dress just so much against wind. Then it would go up sleeves and into pant legs. At the few times when it wasn't windy, there was fog.

"The fog was wetter than on the Bay Bridge," says Al Zampa. "It was

Topping off the north (Marin) tower. The man in the suit, tie and hard hat is Russell Cone.

miserable going out and coming back. It was really damp. You could hardly see. About two o'clock every day, the wind would come up. It could nearly blow you off. You had to lean into it. Get near the tower for shelter, then step out into it again."

Sometimes there were two kinds of weather at once. "We'd be in sunshine on the top, and you couldn't see the water because of the fog."

In sunshine and in fog, in wet weather and in wind, the tower continued to rise until the spidery traveler had been jumped all the way to the bottom of the uppermost tower portal. On May 4, 1934, Cone, Reed, and the McLintic-Marshall engineers, wearing dark suits and in some cases topcoats, along with primitive leather hardhats—the first made mandatory on any bridge job, and which gave the men who wore them a certain resemblance to the mounted warriors of Genghis Khan—rode the elevator some 700 feet up from the tower's concrete base. On a raw, windy steel-plate-and-beam promontory, they raised a snapping, breeze-torn American flag. The first bridge tower, which would require more than a year more of riveting, adjusting, and finishing, had been topped off.

The wind and fog that made life difficult for men working on the tower were also beginning to affect the structure itself. The exterior of the steel members just erected was already beginning to show signs of weathering, to a degree much greater than that on the San Francisco-Oakland Bay Bridge, which was on a less exposed part of the same bay. A second, or field, coat of paint would be needed, in addition to the original shop coat, and this nudged the Engineering Board toward making some decisions about the bridge's final color.

Ammann, the man most experienced in the long-term maintenance of bridges, favored a field coat of aluminum paint, like that used on the George Washington and the San Francisco–Oakland Bay bridges. Admittedly, the aluminum would lose its luster, but it would look less dead than ordinary gray paint. Reed pointed out that the navy, the Southern Pacific Company, and various oil companies had all adopted aluminum paint for their standing structures in the West. In a paper read by Strauss, Irving Morrow restated the original proposal for using an orange-red color for the towers, with deeper shades for the suspenders, cables, and approaches. Morrow had made two studies to display the general plan of coloring. Amman, who conceded that "it was the general view that a bright, attractive color would be desirable for the bridge," expressed doubt, based on his George Washington Bridge experience, that it would be possible to get a red paint that would be sufficiently durable at the Gate. Cone, however, maintained that the paint manufacturer now supplying the bridge insisted that it was possible to make a durable paint of yellow ocher and burnt sienna, to give a red color to a white lead and zinc base.

It had become apparent, first to Morrow, then to Strauss and the other engineers, that the scale of the bridge required a bright, warm color, and that this should be kept simple and not used, as Reed had suggested, to accentuate the portal enclosures, and thus fragment the design.

Since it was clear that the bridge was going to require more durable paint, in greater quantities, more often than had been anticipated, it was decided to establish a test for the final field coat at the site. Steel panels, given the regular coats of paint now being applied to the Marin tower, would be hung on a wooden rack on top of Fort Point. Manufacturers would be invited to submit samples, which would be applied to the previously painted and exposed panels. The panels would be regularly inspected and observations recorded as to color and weathering. Because of the potential amount of money involved in awarding a paint contract that would be, in effect, an annuity for the winning firm, paint samples and inspection records would be stored under lock and key.

Meanwhile, to protect the tower itself, painters would continue to apply the existing reddish primer. This was mostly done by men with brushes, dangling in bosun's chairs, high above the rocks of the shore or the waves of the strait. "The painters used to hang off of the cable," says Peanuts Coble, "though they weren't supposed to. I saw a guy swing out once, and get blown against the tower, again and again. Red lead all over him. They didn't hang from the cables after that."

Strauss was displaying increasing confidence in Morrow, the San Francisco architect he had chosen, originally, in an emergency and, in part, for political considerations. Morrow's strong aesthetic sense, his painterly grasp of color and light, his ability to articulate his views in thorough, well-reasoned, and readable papers, drew Strauss's strong support. "The change in architects," Strauss wrote Moisseiff, "was a very happy one indeed."

There was, as always with Strauss, an intensely personal aspect to this relationship. Morrow did not compete with him. When the architect had come aboard in the summer of 1930, the engineering design of the bridge had already been completed. With no ties to Ellis, Morrow had no reason to dispute Strauss's assumption of credit for designing the bridge. As a local architect, grateful for the opportunity to work on so significant a structure, Morrow would not, like John Eberson, begin making claims about having originated the stepped-off construction of the towers, which, Strauss informed Clifford Paine, who had not been involved at the conceptual stage of the bridge design either, "originated with me, and is covered both by design and construction patents issued to me."

Strauss's satisfaction with Morrow was such that he now proposed that the architect be in charge of lighting design for the bridge, work that was usually done by an electrical engineer and toward which Ammann and Moisseiff both expressed strong reservations. Strauss, as was often his way, trusted his instincts in this matter and let Morrow proceed. As it was, the chief engineer was now dealing with problems enough of his own.

In September of 1934, Bailey Willis finally approached Strauss directly, writing the chief engineer to request a public hearing concerning the safety of the San Francisco pier foundations. To this Strauss agreed, offering to schedule a meeting of the Building Committee for the following week, at which Strauss, Professor Lawson, and Derleth would all be present. When, on September 15, Willis insisted that the construction on the south pier be halted pending the outcome of the investigation, Strauss exploded, informing the elderly geologist, "I am convinced that any discussion with you would serve no good purpose."

Not only had Willis never visited the site, examined the sample cores, studied the boring logs or the sounding sheets or the plans, or met with the engineering staff or the consulting geologist, Strauss continued, "you did not write or consult with me until you were instructed by members or officers of the board it would be necessary to take it up with me." Despite "this rank discourtesy and unethical conduct on your part," Strauss wrote, "I offered you full opportunity to familiarize yourself with the plans, the cores, the records and the site . . . but you have only continued your hostile activities." The plans, Strauss reminded Willis, had been a matter of official record since 1930. Since Willis had had four years in which to express his interest in the project and had not done so, "we are not impressed with anything you might say now."

"If your purpose," Strauss now warned the professor, "is on some pretext to hold up the work and prevent the sale of the bonds, you are wasting your time. The extent of your activities in this direction is an unwarranted interference with a public enterprise."

"In my previous letters to you," Strauss concluded, "I stated my opinion that you were not qualified to pass on engineering matters, and that as far as geological matters are concerned, we are relying on Professor Lawson. That is, and will continue to be, our position."

Willis, if diappointed by Strauss's emphatic letter, copies of which were sent to Reed, the directors, Lawson, the Engineering Board, Paine, and Cone, remained unabashed. Yielding on the construction moratorium, he agreed to come to the hearing before the Building Committee.

"We are having court reporters make a record of all his statements," Strauss gloated to Ammann, "and as soon as this record is complete, we will send a copy to you. In these hearings, we are gradually pinning him down, and I think when we get through, he will be through."

Strauss's confidence that the Stanford geologist's authoritative-sounding criticism of the bridge could be overcome could only have been reinforced by the example of Michael O'Shaughnessy. Since the time of the city engineer's intimidating opposition to the bridge bond issue, O'Shaughnessy's fortunes had declined drastically. The actual construction bids submitted by the contractors in the aftermath of the election, followed by the slow, difficult, but successful sale of the bonds, had increasingly minimized, to a point of irrelevance, O'Shaughnessy's insistence that the bridge could not be built for less than $100 million. His patron, Rolph, was gone after twenty years in the mayor's office and beset with his own problems as a one-term Depression-era governor of California. O'Shaughnessy's enemies, sensing "The Chief's" new vulnerability, now came after him on almost every issue, public and personal. The city charter revision of 1931 had stripped him of his city engineer title and of much of his authority, but he had remained a consulting engineer under the new city Public Utilities Commission, responsible chiefly for Hetch Hetchy.

The great water project, twenty-two years in the building, was at last nearing completion. In the fall of 1934, the first Hetch Hetchy water was scheduled to be delivered to San Francisco, and a great civic celebration was planned, with Secretary of the Interior Harold Ickes making the dedication speech, and featuring O'Shaughnessy as an honored guest.

He was still an imposing figure in San Francisco, representing as he did the spirit that had rebuilt the city after the earthquake, and his presence was requested at most important civic gatherings. Izetta Cone remembers him from a dinner dance the American Society of Civil Engineers staged aboard an American President liner in San Francisco Bay, the great mover and shaker of the past, still distinguished in black tie and tuxedo. "I remember him and Mrs. O'Shaughnessy dancing together. They both must have been in their seventies. They were *enthusiastic* dancers."

The Hetch Hetchy completion ceremonies were scheduled for October 28, 1934. On October 12, O'Shaughnessy attended a meeting of the Federation of City Employees at the Dreamland Arena in San Francisco. Feeling ill, O'Shaughnessy, who had been suffering recurring heart attacks over the past two months, left early and went home. Early the next morning, his wife called their five children to their father's bedside.

"God bless you all," O'Shaughnessy told his family. And he subsided into the last sleep of his vigorous and embattled life.

15

"WE DESCENDED IN A BUCKET . . . TO THE OCEAN FLOOR."

I

Ever since the building of the Brooklyn Bridge in the 1870s—the first long-span bridge built with spun steel cable—the construction of suspension bridges had been associated with men working underwater in pneumatic caissons. Inside these giant steel chambers, sunk to the bottom of a body of water, men would work far below water level, protected from pressure by compressed air, on the detailed excavation and preparation necessary for the foundation of the underwater pier on which the towers to support the cables would rest.

In keeping with the dimensions of the Golden Gate Bridge, the caisson proposed for the San Francisco pier would be the largest ever built. According to the specifications written by Charles Ellis in 1931, the caisson would be divided into five compartments. At the top, above high water level, would be an air lock equipped with a hoist-cage elevator. Below this would be two horizontal cylindrical locks, with capacity for at least sixty men. Between these two locks there would be an airtight refuge chamber with room for a hundred men and provided with an air supply line. At the bottom would be the working chamber, where gangs of men would work much as they had on the base of the Marin tower—building cofferdams, drilling and excavating a level, firm bottom, installing a webwork of reinforcing steel and connecting dowels, and pouring concrete—all done far below the level of the strait, in an atmosphere maintained by artificial air.

The caisson, 185 feet long and 90 feet wide, was built at the Moore Drydock yards in Alameda, where the work was periodically inspected by Paine, and occasionally Cone and Pacific Bridge's Jack Graham and Phil Hart.

As the plan now went, the fender for the pier would be built with one end left open. The eastern or bay side of the ellipse would have concrete poured only to a height forty feet below the surface of the strait. The caisson would be floated through this space inside the fender. The fender wall would then be poured, completing the ring so that work on the pier could proceed in calm, enclosed water. The caisson would be sunk to the bottom by filling its lower cells with concrete, which would also become part of the base of the pier. Inside, the necessary struts and bracing could be fabricated, then embedded in concrete, and eventually the whole caisson could be incorporated into the pier.

By the beginning of October 1934, the fender was ready to enclose this giant box, and on October 8, at four-thirty in the morning of a calm, still, Indian summer day—this was traditionally the season of most agreeable weather in San Francisco—the caisson was towed from Oakland out to the pier site.

It took three tugboats to maneuver the enormous dark box—big as a four-story building and weighing more than 10,000 tons—into place. By the end of the day, the caisson was securely within the ring, moored with two-inch steel cables. That night, however, another unseen Pacific storm, this one originating some 1,500 miles at sea, began sending enormous swells through the Gate, tipping the caisson crazily, causing it to surge to and fro in the fender from fifteen to thirty feet, and to bob up and down up to five feet. At their height, the swells prevented the returning fishing fleet from entering the bay, and waves broke over the trestle and crashed to a height of fifty feet on the bridge's concrete pylons.

Jack Graham and Phil Hart of Pacific Bridge, alarmed at the rocking and bobbing of the caisson, called Cone, who came out in the early hours of the morning to the site. The rearing and plunging caisson had snapped the original mooring cables, and the Pacific Bridge crew was trying to hold the giant box in place with live lines from the cranes and with six different 7/8-inch cables. It was like trying to rope a bull with twine. The caisson was now banging against the fender with such force that it shook the entire structure, and seams were opening up in the steel box itself. The steelwork supporting the working platform was being battered and damaged, and there was danger that the cutting edge of the caisson's bottom might gouge out the west end of the fender.

Cone called Strauss. "I went out to the scene at 5:00 A.M.," the chief engineer later reported, "and directed the Resident Engineer, Mr. Cone, to examine the caisson to ascertain its water-tightness." Cone told Strauss the caisson was leaking in two places. An hour or so later, Derleth arrived, and he, Strauss, Cone, Graham, and Hart discussed the possibility of flooding the caisson and sinking it in place. "This was found to be impossible because of the surging of the water." While the fender appeared secure, the danger was that the leaks in the caisson would widen and sink it within the enclosure, a situation that Strauss claimed "would have been very serious."

Hart and Graham insisted that the only safe way to handle the situation was to remove the caisson by towing it out of the fender and hauling it back to Alameda. It would take a full two weeks, they estimated, to close the open end of the fender, and with the winter storm season approaching and the possibility of swells like this occurring any day, it was simply too hazardous to expect a contractor to continue on this basis.

By this time Reed, whom Strauss had summoned, had also arrived, and it was agreed that, since the method of construction was the option of the contractor, and since the risk was now greater than the contractor was willing to assume, the district had no right to order Pacific Bridge to carry it out. Strauss approved the removal of the caisson, but the water was still so rough that the tugs could not get into position to tow it out until nine o'clock that night.

At one-thirty that afternoon, Strauss wired Ammann in New York:
CAISSON BATTERING FENDER DUE TO EXTREME SWELLS. IF ATTEMPT, THINK IT MAY WRECK BOTH CAISSON AND FENDER. THEREFORE AGREEING CONTRACTORS REQUEST REMOVE IT AND BUILD PIER WITHIN FENDER, WHICH WILL REMOVE HAZARD, SAVE MUCH TIME AND PERMIT SAME IN-SPECTION OF BOTTOM AS NOW. DERLETH, CONE AND REED APPROVE.

JOSEPH B. STRAUSS

What Graham and Hart now proposed, with Strauss's concurrence, was to build the pier without using the caisson at all. The earlier revised plan for the deeper excavation and broadened concrete base for the pier had allowed for eight vertical shafts or wells, to ensure inspection access to the pier bottom. Only a week before, Professor Willis had charged that the berthing of the caisson inside the fender was the first step in a massive, and literal, cover-up. "Should this be done," the Stanford geologist threatened, "the rock will be covered beyond the possibility of further examination of excavation, except at great expense, and the design of the pier will be fixed." Actually there had never been any plan to seal off the bottom from inspection. "Eliminating the caisson," Strauss observed, "involves merely the extension of these wells upward, and provides the same facilities for access to the bottom as before." Professor Willis, should he so desire, could climb down inside a well to the strait's floor and stick his nose against it.

The professor, however, remained unsatisfied and had now carried his attack on the pier to the press and radio. "The change of plan involving junking of the great caisson," Willis commented on October 11, "does not remove the threat which exists in the presence of the slippery fault plane under the pier." The professor, the Renaissance man of doom, was now signing his correspondence: "Bailey Willis, Geologist, Member Earthquake Safety Committee and Junior Chamber of Commerce."

Bailey Willis's prophecies notwithstanding, plans proceeded to resume construction of the pier, using the fender itself as a caisson. Now, however, under the increased pressure—public, financial, oceanic—being brought to bear on this particular aspect of the project, certain strains began to reveal themselves among the engineers.

Before pumping water out of the enclosure, Strauss, backed by his assistant Clifford Paine, wanted to build four concrete cross-walls inside the fender to brace it against ship collision, adding, as well, concrete and steel to the main pier shafts. Cone, the Graham brothers, Hart, and Derleth objected to the cross-walls as unnecessary and claimed they would only add to the job's delays and costs. Derleth, particularly, was also afraid that Strauss, to get his way, would simply go around the others to get the approval he needed.

Derleth telegramed Moisseiff on October 29:

RUMOR TELLS ME STRAUSS MAY ASK YOUR APPROVAL HIS SCHEME WITHOUT LETTING REED CONE MYSELF KNOW WHAT HE IS DOING UNTIL AFTER YOU REPLY. STRAUSS AND PAINE ARE NOT AT PRESENT PLAYING BALL WITH US. THEREFORE BE CERTAIN ABOUT GIVING YOUR OPINION OR APPROVAL UNTIL YOU HAVE COMPLETE FACTS.

Actually, on October 27, Strauss had written Ammann and Moisseiff telling them that the decision to use cross-walls had already been made "because I am unwilling to take the risk that the fender may be struck (by ship) even for the few weeks that it is acting as a cofferdam."

Strauss, sensitive to Derleth's clout with Reed and among the Bridge District Board of Directors, rushed to mend fences with the Berkeley dean. "It appears," Strauss attempted to soothe the slighted engineer, "that I am like the private who claimed that the rest of the company was out of step. However, there is consolation in the fact that this is the first time that the Consultants and I have differed." Strauss assured Derleth that his opinion and cooperation were always valued. "I do not wish," he wrote, "at any time, to have you do other than express your own opinion—whether it differs from mine or not." Perhaps, said Strauss, he had become unduly cautious because of what had happened heretofore. Nevertheless, despite their endorsement by the chief engineer, the cross-walls were disapproved at a meeting of the Engineering Board on October 31.

"I hope," Strauss commented glumly, "that luck will be with us and that no vessels will strike the fender during the unwatering period."

2

On October 28 the fender ring was completed, and by November 4 the blanket of concrete poured within it by tremie had reached to sixty-five feet below the level of the strait. To spread the load over the base of the fender, as well as that of the pier, it was decided to add more concrete to the seal, carrying it up to thirty-five feet below surface level and more than doubling the bearing area of the pier. The total weight of the pier, which would eventually be more than 300,000 tons, would put a pressure on the base of 10 tons per square foot. After the pier was built, water would be let into the space between it and the fender, reducing the pressure from the water of the strait on the top thirty-five feet of the fender's protecting wall.

On November 27 the pier area inside the fender wall was pumped out, and for the first time an oval cavity appeared in the rippled blue surface of the Golden Gate. Just a few feet away from the currents, tides, and waves was an enormous concrete hole that required surprisingly little pumping to keep dry: the wall was remarkably tight. Into this hole, within the ring at the end of the trestle, there now scrambled gangs of carpenters, ironwork-

Inside the San Francisco fender, with the water pumped out—a dry space the size of a football field, surrounded by open sea.

ers, and laborers to build the necessary forms and prepare the ironwork for the base of the pier itself.

While trying to crisis-manage the overall construction of the San Francisco pier with one hand, Strauss was still contending with Professor Bailey Willis with the other. On November 8 Strauss, after analyzing Professor Willis's testimony at the public hearings, his letters, and his articles in *The Argonaut* magazine and in newspapers, submitted a detailed report to the Building Committee challenging Willis on just about every point. The professor, Strauss noted, undoubtedly with a certain satisfaction at taking an academic to task for deficient scholarship, had based his geologic objections to the south pier of the bridge on the assumption that the district had consulted only one geologist, Professor Lawson. Actually, Professor A. E. Sedgewick of the University of Southern California had also served as consultant throughout the design period, and both Sedgewick and Lawson, eminent men in their field (Lawson had first identified and named the San Andreas earthquake fault whose proximity to the bridge so concerned Willis now), were in full agreement as to the safety of the pier and its foundations.

Over the last ten years, Strauss observed, Professor Willis had plotted an alleged fault, adjacent to the pier, in three different locations. He had assumed that sandstone at the pier site was faulted under the serpentine, when the only sandstone near the site was on top of the serpentine behind the anchorage. Willis, said Strauss, had failed to locate the pier within 200 feet of its correct position. The "tilting back and forth" he predicted for the towers indicated an entire lack of knowledge of the true stress conditions. The professor had used exaggerated scale in his diagrams. And Willis's prediction that the rock underlying the pier was so unstable that the structure might slide of its own weight had been refuted already by the building of the fender and base slabs of the pier.

"Summing up," wrote Strauss, "Professor Willis's one-man campaign is apparently a continuation of the same attacks made by the opposition forces upon the Golden Gate Bridge from the moment of its inception." The professor's case, Strauss maintained, rested upon "faulty hypotheses, false assumptions, erroneous measurements, incorrect scales, distorted soundings, and misinterpretations of geological conditions." In his hearings before the committee, Willis had admitted that his principal contentions had been based upon errors and that he was unfamiliar with either the site or the structure. These admissions, Strauss concluded, left nothing to the professor's case "but a lively imagination and an amazing faculty of believing what one wants to believe."

At the heart of Willis's case were the old unanswerable, unanswered questions about the long-term effects of basing an unbuilt bridge on an unseen ocean bottom. One of those questions was now, for the first time, about to be resolved.

On the evening of December 3, three men—Russell Cone, Jack Graham, and Chris Hansen, Pacific Bridge's chief diver—were bolted inside an air-lock at the top of one of the inspection wells. Earlier that day, while gangs of workmen were busy preparing the foundation for the great pier, the lock had been fitted with an air hose, electric light, and a telephone line. At about 8 P.M., air was let into the lock, and the pressure raised to fifty pounds per square inch—the amount required at the bottom to keep it tight and dry—and the water in the shaft was forced out through a blow-hole.

"We descended in a bucket," Cone later reported to Strauss, "107 feet to the ocean floor." At the bottom of the ride, they reached a dome-shaped steel chamber, 15 feet across, "very well lighted with four 50-Watt lamps. In addition, I carried a new flashlight." Inside the chamber, confined beneath 65 feet of poured concrete, the temperature was now more than 100 degrees.

"The foundation bed was exposed and dry," noted Cone, "with the exception of a small puddle about six feet long and two feet wide near one side of the well. The rock at the center of the inspection well was about two feet higher than at the edges. The surface of the rock was hard and firm. It was plainly bed rock."

It had been anticipated that the bottom of the wells would be cluttered with loose rock and debris, requiring an extensive cleanup by gangs of sandhogs. Instead, the floor had a smooth rock surface, resembling a cobblestone street. Inside the stifling chamber, Cone and Graham were grinning and pounding one another on the back. They they got down with flashlights and went over all the rocky nooks inside the bell.

"I walked about and carefully examined all of the area within the inspection well," Cone reported. "The rock was serpentine similar to that exposed in the anchorage excavations last year, except that it was much harder."

Cone found that some concrete particles had run in from the seal, but when he scraped the small area they covered, there was firm rock beneath. "The serpentine fragments lying on top of the rock had the veined and crackled appearance of the hardest serpentine boulders." Cone scraped clean several points around the wall: all appeared to be this veined or crackled hard serpentine as opposed to "talcy" laminated serpentine. His examination had been made on the original bottom, which had been cleaned by the divers at the time the seal was poured by tremie. "Based on this examination," Cone concluded, "my opinion is that the bottom had been satisfactorily cleaned and the material makes a good foundation."

In an atmosphere whose pressure was more than three times that of normal, the men could not remain in the chamber more than twenty minutes at a time. It was now after midnight. Cone, Graham, and Hansen rode the bucket up to the top of the well. Here they waited thirty-five minutes, while a slow release of air brought the pressure in the airlock back to normal. Eager as he was to spread the good news, Cone fully understood the dangers of sudden decompression. Once, on the Philadelphia-Camden Bridge, he had been hit with the bends after exiting too early from a caisson. Cone had gone all the way home to his Philadelphia apartment, where he had been struck with excruciating pain.

"I was pregnant and it was a bitter night," Izetta Cone remembers, "but we traveled back by subway, ferry, and bus to the medicine lock situated at the Camden end of the bridge and got him into the chamber." Cone, in dreadful pain, had forced himself to exercise his arms and legs. A young doctor on duty thought it was amusing to see Cone cavorting inside the chamber. "He was a fool," Izetta Cone says of the young physician. "If Rusty had been the kind of man who says, 'I feel terrible, I'm going to bed,' he could have been crippled for the rest of his life."

The day after they had descended to the floor of the Gate, Cone and Graham resumed their inspection tour and found conditions at the bottom of the other seven wells similarly satisfactory. On December 7, Professor Andrew Lawson, more than sixty years old, was lowered in a bucket to the bottom of well no. 3. "I had the opportunity," the geologist reported to Strauss, "of examining a clean washed surface of the rock. The elevation of the bottom is −107 feet. The rock of the entire area is compact, strong serpentine remarkably free from seams of any kind. When struck with a hammer, it rings like steel."

To put to rest objections concerning the foundation material's soundness, on the part of Bailey Willis or anyone else, pressure tests were made by bracing hydraulic jacks against the roof formed by the chambers and cranking the pressure up to forty-three tons per square foot. Also, at the west end of the pier, diamond drilling was taken down to 250 feet, with sample cores showing no trace of the sandstone that Professor Willis had insisted underlay the serpentine and would "inevitably" cause slippage.

"The foundation bottom in all of the inspection wells has been examined," Strauss reported to the Board of Directors on December 19. "The same excellent character of rock has been found in all." The evidence necessary to forestall any reasonable doubt concerning the foundation of the San Francisco pier seemed to be well in hand. The doubts and delays were forgotten. Work was already under way on the pier itself.

"The west shaft of the San Francisco pier has been concreted to an elevation of plus twelve," Strauss continued, "and concreting is in progress on the east shaft to the same elevation. The anchorage steel has been set and is imbedded in the concrete." The eight inspection chambers and shafts, having served their purpose, were now filled with concrete. The floor of the Gate, revealed to man in dry surroundings for the first time, was now sealed again, forever.

3

With the pouring of the concrete base for the San Francisco pier and the installation of the fascia plating or falsework covering the struts at the portals of the Marin tower, the subtle ingenuity of Irving Morrow's architectural treatment of the bridge began coming into full bloom. The vertical fluting he had added to the portals, which made a playful, savoring use of light and shadow, had been extended to the concrete tops of the pylons and to the base of both the Marin and San Francisco piers. These touches verified the sense of the bridge as a designed and not just a functional object, in service to an idea, and with aspirations to eloquence, perhaps poetry. They also enhanced the almost living quality the bridge was beginning to reveal, of never appearing exactly the same any two times you looked at it. The alternating faces between the tower portals, the curtainlike proscenium clefts at the portal corners, the sculpted lines tapering up the length of the towers, their echoes in concrete at the pylons and the bases of the towers, all these combined with the changing light at the Gate, the low swift fogs, the bursts of brilliant sunshine, the reflection off bay and ocean, to give the sense of a structure that, like an organism, could alter itself from minute to minute in response to the position and intensity of the sun.

People who had watched the towers rising had sensed this feeling, of something being built that had a dimension beyond other things that had been built, and they were profoundly moved by it. "I have been watching very closely the progress of the towers on the Golden Gate Bridge," the sculptor Beniamino Bufano wrote Morrow in January 1935. "In its structural beauty, its engineering and architectural simplicity, and of course its color—like red terra cotta—it moves and molds itself into the general beauty and contours of the hills. It is this structural simplicity that carries to you my message of admiration."

"I have watched the Marin tower from the ferry and the city in almost every kind of weather and light," Robert B. Howard, a Marin County commuter, wrote Morrow in February, "and find it superbly in harmony with the landscape both in design and in color. Now that the south tower is beginning to appear, the beauty of that color of red lead has been brought home to me even more. Couldn't the Golden Gate Bridge be left in red lead or some finishing paint that approaches vermillion?"

"The tone is beautiful under all light conditions," wrote E. P. Meinecke, a U.S. Forest Service pathologist:

> It has the advantage of defining structural lines clearly and yet not obtrusively, and of fitting harmoniously into sky, water, distance and the Marin hills, whether in their spring green or their warm summer tones. If a paint of this value were applied, the Golden Gate Bridge would stand out against its surroundings without danger of ruining one of the most beautiful harbor entrances of the world.

Morrow, responding to this groundswell of shared, stirred emotion, and the verification of his own ideas in concrete and steel, found his confidence growing as the budding bridge unfolded. With Strauss's support, he had begun to assume something of Ellis's role, as the aesthetic conscience of the bridge.

"The magnitude of the structure and its crucial, definitely 'located' position in the landscape," Morrow wrote, "suggest that it should be emphasized rather than played down. The proper color for this is naturally a contrasting one. Local atmospheric effects are predominantly cool—gray during foggy weather, and blue during sunny weather. This points to a red, earthy color for the structure. The color chosen is also admirable for enhancing the scale."

Scale, Morrow pointed out, is one of the most elusive of architectural qualities. When achieved in form, as it had been in the engineering design of the bridge, it would be a great misfortune not to have it sustained by the color. A free use of color in general, Morrow scolded, was an obligation that architects generally had continually evaded.

"When the bridge is completed, and has been observed through an entire cycle of a year," Morrow predicted, "the appropriateness and beauty of the color will doubtless be quite generally appreciated."

In December, Morrow had submitted to Strauss a detailed plan for illuminating the bridge that demonstrated the same ability to work on an enormous scale combined with a remarkable sensitivity to nuance and detail.

Uniform intensity of illumination on the bridge, Morrow observed, must be avoided. "Uniform distribution over a distance of a mile and a half would seem too artificial to be real and would compromise the effect of size." It was the same with the towers. If the lights diminished in intensity as the towers rose, "the gradation will give them a sense of soaring beyond the heights susceptible to facile illumination."

Morrow's diligence and thoroughness, and his ability to articulate his ideas and feelings, his knack of making the abstract concrete, had won him the respect of the Engineering Board. "I am much impressed by Mr. Morrow's presentation," Moisseiff wrote Strauss in February, "and inclined to agree with him that his original recommendation of orange vermillion or the color of red lead shop coat will present the bridge to its best advantage. I feel that such treatment tends toward a decided advance in the coloring of long span bridges."

4

Because of the seven-month delay in completing the San Francisco pier, the steelwork on the south tower of the bridge could be erected using many of the same men and much of the same equipment that had worked on its twin on the Marin side. Once again, the bargeloads of steel sections began arriving from Alameda. Again the sections were hoisted into place by a giant-spider crane, traveling up the ninety-foot space between two great tower legs, and connected, bolted, and riveted together by gangs of bridgemen, grown sure-handed if not cocky in their work, and about as accustomed as it was possible to be to the constant wind and cold at the Gate.

The work went faster than it had on the Marin tower. "On the San Francisco tower, they lined up the sections—80-ton pieces made of cells," says Frenchy Gales. "The crane would let go. Haul off on it. The sections would seat themselves. They didn't hammer them into place with a piledriver like they'd done sometimes on the Marin tower." The sense of teamwork, crucial among ironworkers, who depend heavily on one another, was more ingrained. The men had grown used to each other, as well as the work; those who couldn't cut it had been weeded out.

"People who weren't good didn't last," says Harold McClain. "You'd find a way to make their work harder, till they gave up."

Also, the pace of work had been increased. "There was a higher tempo with McLintic-Marshall than, say, on American Bridge [the contractor on

Looking down toward the fender of the San Francisco tower. Where crossbraces had appeared in other bridges, there would be huge open portals framing the sky, clouds, and fog of the Golden Gate.

the San Francisco-Oakland Bay Bridge]. You did more work, and more was expected of you."

There was a powerful financial incentive in this. When construction on the San Francisco pier had fallen months behind schedule, the bridge directors had ordered McLintic-Marshall to stop fabricating steel at its plants for the tower. The company's contract for the south tower called for it to be finished within nine months. To avoid penalties, the Bridge District had postponed the official construction start.

The Directors now proposed a new schedule, with McLintic-Marshall resuming fabricating steel in November 1934 and construction of the San Francisco tower beginning in January 1935. The tower would be completed by July, within nine months of the resumption of work. McLintic-Marshall, however, had contracted for other jobs, and to give priority now to the delivery of the San Francisco tower steel would cost the firm income. The company officials insisted that even if they did reschedule production now, they could not finish the tower within the nine-month due date and would probably have to request a two-month extension.

To compensate for these interruptions and delays, the McLintic-Marshall representatives suggested that the Bridge District pay the company a bonus for accelerated work, $1,500 a day for the first sixty days that the tower was brought in under the existing deadline, plus $750 for each day saved in addition to that.

It was a stick-up, and Strauss and the directors knew it. Stanton, the blustery San Francisco supervisor, ex-protégé of Tom Finn and Murphy Hirschberg, said the McLintic-Marshall officials were worse than "the fellow who puts you up against the wall and takes money out of your pocket."

Among the directors, who had previously done their share of gouging by calling for new bids to take advantage of the Depression's effects on contractors, there was at least one man who perceived an advantage within this latest problem. Hugo D. Newhouse, a lawyer who was also the board's financial expert, pointed out that, since interest on the district's bond money was costing $4,600 a day, the board, by paying an accelerated-work bonus of $1,500, would be saving $3,100 a day toward the date when the completed bridge would be generating revenue. Newhouse calculated that the directors, by paying McLintic-Marshall $90,000 for shaving sixty days off the work schedule, could save $276,000 in bond interest. The directors enthusiastically agreed to the bonus, and work on the San Francisco tower proceeded at a considerably heightened pace.

In the very familiarity and efficiency of the work, there were dangers.

Men needed to remind themselves from time to time that they were working with hazardous materials, usually at precarious heights. There were injuries, but thus far, amazingly, there had been no fatalities, in a type of work—high-steel construction—where the loss of a life for every million dollars' worth of work was considered average. "Most accidents were guys hit with flying stuff," recalls Frenchy Gales. "My first job on the towers was as a rivet punk. I ran for stuff. You'd work on the towers and hear—*zing!*—when hot rivets would fall and hit." Again the size and scale of the job, the weight and dimensions of the materials, the heights at which men worked, added to the potential impact of injuries.

"There was a riveter named Hungry Pete who heard a noise and looked up—he caught a maul right in the face. Knocked all his teeth out."

As resident engineer, Cone was the man responsible for the safety of working conditions. The rules were strict, and they were firmly enforced. Not only were hardhats required at all times, stunting of any kind on the job could bring immediate dismissal. Still, on a project this large and complex, there were plenty of accidents that were short of fatalities. If somebody did get seriously hurt, Cone went to visit him, usually at San Francisco's nearby Children's Hospital.

"I was on the corner like a point on the very top of the San Francisco side," Peanuts Coble remembers:

They had jumped the rig down. Had a big bar holding the thing in place. The guys in the skip box had just taken these out. The boom was way past that. They had boomed 'em up to get clear. There were two spreaders hooked behind where I was. The hooks came loose, I felt something move. The cable, hooks and all, was coming toward me. When it came by, I grabbed it, swung out onto the load line, and slid down into the skip box with these guys. They was white, I'll tell you. I wasn't afraid of a damn thing, but I had to stand spread-legged to keep my knees from shaking. I had damn near slipped off the very tip-top of the San Francisco tower.

Pushed by a contractor earning a daily bonus for fast work, assembled by men, most of whom had just worked on an identical structure, the south tower had risen to a height of nearly 500 feet by the middle of March. It was completed to its full 700 feet by the end of June, but not before the second part of Professor Bailey Willis's geologic challenge had been answered.

In early June, the San Francisco tower, a limber structure unanchored by cables and still held together in many places by lift pins and bolts, was struck by a fairly strong earthquake.

"I was up on the tower when the earthquake hit," recalls Frenchy Gales:

It was so limber the tower swayed sixteen feet each way. There were a lot of seams, all the way to the top. There were twelve or thirteen guys on top, with no way to get down. The elevator wouldn't run. The whole thing would sway toward the ocean, guys would say "here we go!" Then it would sway back, toward the Bay. Guys were laying on the deck, throwing up and everything. I figured if we go in, the iron would hit the water first."

The following day there was an aftershock that set the tower swaying again. This time, Derleth and Lawson had people from the University of California at the base of the tower, measuring the sway and its effects. The earthquake had caused the tower to oscillate more than would be possible when it was tied down by cables, yet of the slippage, sliding, and landslides, the grinding collapse into pudding stone that Dr. Willis had predicted, there were no signs.

"CAN HE DELIVER?"

I

Joseph Strauss's career as a master designer-engineer was about to emerge, like a butterfly, from its chrysalis. The promoter of some 350 bluntly functional railroad and highway lift bridges was on the brink of a new existence as a soaring, brilliant poet in steel. The Golden Gate Bridge, Strauss informed the bridge directors, had already carried San Francisco to the imaginative forefront among American cities. In the spring of 1935, when Pan American Airways inaugurated regular, twice-a-week service between San Francisco and Hong Kong, newspapers throughout the country ran a photograph of the *China Clipper,* in flight, outbound to the Orient, passing the Marin tower. It was as if these two great innovations summed up the entire spirit of American technology and progress and renewed the westward thrust of the nation's future. Strauss had bested a distinguished professor, Bailey Willis of Stanford, in intellectual combat on the professor's own geological home grounds. There was no longer any mention, in connection with the bridge, of Charles Ellis, who had returned to the academic isolation from which Strauss had originally plucked him.

And yet Strauss's earlier existence, with its caterpillar steps of hundreds of undistinguished bridges, its twists and windings over and through the obstacles of local politics, and his association with lawsuits and fixes, politicians and bagmen, continued to restrain Strauss's full free flight as the mastermind of one of the world's masterworks.

In September 1934, Strauss had filed suit against the Bridge District over the question of who was responsible for the costs of the in-plant and on-site inspection of the various materials used in the bridge. The suit was described as "a friendly action," but it represented a failure to resolve a difference by unlitigated action and put Strauss on one side of an issue, with Reed and the directors on the other. In the course of the suit, it was contended that Strauss, as chief engineer, had written into the contract the very terms whose vagueness was now in dispute. In establishing his claim, which Strauss won, to an additional $262,500 to pay inspection costs, Strauss's lawyer also established that Strauss was not an employee of the

Bridge District but a private contractor, responsible for the supervision of other private contractors. It was a legal nicety that would be flung back at Strauss, to his immense disadvantage, after the bridge was completed.

No sooner was this issue settled when Strauss was struck with another legal action, this one not described as friendly, but in fact originating with one of the chief engineer's oldest and closest associates. On March 19, 1935, the mysterious H. H. "Doc" Meyers, described as a "Los Angeles engineer and former associate of Strauss," filed suit against the chief engineer. According to statements at the time of the suit, Meyers claimed that he had helped Strauss get his job as chief engineer and that, in return, Strauss had agreed to pay Meyers a commission of $110,000, at the rate of 15 percent of each payment Strauss received from the Bridge District. The suit threatened to blow the lid off the various personal and political arrangements among Strauss, MacDonald, Murphy Hirschberg, and certain of the San Francisco supervisors, with the possibility that the construction funds for the entire project might be tied up by investment house officers, wary of scandal, refusing to accept additional bridge bonds.

On March 27, the Bridge District directors met in executive session to discuss the suit and to review their contractual arrangement with Strauss. According to William Filmer, president of the district, when he and Congressman Welch had been in New York before contracts were let for the bridge, they had called on the John A. Roebling Company, the world's largest manufacturer of bridge cables. To the two directors' astonishment, a Roebling official told them that "this man Meyers has been around here saying that he can deliver the cable contract to us. Can he deliver?"

Filmer and Welch had assured the Roebling people that Meyers could deliver nothing.

Meyers's lawyer insisted he had no knowledge of the Roebling matter and that it had no bearing anyway on the present issue, which was Strauss's failure to honor a contract. The chief engineer had already made payments to Meyers amounting to some $45,000, and Meyers was suing to collect the balance, on which no payment had been made since July 1934. Although there were rumors that Strauss's first wife had been brought to San Francisco and would be subpoenaed to testify as to the agreement between Meyers and Strauss and that Annette Strauss, the engineer's new wife, who had also been friendly to Meyers, would also be called and sworn, Meyers's attorney refused to comment.

There was no way that Strauss could win a courtroom confrontation of this kind. Strauss was known, admired, feared, and gossiped about in San

Francisco. There were even rumors that the chief engineer's fragile health was a product of collusion between the "Doc" without a practice and Strauss's younger, attractive wife. Having admittedly made the previous payments to Meyers, Strauss under the circumstances could only be either exposed as a schemer or revealed as a dupe. Choking back his usual zest for combat, the chief engineer took the only course he could and settled. "Strauss, it is understood," reported the San Francisco *Chronicle*, "will carry out the original terms of the mystery agreement . . . which have never been made public."

Strauss's lawyers tried to portray the settlement as a matter of gentlemanly diplomacy rather than successful near-extortion, insisting that Strauss had got his job "on his world-wide experience as a bridge engineer." What seems more likely is that Strauss had discovered the painful truth of the American folk wisdom best articulated by Nelson Algren: "Never play poker with a man called Doc."

These reminders of Strauss's promoter aggressiveness, his willingness to pull strings and grease wheels to get what he wanted, did not sit well with the Bridge District Board of Directors or with James Reed, the district general manager. As a navy man, Reed had been trained to develop an instinctive distrust of men who were unwilling to subordinate their personal interests to the overall welfare of the ship. Strauss, with his penchant for lawsuits, his association with people like the charlatan Meyers, his infrequent appearances on the job, his repeated illnesses with their overtones of malingering, his bohemian avoidance of accepted rules and procedures made him, to Reed, a potential loose cannon, someone who would have to be kept in check to avoid raking the deck of the district itself.

Reed knew he could not challenge Strauss's command of the engines of publicity surrounding the bridge; indeed, that seemed to absorb Strauss's concern more these days than the actual building of the structure itself. What Reed could, and did, do was patiently and methodically gather information that could be presented at critical times to prevent Strauss from simply riding roughshod over the manager's office and to keep Reed from being crushed between the colliding egos of Strauss and the board. In the dispute over inspection fees, for example, Reed had consulted with Ammann as to what was considered ususal and proper in such matters and had asked the New York Port Authority engineer to review Strauss's contract as to the clarity and fairness of its terms. Ammann had told Reed that he "frankly [was] not and have never been satisfied that the contract now in force will secure that high class of work which such an outstanding structure deserves," an opinion that could not but reflect unfavorably on Strauss and on the Bridge District's Building Committee. The threat was implicit and sometimes overt that Strauss, with his hustler-promoter's background, would, if forced to absorb all contingent costs connected with building the bridge, cut corners on materials and inspection in a way that jeopardized the structure's quality. In a board meeting on January 24, 1934, this was exactly what Strauss, who was accompanied by an attorney, had threatened to do. Although Strauss had won the "friendly" lawsuit and the district would have to absorb the additional costs, Reed had carried out what he saw as the highest responsibility of his office; he had assured that the bridge would indeed be a "high class" job, and he had also forced Strauss to go to court to win his point.

There had, in addition, been complaints by Ammann that "matters have been put into the minutes," of Engineering Board meetings, "which are radically at variance with what actually happened at the meeting." These complaints were made not to Strauss, chairman of the Engineering Board, but to Reed, sometimes through Derleth. The feeling was always lurking, surfacing on occasions, as in the dispute over the cross-bracing for the San Francisco pier fender, that Strauss, given the opportunity, would simply run away with the project, leaving the consulting engineers and the general manager, Strauss's concessions to convention, as a collective rubber stamp of approval for anything Strauss wanted to do.

With Clifford Paine assuming more and more of Strauss's day-to-day office and on-site inspection responsibilities, the chief engineer, his physical energies still seemingly sapped since his marriage to his younger wife, was rumored to be turning his attention more and more to flights of fancy: poetry, inventions—it was said that he had devised a machine for washing dishes automatically in restaurants—and proposals for yet more bridges.

The aggrandizing nature of Strauss's personality, his lust for public recognition combined with his penchant for secrecy, his personal remoteness coupled with his insistence that he was "in more intimate contact with the details of the work than any other member of the Board," forced people who worked with him either to band together for self-protection like Reed and the consulting engineers, to accept domination like Paine, or to risk getting fired like Ellis. Only Morrow and Cone, both of whom had come on board after the engineering design of the bridge was done and who possessed skills that were essential now for its completion, seemed to operate with any independence and yet continue to enjoy Strauss's support.

"Your husband," Strauss remarked to Cone's wife, "is my right arm."

Cone himself, still only thirty-nine, was finding the course of his own career unexpectedly changing. Not long after Cone had arrived in San Francisco, Clement Chase, his friend and advisor in the Modjeski firm, had apparently suffered a heart attack while inspecting the Philadelphia-Camden Bridge, fallen from the bridge, and died. Because of Modjeski's advanced age, Chase had become the animating spirit of the firm, and with his passing the Modjeski organization lost some of its sense of purpose and direction. Cone, the man who had been groomed to assume Chase's role, was committed by contract to Strauss to see the Golden Gate Bridge job through. Instead, Montgomery Case, a friend of Cone's, and his original field boss on the Philadelphia-Camden Bridge, became a partner in the firm, whose name changed, almost imperceptibly, to Modjeski, Masters and Case.

Cone, who had established himself in San Francisco and who, along with his wife, had somehow found time for an active social life, began to redirect his long-term as well as his immediate interests toward the bridge. When it was completed, the Golden Gate Bridge would be a functioning enterprise, generating revenue, requiring maintenance, in need of someone with a combination of engineering familiarity and managerial skills to run things, a job that would be in many ways an extension of the sort of work he was doing now. Strauss, with his own firm to run and in questionable health, was not likely to want to continue working for the Bridge District, and there was always the possibility that Reed, who was neither a bridge builder nor a contractor and who had already retired once, from the navy, would eventually move on. If Cone could not assume the manager-engineer's mantle in Philadelphia, he might well be able to do what amounted to the same thing here.

This interest gradually became known to certain of the directors, to Strauss, and, with a certain discomfort, to Reed. Cone, on his part, faced a cutoff date—the completion of the bridge—when the sometimes punishing responsibilities of his job would no longer exist, while Reed began to suspect that the much younger, ambitious resident engineer was trying to undermine him as general manager.

2

While the various personal dramas surrounding the bridge were being played out, the bridge itself was assuming some of the symmetrical balance of its final proportions. The San Francisco tower was completed to its full 747-foot height, a great red ladder matching its counterpart on the Marin shore, and was in the process of being bonded together by a total of some 600,000 rivets. The delays and disputes surrounding the construction of its base, the San Francisco pier, had been forgotten by all save Professor Willis, who still issued an occasional, and now largely ignored, blast against the bridge to regional chambers of commerce and various organizations dedicated to promoting earthquake safety.

Despite broadening support for Morrow's recommendation of a red color for the bridge, the Engineering Board was still divided on the final choice of paint. Ammann, who lived in a house in New York's Westchester County where, with a telescope, he could look down the Hudson at his most recent creation, reported that "we have repainted the cables and suspenders of the George Washington Bridge with aluminum paint, and there have already been very favorable comments in the newspaper." It was the intention of the New York Port Authority to apply the silvery paint to the remainder of the structure the following year. Closer to home, Derleth questioned the use of a red color as being a novelty that people would tire of, or think "gaudy." "Black, gray, and aluminum are colors that have stood the test of time," the engineering dean advised. "Any one of them would suit me, with a leaning toward the gray or aluminum."

Derleth and Ammann, both men who had been associated with other important bridges and who would, in the future, be associated with others, did not feel to the degree that Morrow and Strauss did the visceral sense of the Golden Gate Bridge's uniqueness, its once-in-a-lifetime dimension as sculpture, a work of art. To Strauss, the perpetually aspiring poet, and Morrow, the former Beaux Arts painter, there was something elemental and profound about the red color of the completed bridge towers. As Evan Connell has pointed out, the red of the bridge combines with the blue of the sky and sea, the intense green of the Presidio on the San Francisco side, and the summer gold of the Marin hills to produce a world of primary colors that is immensely refreshing to the eye. The color of the bridge was a debate between logic and emotion, and this time it seemed emotion might actually prevail.

With the completion of the San Francisco tower, there was a general feeling that the worst construction obstacles had been overcome. Work on the bridge, it was now anticipated, could proceed quickly, pretty much coasting to its conclusion. Adding to this feeling was the competitive presence of the San Francisco-Oakland Bay Bridge, now expanding daily toward its completion within sight on the same bay.

Strauss, intensely competitive by nature, had been presented with the prospect of another bridge, on the eastern side of the bay, since the late 1920s. Commissioned by an agency of the state of California, built with state and federal money, the San Francisco-Oakland Bay Bridge was longer overall than the Golden Gate—12.5 miles including approaches—was budgeted at more than twice as much—its total cost would eventually reach $77 million—and, with its two suspension spans plus the longest cantilever span yet built, represented to some people a greater engineering challenge. Strauss had repeatedly found himself in competition with the Bay Bridge people for publicity, of which he won more than his share, and for federal money, which the Bay Bridge, being a government-sponsored project, effectively preempted. There was a certain rivalrous inbreeding among the engineering boards of the two bridges, with Moisseiff and Derleth serving as consultants on both structures—a fact that caused the opportunistic Strauss to schedule Golden Gate Bridge Engineering Board meetings to coincide, if possible, with Moisseiff's Bay Bridge–paid trips West. Ralph Modjeski, for whose firm Cone had got his start and with whom he had aspired to work again, had also been a consultant on the Bay Bridge; moreover Modjeski, an early employer of Strauss, had rejected Strauss's first idea for a cheaper, easily reproduced bascule bridge.

This inbred rivalry extended to the ironworkers involved in both bridges, whose two union locals, No. 377 in San Francisco and No. 378 in Oakland, historically did not get along. The AFL International Association of Bridge and Structural Workers had been riven with dissension in California since 1911, when J. J. MacNamara, the union's secretary-treasurer, and his brother J. B. were convicted of dynamiting the Los Angeles *Times* building in a blast that killed twenty people. At the time of the explosion, the unions were in a concerted drive to make Los Angeles a union shop town, and the *Times*, owned by the vehemently anti-union Harrison Gray Otis, was the leading open shop voice in that city.

In the years since, the union had recovered more strongly in San Francisco than in Oakland, and the Golden Gate Bridge ironworkers were eventually 100 percent unionized, whereas the Bay Bridge workers were a union and nonunion mix. At first, ironworkers from the Oakland local were banned from working on the Golden Gate Bridge, since they were not residents of a Bridge District county; but eventually, as work on the Bay Bridge wound down and the Golden Gate's need for experienced bridgemen increased, Oakland ironworkers were allowed to transfer into the San Francisco local.

Although construction on the Bay Bridge had begun three months later than that on the Golden Gate, it now appeared that because of the delays surrounding the building of the San Francisco pier the Oakland bridge would be completed a year earlier than its Marin counterpart. The inevitable comparisons made of the two bridges' construction timetables raised Strauss's competitive hackles.

"Bridge, bridge," Strauss was heard to grumble, with characteristic brusqueness, to someone who had thrown up to him the Bay Bridge's earlier completion and opening. "That's only a trestle."

In March 1935, one of the last physical reminders of the frustrations and delays associated with the San Francisco pier was scheduled for disposal. The huge junked caisson, towed back to the Moore Drydock yards in Alameda, where it sat for months in the shadow of the Bay Bridge, was now loaded with 800 pounds of dynamite and hauled by two tugs, the *Kodiak* and the *Rustler,* past both the still-building bridges, and out to sea. Aboard the caisson were Jim Graham of Pacific Bridge, Chris Hansen the diver, and Charlie Winters, a timekeeper. The huge box was so unwieldy that it took twelve hours for the tugs to tow it five miles beyond the Gate. It was already dark when the caisson was struck by yet another unexpected Pacific swell that pitched it backward, snapping the towropes and sending the huge jinxed box, loaded with dynamite and with three men aboard, drifting toward the shipping lanes. The caisson lurched, unilluminated, in the dark for six hours, a looming, explosive navigation hazard with a living human cargo. Finally the Coast Guard, using two cutters, the *Shawnee* and the *Ariadne,* managed to get the caisson under control, transfer off the three men aboard it, and attach new lines to the tugs.

The haul was resumed, and the great dark, ugly box was towed some forty miles southwest of the Gate, where it was detonated and sunk.

"ROEBLING BROUGHT CLASS TO THE JOB."

I

With the beginning of the cable construction on the bridge, Strauss was returning to the sources of his own earliest inspiration as a bridge builder, as well as to the dawn of modern suspension span building generally. John A. Roebling Sons, the cable company that had been approached by H. H. Meyers, was founded by the man who designed and built the Ohio River Bridge at Cincinnati, considered at its completion in 1866 the greatest bridge in the world, and a powerful influence on Strauss's boyhood ambitions. Three years after its opening, Roebling began construction on an even greater work, the Brooklyn Bridge, which would take fourteen years to complete and is the genetic forebear of every long-span suspension bridge built since.

On the Brooklyn Bridge, John Roebling, who died of tetanus in 1869 after his foot had been crushed in a ferry accident, and his son Washington Roebling, who after collapsing from caisson disease supervised construction from the confinement of a sickroom for the next ten years, built, on site, cables of steel wire instead of wrought iron. To spin the cable, wire by wire, up and over the towers, from one side of the East River to the other, Washington Roebling developed a pair of carriages, which moved back and forth between the anchorages on a pair of endless hauling ropes, one on each footwalk. Each rope was driven by a reversible winding machine, and the carriages were set so that one was at the near anchorage and the other at the far.

The wire was delivered to the two anchorages on large spools, which were mounted on unreeling machines. The end of the wire was fixed to the spinning wheel attached to the carriage and temporarily made fast. The same operation was carried on simultaneously at the other anchorage. The carriages and spinning wheels proceeded, in opposite directions, from anchorage to anchorage, passing each other at midspan.

This basic method, and the equipment applied to it, had been refined and improved over the years, chiefly by Roebling Company engineers, into something resembling a sophisticated manufacturing production line, operating hundreds of feet in the air. By adding extra spinning wheels and precisely timing the motions associated with setting and adjusting the payed-out individual wires and bunches or strands of wire, the average amount of cable spun had jumped from thirty-three tons per day on the Manhattan Bridge in 1908 to sixty-one tons on the George Washington Bridge in 1929–30. To keep the Golden Gate Bridge on schedule, the existing cable-spinning speed record would have to be multiplied several times over, while stringing a longer cable than had ever been spun before.

The Roebling Company, to service its contract, had opened its own yard near a Marin County landing called California City, today part of the industrial flats of the north part of Sausalito. Here the wire shipped from Roebling's plant in New Jersey was cleaned of its paraffin protectant, re-reeled on larger spools, and warehoused for later delivery to the bridge site. Here also was stored the material for the footwalks, the side-span storm system, telephone and signal lines, and the equipment for the spinning systems at the anchorages—all Roebling responsibilities. Roebling had also imported the company's ablest supervisors, had hired, where possible, experienced bridgemen, and had conducted training sessions for others. The firm imparted a new degree of professionalism to the entire project.

"Roebling brought class to the job," says George Albin:

They taught everybody else about safety. Before Roebling came in, you could wear anything. They furnished rain gear—tried two or three different kinds until they found one that worked best. And web belts with a six-foot manila safety line with an eye on it. You were supposed to use it. Whenever you got yourself in a bum position, if you didn't tie yourself off, you could get fired.

"The Roebling people were the most experienced," Harold McClain agrees. "They were good people to work for."

Roebling was also 100 percent union. "The ironworkers were strongly unionized," says Frenchy Gales. "They had a business agent, a big, tough Irish guy. He told the Pacific Bridge guys, 'You got union men working above you, they're less likely to drop stuff if you join the union.' "

Frenchy joined the union.

In July 1935, the Roebling engineers, a man named Paul Hendrickson most prominent among them, put in place the company's first contribution to the bridge structure itself. At each tower, a 160-ton "saddle," or grooved, half-oval cradle for a cable, was hoisted in three sections to the top and bolted, one saddle to each tower leg, in position at the cables' peak resting points. The saddles rested not on immobile steel beams but on plates designed to roll and move, like giant bearings.

In his original plans for the bridge, Charles Ellis had calculated that the bridge towers, because of the stresses of load and temperature, would tend to bend slightly toward the center of the strait. Ellis suggested that this deflection could be equalized by attaching lines to the tower tops and literally "pulling them shoreward nine inches and holding them there while the cables are being placed on the tower." Roebling engineers, expanding on this idea, came up with the idea of rollers, allowing the towers to move back and forth beneath the saddles as the cables were completed, then freezing the saddles and rollers in place afterward, transferring the movement to the tower tops themselves. The towers and cables could thus respond to changes in stress and temperature, thousands of tons of steel flexing and relaxing, sensitive, like parts of a huge, growing plant, to subtle changes in temperature and weather.

"As the cable changes length," Russell Cone explained it,

> *the bridge itself moves and deflects and the movements are taken care of by expansion joints and articulated bearings at the main tower. It can adjust itself without damage to quickly applied forces of large dimension. This "giving" or elasticity of the bridge gives strength to the whole structure and absorbs strains and stresses. It's almost a living, breathing thing.*

Thus Moisseiff's theory and Ellis's application of it were now being made manifest in steel.

With the cable saddles in place on the two great red towers, work was now ready to begin on the spanning of the Gate itself. On August 2, 1935, for the first time in its history, the entrance to San Francisco Bay was closed to all shipping.

"I'd raise a flag," recalls Frenchy Gales, "to show the cutters outside the Gate was closed."

It was a typical August morning at the Gate, cold, wet, gloomy, with the tops of both bridge towers lost in fog. A full tide was running and there were large, breaking waves. At about 8 A.M., one end of a 1 9/16-inch wire rope was made fast at the Marin anchorage. Then, in a crossing taking about an hour, 5,000 feet more of the rope was played out from a McLintic-Marshall barge, and the wire rope was connected, at its other end, to the San Francisco shore. Next, lines were sent down from derricks at the tops of the towers and connected to the wire rope. At noon, the derricks began lifting the rope from the depths of the strait to the top of the towers. In a process taking about eight minutes, man's first garland had been hung across the Gate.

The wire rope was the first strand of a web of some twenty-five support cables that were spun and hoisted into place by August 25. This was the base for the two footbridges, one for each cable, from which the work of assembling the bridge cables would be done. Each footbridge was to be fifteen feet wide and was built in sections of runners and planks ten feet long. In what was, at least in part, a public relations concession to the bridge's status as the gateway to the Northern California tall-tree country, the planks were made of redwood.

With the rigging of the cables for the catwalks, the bridge's high-wire act officially began. For the shore-to-tower backspans, the flooring for the catwalks would be pulled up in hundred-foot lengths, with men riding the sections up as they rose. Each section would be bolted to the support ropes when it was in place; then the men would ride down and get aboard another section for the next trip up.

For the 4,200-foot main span, between the two towers and over the deepest and windiest part of the Gate, the flooring was lifted in ten-foot sections from barges by derricks to the tower tops. Each section, which had runners grooved at the bottom to fit the catwalk ropes, would be placed on the ropes, then given a shove by the derrick to send it sliding forward from the top of the tower down toward the center of the span. To fasten it in place, a crew would be sent down in a construction car, a box riding on a rope. To run the fastening bolts over the rope and through the planks and tighten down the nuts that held them required men high in the air, working on a steep incline, to assume sometimes monkeylike positions, hanging from one rope with one hand while working on another while fully exposed to wind and weather. It was like moving furniture on a high wire.

"Rigging up ahead of time was dirty work," George Albin remembers grimly. "Putting up cables and catwalks. Trying to make connectors in the fog. Booms as big as a room. You'd holler like hell, and stuff was still coming in on top of you. You'd wonder what the hell you were doing out there—fog, cold, wind in your face."

To stabilize the catwalks, steel truss crosswalks were built between the two footbridges, at intervals of about 500 feet. To reduce wind whipping, webs of storm cables were stretched out from the bridge towers and secured to various points along the catwalks. As a warning to shipping, the storm-cable complex was hung with red and green lights, giving the bridge at night a prematurely filled-out, festive appearance.

Still, despite this complex and concentrated effort at battening down, the catwalks and the men on them remained vulnerable to the near-constant cold and prevailing wind hundreds of feet above the Gate.

"I helped tie down a cable at night," recalls Frenchy Gales:

I was at a party. Me, Jimmy de Cello and Paul Hendrickson went out and walked the catwalk out to the center span. We left at 12:30. It was like walking over waves. The wind was blowing so hard we had to lay down part of the way. We got out to the middle and there was a hut out there. A little house out on the middle of the catwalk. Paul said, "Jesus, a shot of whisky would be good now." He pulled a half-pint out of his pocket. It tasted great.

By the time they got back, it was daylight.

In this manner, assembled in sections that were huge and bulky to the men handling them, yet reduced to two thin soaring and swooping strands by the height and size of the towers, the catwalks were completed by the beginning of October. Handrails were added for safety, and at intervals on each section one of the redwood planks was set on end, as a cleat, to prevent slipping. For the first time, it was possible for men to walk across the Golden Gate.

The first civilian to do it was an official of the Department of the Interior, visiting from Washington. Ted Huggins, a Standard Oil public relations man who had become the unofficial official photographer for the Bridge District, served as tour guide: "There was this guy from Washington with me. I pushed him ahead of me. We were in fog. You couldn't see anything. I wasn't scared. I was wearing those crepe-soled rubber safety shoes."

It soon became common, but never ordinary, to cross between San Francisco and the Marin shore on foot. The views, when the fog permitted them, were almost airborne: the jagged Farallon Islands, some twenty-eight miles out to sea, the distant headlands at Point Reyes and Pedro Point; to the east, the whole sweep of the bay, with San Francisco, Oakland, and the less-tall Bay Bridge in their entirety; and below, the ship traffic now model-like beneath one's feet. It was as if everything had been placed there in anticipation of this view.

There were, from the catwalks, sunrises that suggested the first day of the world, thundering Wagnerian sunsets and afterglows that summoned up suggestions of the last, and with the theatrical drama of a rising and falling curtain, the tumbling cataracts and sudden ghostly disappearances of fog.

Upon this spectacular and dramatically changing setting, the required routines of work were now imposed. A tow rope was added to the west footwalk, like an escalator, to assist men carrying materials up the steep gradient. At intervals along the walks, sheds were erected to store equipment and offer shelter from the wind and cold. There were even portable toilets, hundreds of feet in the air, on a footbridge suspended between two towers. The waste was collected in traps, giving rise to the temptation to open a full trap, as a kind of bomb, on the ships passing below.

The target selected was the *Shensu Maru,* a visiting Japanese freighter. It was, in part, a political gesture: this was a time of deteriorating relations between the United States and Japan; the Japanese, who had seized Manchuria in 1931, were now threatening to invade the rest of China and represented a growing threat to American interests in the Far East. The combined appeal of scatology and patriotism proved irresistible.

The *Shensu Maru*'s schedule was studied. On the footwalks, timing and measurements were secretly calculated. The plan was to open a trap directly over the *Shensu*'s smokestack. On the *Shensu*'s departure day, everything on the footwalk was prepared. The *Shensu* appeared on schedule, steaming toward the Gate in the northern, or outbound, shipping lane. The trap, whose bombardier remains unknown, was opened. The contents missed the stack but hit the ship, giving rise to immediate and outraged protest, but

During cable spinning, the anchorages took on the initial confusion and eventual complex precision of the interior of a watch.

no formal diplomatic complaint. Although the perpetrators were never caught, it was the last larking incident of this particular nature on the determinedly high-class bridge project.

Above each of the two footwalks, overhead tramways were now installed, cross-braced every fifty feet against wind. A hauling rope, running between the tramway support ropes, would pull the carriage from shore to shore as the spinning wheel, four feet in diameter, spun out the main cable wire.

To speed the spinning operation over the longest cable distance ever attempted, the Roebling engineers had designed a simple, ingenious system using two carriages, instead of one, for each cable. A carriage would be pulled out from one anchorage to midspan, where it would stop. A second carriage, starting out from the opposite anchorage, would be pulled out to meet it. The wires on the carriages' spinning wheels would be exchanged, and the carriages would each head back from midspan to their anchorages, spinning two wires in half the travel time it would have ordinarily taken for one. In theory, this would at least double the daily cable-spinning quota.

After the tramway and hauling ropes were in place, guide wires, precisely calculated by Roebling engineers to match the exact elevation of the cables, were fixed to the Marin and San Francisco anchorages. The guidelines were then measured out and strung up and over the towers, with the wires positioned in the bottoms of the saddle seats at the tower tops and adjusted to the exact degree of sag calculated for each cable.

The work was extremely critical; the guide wires would be used to gauge the position of all the wires that were to be spun into each of the main cables. To avoid distortion by heat expansion from the sun's rays, the wires were adjusted at night. Delicate surveying instruments were used to fine-tune the precise position of each guide wire. Behind these measurements and positionings were volumes of all the years of mathematical calculations that had gone into the preparation of the bridge—the measurement of the effect of wind and heat and loads—and the parties responsible were out in force: Clifford Paine, Russell Cone, his assistant Ted Kuss, Walter Joyce, Roebling's resident engineer, Bob Cole, the general superintendent, and Hendrickson and Grover McClain, the assistants. With an immense concentration of attention the guide lines were adjusted point by point to the ultimate main span versine, or sag, suggested by Charles Ellis, 475 feet.

Meanwhile, on either shore, the vast concrete block pits of the anchorages were jammed with a complex of men and mechanisms, suggesting the interior of an enormous pocketwatch. Here were concentrated the motors

Viewed from the footwalks, there were sunrises that suggested the first day of creation, and thundering Wagnerian sunsets that conjured up visions of the last.

that would send out and draw back the spinning carriages, the spools of wire that would have to be rewound, by gangs of men, onto each spinning reel, the concrete-imbedded eyebars, each with its grooved U-shaped shoe fitted to receive the bunched group of wires called a "strand," the scaffolding and stairs, platforms and ladders, pulleys and cables and dangling blocks-and-tackles, with a constant coming-and-going among men and gear that presented, again like a watch, a scene of utter confusion to the eye but was in reality precisely timed and measured. This was how it was at the beginning; things would soon grow even more complex.

The cable spinning would be conducted, throughout, in an atmosphere of opposed and often contradictory tensions. There was the necessity for speed—Roebling must complete the job by July 1936 or the firm would lose money on it—balanced against the permanence of the work: the cables they were spinning must last, in Strauss's term, "forever." There was the enormous scale of the job—tens of thousands of tons of steel spun over tens of thousands of miles, yet everything done with a metered precision that required constant adjustment and could cause the entire line to be shut down to correct a minor mistake. There was the need for almost constant concentration, coupled with the dangers—at these heights, and working with moving machinery—of preoccupation. It was, to some degree for each man involved, a psychological, as well as occasionally a physical, balancing act.

Men reacted to this tension in differing ways. Some, like Harold Mc-Clain, savored the heightened quality it brought to life, the intense involvement, for a relatively brief time, in building a structure meant to last forever. McClain expanded his skills so that he could work, if possible, on every aspect of a bridge and master the experience in a way that let him own part of it, imperishably.

"You couldn't help know it," he admits. "You were building the greatest structure in the world. You knew one thing for sure—there wasn't going to be anything else like it. I was fortunate."

Other men resisted this tension, or internalized it, occasionally letting loose with catastrophic results. "I stayed away from the places where the bridgemen drank," recalls Peanuts Coble, "because bridgemen are mean when they drink. They fight—use anything."

Alone, a continent away from home, working under stress of several kinds, well paid in a hard financial time, yet without any real job security, many bridgemen used the unaccustomed luxury of weekends off to try to equalize the pressure.

"The bars in Sausalito made a fortune," recalls Frenchy Gales. "The ironworkers drank all weekend—'red-hot riveters,' they were known as. The bars would cash checks on Friday night. People would be lined up outside. You couldn't go into a bar or they'd buy you four or five drinks."

Unavoidably, wherever the men went, however much they drank, they carried their work with them.

"Paul Hendrickson would lay $20 on the bar, and not pick up the change till it was gone. Everything would be on him. Paul would show his assistant, Oscar Parker, the next day's job in a bar. He would tell guys all he knew."

Hendrickson, tough, able, respected as a superintendent who would never send a bridgeman anywhere he wouldn't go himself, would die in a fall on a Roebling job a few years later.

On Monday morning, the ironworkers who lived in Marin would make their way, sometimes shakily, to the base of the north tower, where they would catch the elevator to the top and head out on the catwalks to work.

"There was a first aid station on the Marin side," says Gales, "with an intern in it. He kept a bottle of Canadian whiskey there. If somebody looked rocky, he'd give him a shot and send him up."

2

"On Monday, November 11," Strauss proudly reported to the board two days later, "the first wire of the west cable was pulled across from the Marin side." The chief engineer had made one of his now-rare personal appearances at the bridge site for the occasion. The individual wires, each made of acid steel, were some 0.196 of an inch in diameter, narrower than an average pencil, yet so strong a man couldn't bend one double. One end of each wire would be fixed at the anchorage. The other end would be looped around the spinning wheel, which would be hauled by the carriage out on the tramway up and over the tower, then down to the center of the main span. Actually, each spinning wheel would carry two wires. The top or "live" wire was spun out continuously; the bottom or "dead" wire, fixed to the anchorage eyebar, would be adjusted from point to point.

"You'd get to a certain point," says George Albin. "The guy at the tower top had a come-along machine," a more sophisticated sort of winch, "to make the cable go forward or back. The wheels had two bights [loops] on them to start. When they got to the main span, I would pick up the power. We'd bring the wheels in, stop 'em, change bights so they would continue across. Marin would go to Frisco and vice versa. You'd keep your finger on the button for 200 feet."

On the return trip, the live wire would become the adjustable dead wire for the next pass. At the middle of the main span, the dead wire would be adjusted by a man remotely operating a come-along, complete with gauges and balance weights.

"You're out on the catwalk with a good view of the wire," says Harold McClain. "You have a set of buttons in your hand which works the adjusting machinery. You'd adjust the wire visually. We were in telephone communications with the shoreline when we were all hooked up and ready to go."

While this fine tuning, the careful tightening or slackening of the dead wire, was going on, the neighboring live wire would still be spinning. From time to time, as the tension in the dead wire was being adjusted, the spinning live wire would get snagged in the spinning wheel. To free it, men would have to climb up and over the catwalk and balance, teetering out on the tramway ropes, hanging by a thread above the Gate, to disentangle the continuously running cable from its wheel.

When the individual wire came to an end, it would be spliced to the next wire. The splice would be made wherever the end appeared, sometimes at midspan. "It was like splicing a piece of rod you were going to put into

GOLDEN GATE BRIDGE

SADDLE OF SAN FRANCISCO TOWER

CABLE SPINNING

SPINNING WHEEL IN SADDLE

CABLEMEN SETTING WIRES IN SADDLE

TOP OF SAN FRANCISCO TOWER

a turnbuckle," says McClain. "Each adjusting crew carried a splicer. When we cut wire, we cut it with a hacksaw on each side so it would be even." The entire trip, allowing for reeling, adjusting, splicing, and transferring, was averaging about fifteen minutes, per wire, from anchorage to anchorage. To meet the schedule necessary for Roebling to avoid losing money on the job, this simply would not do. Once the bridgemen had grown familiar with the sequence and the exacting and complicated series of steps had assumed some of the qualities of a routine, the Roebling engineers added, to each carriage, a second reel—a practice that had been inaugurated on the George Washington Bridge. Each carriage would now haul four wires at a time instead of two, with a corresponding increase in the complexity of the work.

When a specified number of wires had been spun, they were bunched into a "strand" and tied off with flat metal bands. For the first time in cable spinning, the number of wires per strand would be varied. Roebling engineers, working with a ten-foot-long scale model cable, had found that when the size of the strands was altered the strands could be formed into a more

While still under construction, the bridge had assumed the status of a landmark, featured in news photos and postcards.

perfect circle. When the strands were varied, from 256 wires per strand up to 462, they formed a finished cable that flowed more naturally, with less resistance, once it was in the saddle than in the slightly hexagonal shape they'd had to settle for on the George Washington Bridge.

When each strand was tied off it would be lifted out of the way, to be adjusted into its final position at night.

"Each strand of wire was tested at a certain time at night," says Gales. "A hydraulic hoist with a 240-ton limit pulled it. They'd lift it out of the saddle and get a proper loop over the ocean at a certain temperature. Sometimes we'd be out there all night to get it right."

The wires and strands made the effects of temperature apparent to everyone working on the bridge. "You could see the expansion and contraction in the main cable," says George Albin. "It would be high in the morning, when it was cold, and sag at midday when the sun came out." The adjustments, made at this time, were especially crucial, because the final cables would be too large for adjustment. It all had to be done with the wires and strands.

The strands, although only a small fraction of the size of the final cables, were individually heavy and could be as dangerous as an enormous, cracking whip. "We worked between the strands," says Gales, "even though you weren't supposed to. A guy let go, and I got hit, the cable took the hat right off my head. I had to get out of the way of the whip—it would go all the way down the cable. I had a sore neck and head for a couple of days."

At each anchorage, the gathered strands, after passing through the archway guide of the giant concrete pylons, would fan out again toward the strand shoes, which fitted them to the eyebars and anchored the cable, strand by strand, to the shore. The load, compressed in the cable as it went up over the towers and across the span, was thus spread out again among the huge concrete block weights at either shore.

The entire process was like two high-speed production lines, running simultaneously, on the west and east catwalks, across a great gorge, requiring constant monitoring and tinkering, with every possible attempt being considered that might increase the speed and efficiency of the work.

"We used to splice each line by hand," recalls Gales. "A guy rigged a machine to do it. Some guys came out on the bridge and asked me what I thought of this machine. I told them I thought that some engineer had been unable to sleep and instead of playing with himself had thought up this machine. They left, and Paul Hendrickson told me, 'That's the guy who invented it!' "

"I HAVE FELT ALL ALONG THAT THE ORGANIZATION WAS NOT AS IT SHOULD BE."

8

I

Between himself and the world at large, Joseph Strauss had always required intermediaries. In person, often on the road, trying to juggle appointments, to prepare for meetings or raise money, he was often preoccupied, abrasive, unfeeling, abrupt. He had always been a man in a hurry.

"My other stenographers wouldn't work for Mr. Strauss," recalls Ruth Natusch, who ran a secretarial service at San Francisco's Palace Hotel when Strauss, in the days when he was still promoting the idea of the bridge, was staying there. "They were afraid of him." Instead, Miss Natusch took dictation from Strauss herself.

"I worked long hours. I often worked from seven-thirty or eight in the morning to late in the evening. Mr. Strauss dictated, while I took shorthand. It was a very pleasant relationship. Mr. Strauss didn't mean to be unkind. It was just that everything was against him. He was *very* determined."

Strauss's obsessive single-mindedness, his possessive secretiveness about his work, led him to be rude to people, especially, and in a potentially expensive way, to the press. "There were a lot of engineers living at the Palace then," says Natusch. "There were big projects going on in town. There were always a lot of reporters hanging around. They would come by and knock on his door. He'd tell them, 'Get out—can't you see I'm busy?' You can't expect a genius to be the same as an ordinary person."

To smooth things over Strauss, who had the reputation of never apologizing to anybody, needed a go-between, an envoy to serve as his emissary to the outside world. At first, and for a long time, "Doc" Meyers had filled this role. "He used to heal over things for Strauss. He was a nice person, sort of a hanger-on, with a lot of braggadocio and no visible means of support. Yet he lived at the Palace."

Charles Duncan, the former outdoor advertising salesman who became

Strauss's public relations counsel, also assumed some of these diplomatic responsibilities, but as time wore on and Strauss's health continued to decline, he came to depend more and more on Clifford Paine. After Strauss's falling-out with Meyers, this dependency, professional and personal, increased.

"Mr. Paine," says Natusch, "was a good, genuine person who was a great help to Mr. Strauss."

Paine was modest, unassuming, hardworking, able if not conspicuously gifted as an engineer, and therefore not a competitor; yet, like Meyers, financially dependent on Strauss; in many ways, the ideal intermediary. Unlike Strauss's previous buffers, Paine could carry Strauss's wishes all the way out to the job, which indeed he did.

"Clifford Paine was all over that bridge," says George Albin. "Especially during the spinning of the cables. He was up and down that catwalk, checking things, I'll bet he did twenty miles a day. He used to spend a lot of time at midspan with a young fella from Columbia University. They concocted timing, adding of strands. From what I've read of Clifford, he was the one who really worked the whole thing through."

Wherever there was prolonged detail work to do, or the overseeing or inspection of materials, Paine could work from his strengths. When a finished report was necessary on the findings of the bridge paint tests, Paine prepared it. Problems with quality control at the Roebling plant? Paine would see to it. Paine was thorough, patient, methodical, phlegmatic, an ideal complement in many ways to the mercurial Strauss.

The chief engineer was now showing himself about once a month, at the bridge district board meetings, where he would submit his report and often have little else to say. In an eerie parallel to Washington Roebling, he was spending more and more time confined to his Nob Hill apartment, where he observed the cable spinning on the bridge through a telescope.

Sometime during the summer and fall of 1935, the relationship between Strauss and Paine subtly shifted. Strauss's dependency on Paine became greater than Paine's need for Strauss, and both men were made to realize it. This may have been the time when Paine resigned from the Strauss firm, or at least threatened to. The outcome, in any event, was that Strauss, no doubt under heavy pressure of some kind, made the greatest concession of his career: in October 1935, his firm was renamed "Strauss & Paine, Incorporated," with Clifford Paine as executive vice-president. The loyal sidekick had become co-topkick.

There were, in this relationship, obvious parallels to a marriage. Reed, the general manager of the district, later referred disparagingly to Strauss and "his boy friend Clifford," but in no sense was this a homosexual attachment. Rather Paine was the familiar corporate figure, the top man's loyal go-between and interpreter, assuming most of the daily complexities and frustrations of the boss's job, sheltering him from the world while explaining him in turn to the people he may have abused or offended, allowing him to indulge in the lapses of manners and feeling that are accepted as justified to the degree of one's belief in the existence of the modern state of grace called "genius."

While Strauss observed the distant cable spinning through the window of his apartment, and Paine, his alter ego, patrolled the catwalks to keep an eye on things close up, work also continued on the less dramatic approaches to the bridge, the mounting for which the great suspension span would be the jewel.

On the Marin shore, preliminary pile-driving work had begun for a low viaduct leading from the bridge to the highway now being constructed by the state up and over the Waldo Grade, through a tunnel in the Marin hills. On the San Francisco side, meanwhile, the clearing of buildings in the Presidio had advanced to the point where work could begin on the pedestals and abutments for the high viaduct, which would carry traffic upward, onto the San Francisco entrance to the bridge. The various procedural delays surrounding these structures, particularly those on military property, aroused the politically sensitive directors' fears of cost overruns. Even though the overall job was still well within its budget, the directors decided to make a show of economizing by cutting down the dimensions of the toll plaza.

"I remember Morrow's face when he returned from the meeting where he was told he had to cut," recalls Herb Johnson, a young architect on Morrow's staff. "He was livid." Johnson was assigned the problem of determining where the size of the buildings could be materially reduced, while retaining, if possible, the architectural proportions as originally proposed.

"There had been an elaborate meeting room planned for the directors as part of the toll plaza complex. I peeled that off first. Then I peeled off the garage for their cars."

Morrow, it appears, had learned something about politics and infighting from Strauss.

The Engineering Board also suffered, to a degree insufficiently appreciated, from the absence of Professor Willis as an effective adversary. Whenever tempers flared or egos were wounded, the Stanford geologist had always served as a unifying figure, a catalyst who could dissolve, in necessary opposition to his reckless conclusions, the most insoluble of differences.

Now, with Willis silenced or ignored, and Strauss increasingly withdrawn, differences of opinion between members of the board could harden quickly into adamant professional stances.

On December 18 Derleth, visiting the bridge site, questioned, in the presence of Reed, some of the foundation footings at the proposed site for the toll plaza. When asked how much it would cost to take new borings and relocate the plaza, Derleth suggested a rough figure of around $250,000. These thoughts Derleth passed on to Strauss in a letter on December 21. Strauss, however, arriving for a Building Committee meeting on December 20, was caught completely by surprise when the issue arose. All his old mistrust of Derleth as the directors' plant within the Engineering Board was revived.

"I regret that you did not call me on the phone and discuss the matter with me informally first," Strauss complained to the engineering dean on January 4. "You have heretofore very frequently maintained that technical problems should not be imposed upon the directors. This should hold true in this case more especially in view of the fact that I had the problem in hand."

Clifford Paine, Strauss went on to say, had notified the chairman of the Building Committee that Strauss was investigating the effect of additional borings on the structure. Strauss particularly objected to Derleth's giving the directors a guesswork or "horseback" estimate. "I believe the Board's position in these matters has been that they must be guided in technical matters by the findings of the engineer, since they are not in position to determine by themselves what is and what is not a correct and proper cost."

The information about subsoil borings was now "before Mr. Paine," who, presumably with Strauss, would investigate and make a report to the Board of Directors.

"I am writing this letter," Strauss concluded, "in an entirely friendly spirit and in pursuance of my desire that the teamwork which has prevailed through the difficult stretches of the road may continue to the end, which is relatively so near at hand."

Still, it was increasingly apparent that Strauss, growing ever more withdrawn, was looking more and more upon Clifford Paine as the one man he could rely on, confide in, and trust.

2

Even as it was being built, the bridge was taking on an existence of its own, in ways that sometimes alarmed and surprised the men responsible for it. The north tower, which had been completed for a little over a year, was already so damaged by wind and fog that it would require an estimated $24,000 in repairs, mostly for repainting. Deterioration at this rate, from exposure and weathering, had been anticipated by no one except Professor Lawson, the Berkeley geologist, who had warned in his original report that the bridge might have to be replaced once or twice in a century due to the corrosive effects of salt air. Even Dr. Lawson, however, had not anticipated that the bridge would display such extensive damage this quickly. There was no precedent in steel construction for anything like it. Even on the Bay Bridge, just five miles farther inside the bay, the effects of weathering were nothing like what had appeared at the Gate.

For one thing, the Golden Gate Bridge, in addition to being exposed to the unbroken winds originating in Pacific storms, was situated so that it took these winds broadside. There could be no accommodation, no angled shrugging off of the impact and gradual wearing effect of these coastal winds. More crucial than this was the fog, which sent millions of gallons of moisture tumbling all around the bridge, hurtling at freight train speed through the only break in the Coast Range and coating every steel plate and rivet head on the surface of the towers with a layer of salty dew. To prevent the kind of corrosion prophesied by Dr. Lawson, the bridge directors and Engineering Board were now beginning to realize, no coat or color of paint in the world would suffice. Instead, the bridge was now teaching its builders and managers that for such a structure to survive at this site would require that it exist in a state of permanent repainting, with a crew no sooner finished with painting one end of the bridge than beginning on the other, in an endless cycle of chipping, scouring, sandblasting, spraying, and brushing that suggests one of the endless, repeated tasks of Greek mythology, or the haunting frustration of a recurring nightmare.

The bridge was teaching other lessons to its builders. As the structure grew, it required increasing supervision, not from a distance, but in the field. The expansion in men and material, the various comings and goings, the continuing orchestration of a work that now included repair, finishing, and maintenance as well as the overseeing of ten different contracts, placed geometrically increasing responsibilities on Cone, who was short-staffed as it was.

With the start of cable spinning, the quality and on-time delivery of wire became so critical that Ted Kuss, Cone's assistant, was put in charge of wire inspection, full time. The Bridge District was now left without a backup for the one man charged with the on-site responsibility of building its bridge. Reed, growing anxious, wrote Strauss but received no news of a replacement from either Strauss or Paine. His concern increasing, he contacted Ammann, Moisseiff, and Derleth.

"Chief Engineer Strauss has not informed me in response to my inquiries as to whether he proposes even to replace Mr. Kuss," Reed complained to the other engineers. "It appears to me that the interests of the district would be seriously hazarded if anything should happen to Mr. Cone which would incapacitate him, even temporarily."

Reed asked the engineers what their judgment was on this matter. Ammann shared his concern, emphatically. "I have felt all along that the organization on the Golden Gate Bridge was not as it should be," he wrote Reed on February 7, 1936, "and in particular that the resident engineer in charge had inadequate assistance." On the Triborough Bridge, Ammann pointed out, the average force for the construction period was "probably around 50 engineers and 50 inspectors." The Golden Gate Bridge did not have even one resident engineer for each contract.

"The situation is such," Moisseiff wrote to Ammann, "that Strauss has inadequate engineering supervision on the Bridge."

Cone was in the overworked manager's classical quandary: he needed an assistant, yet he was so busy he didn't have the time to recruit or train one. The question of a replacement, brought up by Reed before the Board of Directors, was referred to the Finance and Auditing Committee to decide who would pay for the added help. There the matter lingered while the work went on and the district, in effect, gambled everything on the continuing good health of one overworked man.

By using double spinning wheels and transferring the wires from carriage to carriage at midspan, the Roebling crews were now spinning cables at a rate comparable to that achieved on the George Washington Bridge. Counting both cables, sixteen wires were now being strung across the Gate at a time. To help the adjusters sort things out, the wire on one of the two spools feeding each carriage was sprayed with a distinctive color. Accepting this colored wire as a "given," the adjusters were able to keep the other wires identified by placing them in their proper guides and positions as the wheel passed each control point. Even with these assembly-line efficiencies, however, there was some doubt as to whether the deadline could be met; on the Golden Gate, after all, with its significantly longer span and taller towers, Roebling was spinning one and a half times as much steel as on the George Washington Bridge.

It occurred to C. C. Sunderland, chief engineer of Roebling's bridge

department, that the carriages could be adapted to accommodate three spinning wheels each, just as they had been adapted to carry two. The raw cable-spinning crews, working under Roebling supervisors, had matured quickly. Men had grown used to rereeling, positioning, adjusting, and splicing four wires at a time, then eight, then sixteen. Why not take the process one step further and spin cable twenty-four wires at a time?

Working weekends with Bob Cole, Roebling's general superintendent, Sunderland added a third wheel to each tramway carriage that rode, rather like a side car, outside and between the other two reels. Slowly at first, the men adapted to handling, adjusting, and transferring six wires at a time.

"It wasn't something you could pick up quickly," says George Albin. "We learned how to do it by building all that stuff into place. You had to take your time, especially at midspan. They had a deck above the catwalk, adjusters working down below. There were six guys up there, three for each set of strands. It would be easy to get screwed up. We had a system."

The Roebling engineers and supervisors understood how to work with bridgemen. By stressing safety and preparation, supervisors working alongside the men, they were able to strike a balance between support and challenge, discovering, as they went on, new capacities among the work crews and in the bridge itself.

"You'd have five men on a gang, including a foreman. You had an engineer with you. At midday, the towers would bow out when the sun would warm them. We'd have a ratchet-wrench. Put a clamp on the cable, and actually take the bend out of the tower. They had a house at midspan with clocks and meters to time everything. Every move they made, the engineers were tickled that things were working out."

At mid-span, the wires on the carriages from either anchorage were interchanged, thus cutting the carriage trip in half and doubling the wire that could be carried.

To compact 122 strands with some 25,000 wires into a single unit, Roebling used radial squeezing machines, applying more than 4,000 pounds of pressure per square inch.

With three wheels working, in split trips, on each side of the bridge, the cable-spinning total had jumped to an average of 271 tons a day, or more than four times the peak output on the George Washington Bridge.

The cable-spinning crews were now completing one "set" of four strands for each bridge cable every twenty-four working hours; there would be sixty-one strands in each cable. Engineering strategy had now moved beyond the speed and quantity of wire spun to the finishing process of compacting and wrapping the completed cable.

When all 122 strands were in place, each cable would be bunched in a rough hexagonal shape some five feet across. To compact this misshapen bunch into a smooth, round cable, with some 25,000 wires acting as a single unit, Roebling would use large radial squeezing machines that were to be fitted around the collected strands and pressed together by a circle of hydraulic jacks, applying a force of more than 4,000 pounds of pressure per square inch.

"The squeezing machine that compacted the cables had so much pressure," George Albin would later recall, "that they gave you special glasses to save your eyes if a line broke loose."

When each cable had been compacted into a perfect circle some 36⅜ inches in diameter, the cable would be banded every fifty feet by clamps of cast steel, which would also serve as saddles for the cable's "suspenders," or hanging wire connecting ropes. These cable bands were being cast for Roebling by the Otis Elevator Company, which was having problems with them.

Herbert J. Baker, the chief inspecting engineer retained by the Strauss organization to oversee the quality of materials as produced in the Eastern plants, had found a number of these bands to show signs of unauthorized welding and to display cracks in the suspender rope grooves. He rejected a number of them. The remainder were shipped west and arrived at the bridge site, where Paine rejected them all.

"It is unnecessary for me to repeat here the exceedingly unsatisfactory history of the castings which you have offered us," Paine, who could be tough when he had to be, wrote to Walter Joyce, Roebling's resident engineer. "Nor is it necessary for me to remind you of my warnings to you over three months ago and my advice at that time that you arrange immediately for other sources of supply." Under pressure from Baker and Roebling, the Otis foundry had changed its casting procedure, but only after a delay of between two and three months. This delay, said Paine, meant that Roebling now faced "almost certain failure to meet the requirements of your contract with respect to completion date." There would be damages to the district, "much in excess" of the ordinary liquidated damages provided for in Roebling's contract.

Charles Jones, assistant chief engineer in charge of Roebling's bridge department, protested Paine's sweeping rejection of the entire shipment of cable bands. "It was our understanding," he replied to Paine, "that your inspecting engineer would accept castings of this group which, upon exploration, were found to be sound castings. We are now informed that the entire group is rejected by you."

Jones, who had recommended in April that a board be set up to establish some standard of acceptability for the bands, complained of "the handicap under which a manufacturer works when the final decision as to acceptance or rejection is in the hands of one person, and that person some three thousand miles away." Through Roebling's own tests on the rejected castings, Jones maintained, the Roebling engineers had convinced themselves that the so-called injurious defects did not, in fact, impair the strength of the cable bands.

"You may be interested in knowing," he chided Paine, "that under a load on the bolt housing equivalent to approximately twice the ultimate strength of the cable-band bolts, rejected castings which we have tested did not fail at the location of the injurious defects, but rather in the shell of the band above the bolt housing."

Nevertheless, Paine held the Roebling engineers to the language of their contract, which specified, "The Engineer shall be the sole judge of the materials' quality and suitability for the purposes for which they are to be used."

Roebling's management refused to accept Paine's rejection of "previously accepted material." On August 10, in a letter hand-delivered to members of the Bridge District Board of Directors, Jones requested an extension of the Roebling contract, as well as payment of some $250,000 in compensation for additional costs incurred because of the rejections.

"As you know," Jones wrote the directors, "our work has been practically shut down since June 15, and cable band erection could have commenced June 16 had it not been for the unwarranted rejection of bands which were available in the field from April 11 until May 27, and which were not rejected until between June 6 and June 24."

Paine, in a letter to Strauss—whom even Ammann and Moisseiff addressed as "Mr. Strauss"—that began "Dear Joe," said that the bands now being delivered "were the good sound castings we have a right to expect." They were being passed just as fast as Roebling offered them. The previously rejected castings, however, "are not acceptable and Roebling knows it. They are trying to work them off on the district." The letter was signed, "Clifford."

Strauss, responding to the Roebling letter, wrote the directors on August 19. Charging Roebling with the responsibility for the delays, he quoted a statement from Sunderland to Jones describing one of their subcontractors entrusted with producing the castings: "It is very disgusting to have to work with such an organization inasmuch as they do not seem to be able to get organized, and the castings they produce are so un-uniform in chemical properties."

Strauss was backing Paine all the way. "In my opinion," he concluded to the board, "the criticism directed against the Engineer was totally unwarranted and out of place." Should Roebling file the suit against the district that its lawyers now threatened, then "the District would not be jeopardized in any respect by taking sufficient time to gather all the facts to make a proper defense."

Whatever actual hazard the flawed cable bands may have represented, Paine, whose quality concerns were no doubt genuine, was also using the situation to make a statement. "The Engineer" specified in the contract—the man who ultimately must be pleased—had always previously been the Chief Engineer, Joseph Strauss. Now it was Clifford Paine.

These differences about the finishing of the cables themselves could not diminish Roebling's pride in the achievement of spinning them. On May 20, 1936, nearly two months ahead of schedule, the last pin was inserted in the last strand shoe in the opening of the last anchorage eyebar. The spinning of the world's largest, longest bridge cables was completed, at a rate four times faster than had previously been considered possible.

"The main cables are not only made of the highest-strength bridge wire furnished on any bridge to date," Roebling's president boasted in a letter to the directors, "the cables themselves were erected and compacted in record time, and we believe that you have in your structure the largest, longest, strongest, roundest, and the most compact cables ever erected."

To celebrate the completion of cable spinning, Roebling honored its own. The firm rented Paradise Cove, a park on a quiet Marin County shore behind Tiburon, and threw a barbecue for the bridgemen. With the tensions of meeting the deadline gone, the violence that often erupted among bridgemen when they drank never appeared. "Usually among these guys there were a lot of fights," Gales observed. "People usually were keyed up. This time, there were no fights."

With some men's work on the bridge already finished and the other men facing the ultimate conclusion of their own jobs and the associations, teamwork, and friendships that went with them there was a certain valedictory quality to the Roebling barbecue. It didn't make it any less of a party.

"They had all the beer you wanted—but they still ran out. Had to get another load." The party went on for three days.

"We had a tug-of-war on the beach. The pushers [foremen] and personnel on one side, riggers on the other. The guys on the upper end all let go, and the other guys all went in the water. It was good-natured stuff. We had a good time."

19

"YOU HAVE TO HAVE A LITTLE BIT OF FEAR."

I

Of the various construction records established at the Gate—the tallest towers, the longest and largest cables—what remains the most impressive achievement is that more than three years after the start of construction there had not been a single death. There had been injuries and plenty of close calls, but despite the size and complexity of the job and the natural hazards of the site, the bridge had thus far generated no obituaries. High-steel construction without fatalities was something unprecedented at the time.

"In those days, a man's life wasn't worth a nickel, anyway," recalls Lefty Underkoffler, a bridgeman on both the Bay and Golden Gate bridges. "They figured one life for every million dollars on any job, no matter what type of work it was." In a time when suits for negligence or personal injury were unheard of and men in the world outside were still standing in line for jobs, it was simple, practical bookkeeping for contractors to skimp on safety considerations in order to cut construction costs. The family of a fallen bridgeman would usually consider itself fortunate to be reimbursed for funeral expenses. According to Underkoffler, twenty-three men had been killed in falls on the San Francisco-Oakland Bay Bridge.

At the Golden Gate, for the first time, strict safety standards were made a policy on every aspect of the project. There were several reasons for this. One was the prominence of the job: so much public attention was concentrated on the unbuildable super-bridge at the Golden Gate. The sensitivity to criticism of the structure's safety in terms of earthquake, winds, and navigation was so high that Strauss, image-conscious long before the term became common, was determined not to be remembered as another man who built a great work at a terrible cost in human life. Added to this were the strict standards insisted upon by Cone and Paine, the one a stickler for on-the-job safety and the other a demanding materials-evaluator who required that the components bear stress tests many times what would be expected on the bridge. Perhaps most influential of all, however, were the

Roebling engineers and supervisors, the most experienced bridgemen of all, who approached the riskiest high-steel work with a minimum of daredevil heroics and who made safety a matter of pride-in-craft.

It was Roebling that worked 100 percent union, ran a training school for apprentice bridgemen at Fort Point, experimented with foul weather gear and safety belts, made hardhats mandatory, and cautioned its workers not to extend themselves beyond their capabilities.

This convergence of interests produced a working concern for safety that was sometimes carried to extremes: not only were on-the-job stunting and drinking causes for immediate dismissal; experiments were made with anti-dizziness diets, hangover cures, and antiglare goggles. Yet whatever the excesses of this continuing vigilance, no one, thus far, could dispute the results.

Bridgemen profess to be unintimidated by working at great heights, yet the knowledge that they are vulnerable to falls, that men just like themselves, often friends, have died because of the most minor mischance, is inescapable. On the Brooklyn Bridge in the 1870s and 1880s, Washington Roebling had hired sailors, men familiar with working at hazardous heights, accustomed to using one hand for the job and one for the ship, to rig, spin, and adjust the cables on the first of the great spun-steel suspension bridges. Bridgemen have been part mariners ever since, working aloft above water, exposed to wind and weather, clinging for support to the structure they maintain. There is a shipmate sense of mutual dependency and transience, a fo'c's'le intensity of friendship and dislike, a roistering shoreside camaraderie combined with an intensely individualistic pleasure in reading,

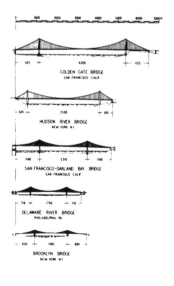

The possible length of an individual suspended span was dramatically extended by the Golden Gate Bridge. In the years since, it has been advanced only sixty feet more.

storytelling, poetry, or crafts. Only here, when the voyage has ended, the men, not the ships, sail on.

"I worked on the Delaware River bridge," says Lefty Underkoffler. "I went in and applied for a bridgeman's job. Didn't know my ass from a hole in the ground. The foreman kicked me off that job. Same thing happened for six or eight more jobs. Finally one bridgeman told me, 'You're no bridgeman, but you've got guts. I'll teach you everything I know.' " The two men worked together, as a team, for twenty years.

"You couldn't visit or bullshit," recalls Al Zampa, an ironworker on both bay bridges. "Had to keep working. The pushers acted like they owned the company. Holler and scream. You don't hear that anymore, but in those times you did."

Always, there is the sailor's common bond of danger, with its fine-tuned awareness of life, and the pangs and insights of the realization of one's own mortality.

"I was walking the catwalk," recalls Alfred Finnila, "and I slipped my watch into my pocket. It was a gold pocketwatch. I missed the pocket, and the watch went down my pants leg and kept going all the way down into the water. It gave me a funny feeling."

"People freeze up there." says Al Zampa:

They hang on—they won't fall, but it would take three or four of us to break 'em loose. We'd put a line on 'em and let 'em down. They were mostly inexperienced men—they think they can do it and they can't. They say, "Don't look down." I never avoided looking down—it didn't bother me. I could look right down in the water, and see big fish down there.

You have to have a little bit of fear—not too much—way back here, in the back of your head. You can't daydream—or you'll take chances.

"I was boarding with this widow lady in Sausalito," says Peanuts Coble. "And I asked her if she'd take another boarder." The new boarder was Ed Reed, a full-blooded Indian who boxed professionally on weekends.

Ed was no Emily Post. We were eating one night, she and her kids at the table. The widow said, "Don't you feel proud that some day your son or daughter will look up at that bridge from a cruise ship and say, 'My father helped build that.' " Ed said, with a mouthful of food, "Just so the little bastards don't say I fell off of it."

Reed was killed in a building-truss construction accident some years later.

In June 1936, Strauss announced the most elaborate and expensive safety device ever conceived for a major construction job. An enormous net, like a circus net, would be hung beneath the suspended steel structure of the bridge, extending outward with the structure itself as it was built. Anyone who fell while working on the span—a crucial time in the construction of any bridge since men were constantly building outward instead of up, as in the towers, or back and forth as when spinning cable—would presumably land, shaken, but safe, in the net.

The net was not a new idea. According to Underkoffler, the ironworkers' union had been fighting for safety nets for years with contractors who considered such an investment an unjustifiable extravagance. Strauss and Paine did not view it as such, although who first suggested using a net on the Golden Gate Bridge is not clear. The manufacture of the net itself was contracted through Roebling, however, and the firm had initiated most of the significant safety measures already employed at the bridge. No doubt it occurred to Strauss and Paine that the cost of the net—something over $125,000—might be made up in accelerated construction time by men moving more freely over a protected space. Also, when the job was finished, the net might be resold to another contractor. Whatever the reason, the high-iron men at the Gate would now enjoy at least the minimal safety protection customarily offered circus performers.

The net was made of half-inch manila rope, woven into six-inch squares, a webwork more than substantial enough to sustain the impact of a grown man after a sixty-foot fall, yet which was reduced, by the dimensions of the bridge and the breezes at the Gate, to a fabric that, from a distance, seemed to take on the filminess of gauze. There would be four nets, actually, two extending outward from each tower, one toward the shore, and one toward the strait, as construction progressed in four directions at once. Upstaged to a degree by the approaching completion of the Bay Bridge, Strauss undoubtedly felt a certain pressure to accelerate construction now. The exasperating frustrations surrounding the building of the San Francisco pier and the current delay caused by the dispute over cable bands could be sublimated by the speedy completion of the graceful suspended structure.

By July 30, 1936, the compacting of the cables was finished, and the signs of the complicated airborne production line of cable spinning had all but disappeared. For the first time in over three years, activity at the anchorages had quieted, and the complex points underneath the twin pylons at each shore, where the cables splayed out again into the strands that were fixed to the strand shoes at each anchorage, were now completely hidden and protected by concrete housings.

With the cables ready for banding, the cable bands themselves were still in dispute. While Strauss, Paine, and the Roebling engineers fumed and stewed, Cone and the crews on the bridge had to let construction wind to a halt. "In ample time," Strauss complained to the directors, "I warned the Roebling Company to get their preparatory work underway, but they took the position that we would not be ready when we said we would, and so misjudged their time." Roebling, Strauss protested, had gone ahead and made castings without furnishing Paine the necessary plans to make an accurate check. "This, in my judgment, is the original cause of the trouble with the cable bands."

To accept castings that, for any reason, did not meet specifications, was completely contrary to engineering policy on the bridge. "The risk involved is too great." The same was true regarding the use of liners or shims to compensate for the differences in diameter. "On a bridge, particularly of this magnitude, patchwork should not be permitted."

The Roebling engineers, now delivering bands of specified quality as fast as possible, continued to maintain the acceptability of the rejected castings. "It is a fact, well known to all engineers and foundry men," Charles Jones protested to the Bridge District directors, "that in castings of like size and distribution of metal, there is no such thing as a one hundred percent steel casting." The contract had specified only that the castings be free from injurious defects. Yet "the Engineer has reputedly refused to define an injurious defect."

To underscore this point, Roebling's president, on August 10, sent Reed and the district a bill for $8,029.27, "to reimburse us for the unnecessary and added expense incurred by us because of the unwarranted and temporary rejection by your Engineer of some 2,160,000 pounds of Bridge wire on April 29." The wire had subsequently been accepted by Paine on May 7 and incorporated into the cable.

The total loss to the contractor because of this disruption "on an otherwise smooth-running job" could not be estimated, contended the Roebling executive, "nor can the damage which resulted from articles appearing in the San Francisco newspapers and the unfavorable radio broadcasts to the effect that the contractor had shipped defective wire for the cables."

With Strauss and Roebling both trying to delegate blame, it was the end of August before cable bands acceptable to all parties arrived, in quantity, at the bridge. Construction had been delayed some six weeks, and the anticipated completion date for the bridge had been pushed back to July 1, 1937. At the bridge site itself the lull had been like the gathering of a last, exasperated breath. Now it was time to begin the stretch run.

2

At intervals all along both cables, some 254 bands were now clamped around the smoothly rounded cable and bolted into place. Each cable band was grooved to serve as a saddle for two of the wire suspender ropes that were to connect the cable with the suspended structure. The suspenders, grouped in twos to share stress, precut to varying lengths and looped over the saddles, would, like wires in a piano, be tuned to Ellis's original calculations. Through them, the harmonics of loads and stresses on the suspended framework and roadbed would be transmitted to the cables and along them to the towers and abutments, thus shrugging off the impact of high winds or heavy loads.

The length of the suspenders, which dangled now just like those from a pair of pants, ranged from nearly 500 feet near the towers to less than 10 feet at each cable's deepest sag. When the four ends of the paired suspenders were bolted together, they were prepared to cradle the steel chords, connectors, and beams that would support the bridge's roadbed.

To keep the load on the cables evenly distributed, the steel structure, as it was built and connected to the suspenders, would have to proceed at the same pace from each tower. In practice, this would mean that work on one side of the bridge would have to be slowed or halted so that work on the other side could catch up.

Once again, barges began arriving at the bases of the towers, this time carrying not honeycomb steel sections but the varying rectangular lengths of steel frame construction like pieces in an enormous Erector Set. There were the enormous lengthwise girders, with the uncharacteristically flimsy name "stringers," to be used as strengthening members; the hollow, boxlike beams called "chords" that would form the outside part of the support structure or truss; the heavy members that would serve as floor beams; and the various connecting pieces and plates that would form the crosshatch of support, sustaining the structure's main elements.

Each of these pieces would, like the wire, be subjected to close inspection on arrival, then be stored at a tower's base. Once again, the steel would be lifted up on the tower by cranes, one at each leg of each tower at the level of the roadbed, each operated by a lifting gang. When it reached road level, the steel would be carried into position by another crane, a traveler that moved outward as the structure progressed, just as the enormous spidery traveler had earlier moved with the towers up to their tops. Four of these travelers would be working in four directions at once; as they moved outward from the towers, cars would be used, rolling along on stringers, to carry the steel out to the travelers.

The gangs connecting, bolting, and riveting the steel together were the same that had worked on the towers, the same men mostly, augmented now by men released, now that work had wound down, from the bridge between San Francisco and Oakland. There were no red lead paint fumes to contend with, but there were was still enclosed spaces. In order to connect up the boxlike rectangular chords of the suspended structure, men up to a certain size would have to climb inside a chord and work in the dark. Other men would be hooking up or positioning steel truss sections while balancing on the smooth, often slick surface of steel beams. In addition, the men were now exposed to the full force of the winds, rain, and fog to be encountered some 230 feet above the center of the Gate, and to the sense of extending themselves, a little farther each day, out over the stretch of water the bridgemen, most of whom had first worked in the East, referred to simply as "the river."

The stiffening trusses, the steel crisscross supports that would hold up the sides of the roadway, were hoisted up and out to the structure in twenty-five-foot sections. At the end of each section, holding it in place and connecting it with the next section, was a steel post. Every other post was attached to the suspenders that hung from the cables. Thus the bridge towers and spun cables helped support the structure even as it was being fit into place.

Every fifty feet, as the truss to either side moved out from the tower, a ninety-foot cross-beam would be derricked into place. On top of this, once enough sections were connected, the derrick would lower the stringers. Connecting gangs, working with the derricks, would align the steel pieces and bolt them into place. The riveting gangs, following in the traveler's wake, would bind them together with hot steel.

"We were connectors," recalls Al Zampa:

I was the right size. I could get in there {crawl inside the hollow-beam chords} and buck up a piece with a jack.

We could see pieces coming up on the crane and know if they were right. We'd say, "Send it up work-ways"—not end-for-end—so we wouldn't have to change it.

262

We had to carry planks which were used covering the beams. The steel itself wasn't hard, but it was difficult to walk the beams carrying planks. You were high, and in the wind.

There was a four-by-six panel board called a float, which you could swing out on, and the wind was so strong it could blow it out and hold you out there, until somebody pulled you in.

You had to be surefooted like a mountain goat, and hang on like a monkey. And you were under the iron all the time.

As the structure was being extended, four ways at once, the net was being extended, below and just ahead of it. Hanging from a frame that was ten feet wider than the bridge, and that rode out some fifty feet in front of it, the net moved by a system of rollers and clamps, just in advance of the work. Billowing up toward the trusswork and beams in the wind, casting its own spiderweb patterns of light and shadow, the net took on an existence of its own that added another dimension to the job. Inside, it was like a latticework tunnel, with a trampoline floor. Most things dropped by the workmen would get caught in it. "The punks [young apprentice ironworkers] would get a kick out of jumping into the net and retrieving stuff," says Underkoffler, "lumber, tools, cans, anything larger than six inches. They'd jump down in there for kicks." About once a week, the net had to be cleaned out entirely.

"There's no doubt the work went faster because of the net. It gave men a little security."

The safety net: nineteen men fell into it while the suspended structure was being built.

It had also begun to save lives. "I was on the Marin side," recalls Al Zampa. "It was a wet and slippery morning, and I had just started moving. I stepped out onto a stringer and missed. I flipped three times and fell into the net." At the time, the net bellied down onto the rocks at the Marin shore. "I bounced off the rocks. Broke four vertebrae. I was out twelve weeks in a body cast." Zampa was the third man to fall into the net while working.

"I was cut from $25 a week to $12.50. I couldn't work for two years. They told me I'd lost my nerve. But I went right back to ironwork."

The net on the Marin side was afterward drawn up tighter.

From a distance, in the fall of 1936, the bridge was perhaps at its most interesting so far, with its final proportions apparent, yet enough of the intricacy of its works still visible to give everything scale. At the top of each tower were two great stiff-legged cranes, still being used to hoist materials up for finishing work on the cables, yet reduced now by the size of the towers to tiny and rather stunted insect antennae. Just below the cranes, at the topmost portal of the towers, the steel-beam cross-members were still visible, not yet covered by the fascia plating—the aluminum falsework with its vertical fluting that gave the towers overall a streamlined, sculptured look with no hint of the structural muscle beneath. From the tower tops the cables—banded and dangling suspenders but with the strands still exposed, not yet sheathed with a covering of wound wire—and the twin catwalks swooped down and then up together like a giant roller-coaster, with men along the catwalks visible, antlike at a distance, only through fieldglasses or a telescope. Below this was a steel structure, reaching outward from each tower in a confusion of cranes and nets and steelwork and metal-on-metal overwater riveting noise. The two ends of what would be the center span—the largest single stretch of bridge steel yet erected—both appeared to bow up at the ends, the result of the derrick crews trying to maintain an even load on the cables at all times.

The bridge itself resembled a great gate now, an entry portal, open only temporarily at its center, soon to be closed by the steelwork now advancing more than a hundred feet each day.

On October 6, at a meeting of the Engineering Board in San Francisco, O. H. Ammann, the last holdout in favor of an aluminum paint for the final color of the bridge, formally proposed that "inasmuch as the board has examined in the field as to color and condition the last coat of paint applied to the towers nine months ago, and has found the paint not only in excellent

preservation, but also of suitable color, it therefore approves of this for the final coat over the entire main structure." His motion was seconded by Derleth, and carried.

The paint, Dulux International Orange, was actually the earth tone advocated by Morrow. "It was the color he had always wanted," says Herb Johnson. As a young man Morrow had hiked frequently in the Marin hills and the red-orange rock of the slopes ascending from the Gate had suggested to him a color which, to his painter's eye, might appropriately connect a massive work of man with this particular natural setting. The bridge that, among its other singularities, was to be the most-repainted bridge in the world would at least have a color that was all its own.

Even as the steel structure was still being erected, Cone, sensitive now to the dangers of exposure at the Gate, had painters, hanging in bosun's chairs, already slapping a protective coat of the approved paint on the newly joined girders and beams.

By the fall of 1936, the bridge had assumed its most interesting shape yet—a great gate about to be closed, at the border between ocean and bay.

265

From his office in the engineering shack at Fort Point, still tied directly by the trestle to the San Francisco pier and tower, Cone would be in telephone contact with his assistants and the other inspectors overseeing the on-the-job work on each of the ten different contracts. This now typically ranged from riveting on the San Francisco tower, to steel erection on the suspended structure, to detail work around the pylons and anchorages, to caulking and adjustment of the cable bands. The crews on these jobs ranged from as few as 4 to as many as 257 men. It helped, in keeping track of this army of men and equipment, working simultaneously on, above, below, and to both sides of the Golden Gate, that Cone was assisted by men who were a long way into their relationships with him. Both Ted Kuss, Cone's chief assistant, and Ed Davenport had worked with Cone on the Ambassador Bridge at Detroit; Cone and Davenport had studied structural engineering at Illinois in the department built by the legendary Professor Hardy Cross. The level of trust and mutual respect was high, and Cone was able to keep a close watch on things without having to be the heavy-handed overseer. It didn't matter that, for a man charged with such responsibility, he was still young. "He was just doggone good," says Davenport. "He knew his work."

While the inspectors, and to some degree his assistants, could concentrate on a single aspect of the job, Cone was the man who had to coordinate them all, to schedule the arrival and departure of men and material and equipment so that they would not conflict with each other or lie wastefully idle. He was the one man who had to keep the entire project on his mind at all times.

Cone was also in regular daily contact, either by phone or in person, with Derleth, Reed, Paine, or Strauss. Several times a week there would be meetings, lunches, conferences concerning the strategic future of the bridge —projected tolls, the possibilities of rapid transit—and often in the evening obligatory social engagements with the brokers and bankers and newspaper proprietors whose support was still crucial to completion of the bridge. There were also occasionally special functions like the dinner given by Filmer in February 1935, celebrating completion of the San Francisco pier.

The unending and overlapping demands of the job, professional and social, could not help but take their toll on Cone, who had not had a vacation—as both Strauss and Paine had—in more than three years, and who rarely even enjoyed a day off. He was not spending enough time with his family, and shortly after the bridge was finished Cone's marriage would end. He would eventually remarry, his second wife a woman who had worked as his secretary at the bridge site. Like Strauss's health and Ellis's career, Cone's marriage can be counted among the bridge's unreckoned costs.

3

There were now more men working on the bridge, with construction proceeding at a faster rate, than ever before. With the intensified concentration and pace of work came an increased accident risk.

"They had tracks running out on the deck," recalls George Albin, "whirly-rigs—a lot of big steel. There was a lot of stuff dangling out there."

On October 21, 1936, the project's perfect safety record was overtaken by the actuarial odds. Kermit Moore, a twenty-three-year-old bridgeman working on the suspended structure, helping lay the steel deck for the roadway, was part of a crew bossed by his brother-in-law, a man named Jack Turnipseed. The crew was working some 1,400 feet out from the Marin shore when a pin in the traveling crane that was laying the steel for the decking pulled loose. With a loud noise, the entire crane collapsed, and the men working around it leaped and scrambled out of the way. The boom of the crane toppled out over the west cable and fell into the strait. Moore, climbing off a girder that the crane had been picking up, was hit with the supporting leg of the collapsing crane, which crushed him between it and the girder. "Jack's brother-in-law just froze," recalls Underkoffler. "The thing came down and decapitated him."

A second man, Miles Green, was hit by a cable whipping out from the falling crane and knocked off the bridge into the net; he was uninjured. In a moratorium traditional among bridgemen, all work was halted, and the men were sent home for the day.

A combination of circumstances, among them the height, size, and complexity of the work already accomplished on the bridge, now contributed to what was, compared to the tranquil first three years of construction, a rash of serious and near-serious accidents. The bridge was now of such a size that it could occasionally get in its own way. Virtually all of its work was now done at heights, by men working without handrails, often working along with heavy equipment. There was simply more going on, throughout the project, than there had been at any time before, with a corresponding incidence of slips and stumbles, falls and occasionally breaks.

Men continued to fall into the net, at a rate of more than one a month. After Zampa's broken back, no one seemed more than shaken up by such a fall, and a sort of larking elitist spirit arose among those who had made the trip, like fellow survivors of rides over Niagara Falls in barrels. In December the veterans of these various half-falls formed the "Halfway to Hell Club," whose sole membership requirement was an accidental fall into the net. There was, at the same time, the sobering example of the risks of working without a net, both on the Bay Bridge and, closer to home, at the Gate itself.

In November 1936, Dean Kinter, a riveting inspector, was examining some of the rivets driven into the high viaduct on the San Francisco approach to the bridge. Kinter, working where there was no net, absorbed in the close work of measuring the space around rivet heads with a feeler gauge, stepped backward and fell more than forty feet onto the government ground of the San Francisco Presidio. He suffered a fractured skull and a broken back.

Less than two months later, a three-man crew was at work dismantling a work tower beneath the steel arch "bridge within a bridge" that Ellis had designed to serve as a support on the San Francisco shore while passing over, and thus preserving, Fort Point. Again, the men were working without a net; they were on a 120-foot tower that was in the process of being dismantled when the tower was hit by a sudden gale. The tower swayed and then began to collapse, blown inward by the wind against the squat brick fortress. Two men were pitched over the fort wall and into the courtyard, where they landed bruised, but with no bones broken. The third man, Jay Hollcraft, fell onto the walled roof of the fort, where one of the falling tower beams landed on him and broke his back.

Later the same month, a seven-ton locomotive being used to haul concrete for the bridge's roadway broke loose while approaching the Marin shore, plunged off the north end of the bridge, and crashed, after a fifteen-foot drop, against the Marin anchorage. The engineer, who had yelled to the men working nearby to get out of the way, had leaped off the locomotive before it struck and avoided injury.

These incidents, crowded now into the completing stages of the bridge, were like construction costs that the most safety-conscious engineer could not evade. They had occurred after Strauss had made his most expensive investment in safety yet, the $125,000 net, and they suggested that the bridge's exemplary safety record was a debt, long deferred, upon which payment would ultimately be exacted.

"I FELT A FUNNY SHUDDER . . ."

I

By November 1936, the "friendly" lawsuit for additional payment that Strauss had brought against the Bridge District, and won, had reached the dunning stage. Strauss's lawyer, John McNab, wrote a letter to Filmer, summarizing the history of the dispute, and urging the directors to begin making payments on the $262,000 awarded Strauss or risk paying an interest charge amounting to some $50 a day. The directors had delayed apparently in consideration of an appeal, which would have been a violation of the "friendly" nature of the action: there had been an unwritten understanding that both sides would accept the first court's decision as final.

The dispute recalled the circumstances surrounding the bridge's original design under the pressure of the 1930 bond issue and shed some additional light on the sort of stress that Ellis had worked under then and that the Strauss engineering staff, and particularly Clifford Paine, had lived with ever since.

Strauss's suit for additional payment had actually been two suits, one for additional money to pay inspection costs and the other to cover the expenses of redrawing an entirely new set of engineering plans. Under the terms of Strauss's original contract, it had been assumed that the detailed plans for the bridge would be worked out, as was customary on other projects, one unit at a time. As work on the bridge advanced, engineering work on the next succeeding portion would be completed, thus spreading the detailed working-out of the bridge, and its costs, over the four-year construction period.

At the Golden Gate, however, because of the bond issue and the intensity of the opposition to the project, during which authorities like O'Shaughnessy had estimated that the bridge could not be built for less than $100 million, the Bridge District directors had pledged that, if the bond issue were approved, the entire project would immediately be submitted to general bids. If the bids did not come within Strauss's $35 million estimate, no further work would be done without another public vote.

At the first meeting of the Bridge District board following the success of the bond issue, the directors had passed a resolution ordering Strauss to prepare all plans for the bridge, complete enough for lump sum bids. They also instructed MacDonald, still general manager at the time, to see to it that this was done at once.

Derleth, Moisseiff, and Ammann had protested to the board against the inadvisability and wastefulness of doing so much detailed engineering planning in advance of any sort of actual construction. Nevertheless, MacDonald, under pressure from the board, insisted on it. Acknowledging to Strauss that it was all extra work and a complete departure from the chief engineer's contract, MacDonald promised that the additional effort would be paid for whenever the bonds were sold and the Bridge District had some money.

It was under these pressures and deadlines that Ellis, assisted by the Strauss engineering staff in Chicago and San Francisco, produced in seven months the detailed engineering plans, including stress analyses, material specifications, and cost breakdowns, for ten different bridge contracts. The total lump sum bid had come within Strauss's estimate, and the directors' pledge to the voters of the district had been fulfilled.

With the start of actual construction, however, these engineering plans had, as in any other major construction project, to be redrawn. The realities of the site, the timetable, the availability of manpower, equipment, and money caused the original detail plans to be modified and, in some cases, like the San Francisco pier, scrapped entirely. The work was changed by the process of doing it.

Strauss's lawyer maintained that making these changes had required the work of Clifford Paine and thirty-two other engineers. He contended that the value of the extra work amounted, in Strauss's estimate, to 1 percent of the cost of the bridge structure—not counting approaches and abutments —which was $27 million. The jury, despite a cross-complaint, agreed, and Strauss had won a judgment. Now the directors, reminded by Strauss's lawyer of their promises and his client's efforts to fulfill them, decided that it would be in their best interests to remain friendly after all. Abandoning the idea of an appeal, the directors reluctantly began to pay up.

The man most responsible for the directors honoring their original pledge, Charles Alton Ellis, was by now an established and respected member of the engineering faculty at Purdue. "I make my living playing bridge," Ellis quipped to an interviewer at the Civil Engineering School in what may have been, in part, an attempt to fend off with irony the bitterness of dealing with bridges now only in theory. An old college baseball catcher, Ellis was known to sneak out occasionally during office hours to catch workouts of the Purdue varsity. His students called him "Uncle Charlie," although never to his face. Ellis, however, was well aware of the nickname.

"Once he put in a call either for a student or one of his assistants," Professor Marion Scott remembers. "The person taking the message said, 'He's gone to see Uncle Charlie. Who shall I say is calling?' 'I guess you can say that this is Uncle Charlie,' Ellis replied."

Through contacts like Derleth, now a fellow academic engineer, and Beggs, the professor at Princeton who had conducted the stress tests on the bridge tower model, Ellis maintained a lively and often gossipy interest in the Golden Gate Bridge and the personalities surrounding it. At some point he entered into a correspondence with Reed, the district general manager, who frequently felt the need to check statements made by Strauss with other engineers who had been associated with the planning of the bridge. Bound by a shared dislike of the chief engineer, the two men were to reveal a war veteran's relish for exchanging stories of the difficulties suffered or caused by an eccentric, overbearing, and occasionally arrogant commander.

Ellis, while accepting a decrease in salary and a certain shrinking of his ambitions, at least could enjoy the security of a continuing job. His former colleagues at the bridge were not all so lucky. With the completion of the bridge, the character of the enterprise would change entirely. The emphasis would shift from construction to administration, with all but a few people on the building side rendered superfluous. Even Reed, anticipating that the salary of the general manager would be cut once the bridge itself was finished, was now privately planning to leave.

Cone, whose present burden of responsibilities must have given him mixed feelings about being out of a job, had begun entertaining thoughts of succeeding Reed. When hired by Strauss, Cone had in fact been acting as general manager of the Tacony-Palmyra Bridge in New Jersey, and he was familiar with the administrative side of toll bridge operation. He had begun making it known that, when construction was completed, he would be available for additional duties at the Gate.

Strauss and Paine now had an entire San Francisco office staff to keep employed, and the firm had already contracted for and built another of Strauss's patented bascule lift bridges on Third Street, just a block away from the original bascule bridge Strauss had built for O'Shaughnessy in 1917. The firm had achieved an enormous step upward in stature by what had been accomplished at the Golden Gate, and both Strauss and Paine were anxious to keep operating in the big-ticket arena of long-bridge construction. In whatever time that Paine's responsibilities and Strauss's health permitted, the partners were exploring the possibilities of building another major bridge across San Francisco Bay, ideally at the last remaining major ferry point, between San Rafael in Marin County and Richmond in Contra Costa County, north of Oakland. There was some skepticism about whether the communities around the bay, which had supported the efforts to build two great bridges in the last five years, could possibly be roused to the enthusiasm required to sustain a third.

2

With crews rigging, connecting, and riveting steel in four directions at once, encouraged to move quickly by the underlying assurance of the net, their efforts multiplied by hoisting derricks and traveling cranes and hauling cars that rolled the steel out on rails, work on the suspended structure advanced quickly through the fall of 1936. From a distance, amateur superintendents could observe the progress on the main span: each day, from two directions, the Gate was moving toward its closing. By November 18, the two expanded half-spans were ready to meet.

There was a small ceremony, limited by the size of the crowd that could be accommodated on the bridge, with Strauss himself the star, white-haired and frail-looking, accompanied by a delegation from the bridge board and various city officials. Somehow it had been decided that Strauss himself would maneuver the last piece of the main span into position. In the middle of the afternoon, to allow for the sun's full expansion of the steel sections, Strauss was assisted to the controls of one of the main span's traveling cranes. With the chief engineer working the levers, the last steel chord was hoisted, swung, lowered, and connected in place. The first pass of the bridge's suspended structure had been completed.

It was, as yet, just a skeleton, yet the length of the single center span, with its joined stretch of net still beneath, had given the bridge the elongated, stretched-ribbon quality that until now had existed only in the Maynard Dixon painting and in the architect's drawings. With the suspender ropes now connected and taut, the superstructure now took on something of the look of a great red harp, hung in the western sky. Yet the span, as long as it was, remained in proportion with the tall, stepped-off towers, so that the actual size of the bridge, and the dimensions of the distance that it covered, were not apparent until a good-sized ship—a large freighter or passenger liner, or a navy battlewagon—steamed beneath it. Only then, with the shadow of the bridge passing slowly bow to stern, could be gauged the almost airborne height of the roadbed, high above the lightest-laden ship's tallest mast; the mile-wide strait it spanned, with its ocean swells and rivermouth complex of currents and eddies; and the mass of the pier, foundation, and legs of the San Francisco tower, which could make the mightiest warship passing it seem transitory and frail. All this added to the depth and drama of the site, giving every passage beneath the bridge an eventful quality, as if an arrival or departure was not just being marked but its significance measured in time.

To the profound and subtle thought he had devoted to the color of the bridge, Irving Morrow added a concern, equally deep, for its nighttime illumination. In the spring of 1936, along with his analysis of the site, dimensions, and significance of the daytime bridge, Morrow submitted to Strauss his recommendations on lighting, which in another way demonstrated Morrow's unusual combination of analytical ability and aesthetic sense.

Noting that it was common illuminating practice to seek both high intensity and uniform distribution of light, Morrow proposed, for the bridge, the reverse: a system that combined relatively low intensity with constant gradation of light. "Virtuosity or display have no place here," Morrow observed. "The object is to reveal aspects of a great monument which are unsuspected under the conditions of natural, or day, lighting."

Morrow approached the lighting of the bridge in the spirit of a painter addressing his subject. Artistic quality and increased verity would be achieved by stressing significant points and leaving minor ones to the imagination. The picture aimed at would be this: the towers, at their bases, would be "enveloped in a mellow glow," with the tone gradually disappearing into the night. The cables would be suggested by diffusion from the roadway lighting and from the towers. The anchorages would be merely suggested in position, and not detailed in form, while the approach viaducts would gradually disappear into the night. "Threading through all this," suggested Morrow, would be "the continuous horizontal line of diffusion from the roadway lighting."

Morrow proposed to position lights of varying intensity in such a way that both light and shadow would be used to emphasize the size and scale of the bridge. It was a nighttime adaptation of the way he had used vertical fluting to accent the height and grace of the towers and to make the most

Traditionally, electrical engineers had designed bridge lighting. Strauss, over protests, entrusted the job to Morrow —who gave the bridge as distinctive a look by night as by day.

of the sun's play of light as it moved, almost perpendicular to the bridge, varying course only slightly each day.

For the roadway practical considerations, particularly fog, dictated a strong, continuous band of illumination. Yet here, too, Morrow insisted, the light could have decorative value. The roadway illumination could be a bond of light connecting all the elements of the bridge, and indeed the lighting eventually chosen—sodium-vapor lamps that cut through the densest fogs while bathing the bridge and everything on it in a strange, trans-forming orange glow—seems to have achieved exactly this effect. The strong roadway light, combined with the more subtle illumination of the towers and anchorages, give the bridge, even in darkness, a realized quality, a sense of having been subjected to a greater degree of thought and care than that given to other structures. In fulfillment of Morrow's proposal, the bridge at night, whether seen from a distance or close up by people passing over it, even in fog, inescapably declares its own uniqueness: the thrill of fundamental aspiration, enduringly achieved.

3

The traveling cranes that, meeting at the center of the main span, had joined it together, were now backed away from each other as they hoisted into place the stringers, beams, and girders that would complete the steel structure for the bridge's roadway and sidewalks. On this second pass, which took about a month, real rails were laid onto the stringers so that the locomotive could pull cement by the carload out onto the two side spans and the main span, pouring pavement at the same rate that men might pave a highway. When the uninterrupted stiffening truss and grid of floor steel was completed, the Pacific Bridge concrete crews returned to the job, welding into place the steel reinforcing rods and building the steel and wooden forms necessary for pouring cement.

As work on the bridge moved toward its conclusion, the complexities of scheduling and coordinating it increased. Until the roadway was poured, the cables would not bear the full weight of the suspended structure; until the cables were stretched full—they would eventually expand twenty-one feet longer when fully loaded than when spun—they could not be wrapped; until the cables were wrapped, the footwalks could not be removed. And until the footwalks and ropes were removed, work could not be started on the upper portions of the pylons, portal enclosures, and tower tops. The fact that this diverse and complicated work was being carried out during the uncertainties of what was turning out to be a very wet winter added to the confusion, as did the requests by various contractors for extensions or for additional money to cover extra finishing-up work. And the impatience, on the part of the directors, for some sort of definite date when the bridge would stop costing money and begin earning it.

The pressure for sorting out these matters fell increasingly on Cone. He was the man on the site, usually wearing his leather safety hat, a business suit, and a pair of scuffed shoes he kept at the field office, whom people came to for answers: the interpreter, who had to put the engineering realities in terms understandable to lawyers and financiers and make the concerns of the Bridge District's management clear to engineers and conractors.

The bridge itself continued to provide surprises. Because no one had anticipated how much painting the steelwork was going to require, a way had to be found to provide continuous access to the exposed underside of the suspended structure. Cone and his overworked inspection staff, working on a tight budget, designed a traveling scaffold that would roll on a running rail carried by one of the floor stringers. The traveling platform could be used for inspecting the undercarriage of the span as well as repainting it.

Strauss pointed out that the platform, hanging below the steel structure of the bridge, would infringe upon the strict navigational clearance that had been insisted upon by the navy. He also warned that, because of the pioneering nature of this installation, particular care must be taken that it did not overload the stringer, causing eccentricity, and that special care must be given to avoiding distortion, binding, or derailment. Strauss also noted, in an observation that was soon to assume greater significance, that "due to the grade of the bridge, the matter of brake power should be carefully developed in order to prevent accidents." The Golden Gate Bridge, because of the length of its main span, is cambered—it bows up toward the center of the span, where it comes closest to the ultimate sag of the cables—then slopes away toward the two towers. Any rail-mounted structure moving on wheels along this doubly declining plane had the potential of breaking loose, gathering momentum, and crashing.

While this traveling inspection carriage was still in the planning stage, a pair of moving platforms, suspended from the underside of the bridge, were actually in use. In order to strip wooden concrete forms from beneath the poured roadbed, the Pacific Bridge engineers had built two temporary platforms that rolled along hanging from floor beams. The platforms, each sixty feet long, which crews of carpenters and laborers would stand on, reaching up to erect or remove the temporary wooden pieces overhead, were moved along by cranking a hand winch and held in place by clamps that gripped the center of the beam. Because of the camber of the bridge—its overall convex shape—it was necessary for men to go ahead of each platform, as it moved, and make sure that it was clamped securely at several points, to avoid rolling away.

The stripping platforms were an improvised expediency, jury-rigged devices uncharacteristic of the degree of concern that had gone into every stage of the bridge's construction thus far. They looked ungainly, and several supervisors and engineers had expressed reservations about the Rube Goldberg way they worked: the clamps, for example, were held in place by aluminum plates and what looked to be undersized safety bolts. Still, the job of form stripping was peripheral to the actual construction of the bridge, finish-up work, with the goal of completion tantalizingly in sight; and there was always the assurance that even here, hanging from the undercarriage of the bridge, the men were working above the security of the net.

The net itself was considered such a success that the manufacturer, the J. L. Stuart Company, had patented it, and Ammann's assistant general manager, Billings Wilson, had inquired about buying a portion of it for the New York Port Authority. "We made some tests recently," A. F. McLane, manager of erection on the Golden Gate Bridge, wrote Wilson on February 16, 1937, "and found that the ⅜" Manila rope was still as good as the theoretical strength furnished by rope manufacturers."

The following morning, a crew of eleven men was in the stripping platform near the north tower, removing wooden forms from the Marin side of the main-span roadbed. Two more men were in the net below, cleaning out debris. The platform has just been moved, and stopped, when there was a sharp crack. Because of stress on the cast aluminum side plates of the platform's trolley, the casting, near the center of the scaffold on the west or ocean side, had failed. The entire platform structure, weighing about five tons, tilted suddenly.

"I felt the stripper give a funny shudder," Slim Lambert, foreman of the stripping crew, later recalled, "then it lurched to one side. I shouted and without waiting I jumped. I must have acted instinctively, because I don't remember thinking. I landed in the net. A moment later, I heard a sound like thunder as the stripper ripped from its hangers."

One set of scaffold wheels had slipped from its supporting rail; now the immense weight of the sixty-foot platform, canted at an angle, jerked the remaining two sets loose. The whole mechanism—platform, rods, wheels, and scaffolding—now started downward, "like an elevator," the San Francisco *Chronicle* reported the following day, "without a counterweight."

The net held the five-ton stripper only for an instant; then, with an enormous "rumbling, ripping sound" the support ropes pulled loose from the undercarriage of the suspended structure and from the outriggers extending beyond it. "The net cracked like a machine gun," reported a bridge painter. "Then," added another, "it ripped like a picket fence splintering."

In an enormous, cascading spill, more than 2,000 feet of safety net, along with its proudly patented metal intersections, fell some 220 feet— the height of a twenty-two-story building—into the strait, where it landed with an impact that raised whitewater waves in the outgoing riptide. Twelve men, all but one of the crew that had been working on the platform and in the net, went spinning and clutching downward. The men were reduced to specks by the dimensions of the net and bridge and seemed to float down as the net, buoyed up by the strong wind blowing through the Gate at the time, was rended in an exaggerated flapping action resembling a great, descending flag.

"The net sagged slowly," Lambert recalled, "so slowly that it seemed

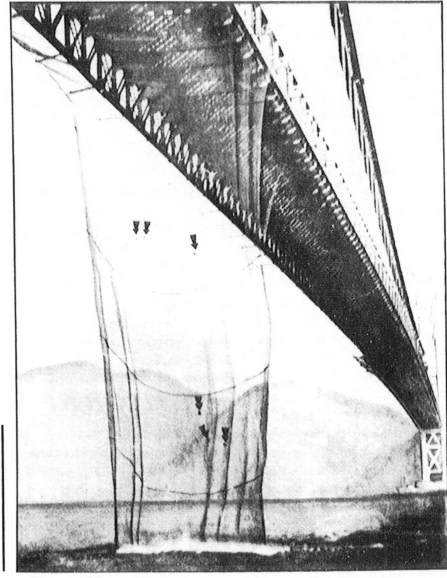

February 17, 1937: the net breaks. The arrows indicate falling men.

like a slow-motion picture. Men were screaming. The net tore like tissue paper."

The thirteenth crewman, a laborer named Tom Casey, was dangling in the air, hanging above the strait from a beam in the bridge's suspended structure. "I felt the scaffold start to go out from under me," Casey told the San Francisco *Examiner*, "and I jumped. I grabbed a beam and hung on. I just hung on and kept hanging on."

The snapping of the net, hanging from what now was the fully suspended taut structure, shook the entire bridge. "I've never been more scared in my entire life," recalled Edward Riley, a paint foreman working about a hundred feet from where the scaffold had collapsed. "I looked around. I didn't know what was happening. I thought something was crashing down from above. Then I glanced toward the water and saw the net falling."

John Marders, who was stripping scaffolding on the deck, was brought to his knees by the shock of the net ripping away. "I got so excited, I kneeled on the bridge floor and started reaching down. I guess I was out of my head, making a futile gesture to save those poor fellows."

To Albert Todd, a paint sprayer, the scene was framed like a picture by an opening in the bridge floor. "The jumbled mass of men and twisted rope, the contorted faces, slowly getting smaller. It was horrible."

"As the net was falling, I could hear faint, babylike cries," recalled Tex Lasiter, a steelworker on the suspended span. "When the net hit the water, the men seemed like little blots of ink on the surface. The net was like a sinking raft, with the tiny men entangled fighting to get free."

Riley recalled some of the men getting entangled in driftwood, swept down on the river tide ebbing out through the Gate, their faces "ghastly" in the brownish green water.

Frank Dowling, a timekeeper, recalled two men clinging to steel beams as they fell into the water, as though they were hanging onto the steel for safety. "Two other men were clinging to the net and struggling, as though trying to climb up the falling ropes."

"All of us who saw the crash were scared stiff," recalled Alex Johnson, a painter. "We thought the bridge was coming down around our ears. I could see two men in the water. One of them was swimming and the other was hanging on a floating timber. We began running around the bridge deck, looking for life preservers to throw down, but we couldn't find one. Nothing but fire extinguishers."

The man swimming was Slim Lambert, the foreman, who had somehow survived the fall. "As I was falling, a piece of lumber fell on my head. I was almost unconscious. Then the icy water of the channel brought me to."

Lambert, only twenty-six and a strong swimmer, was trying to get free of the net and the wreckage, both of which were now going down fast.

"As I drifted with the tide, I saw three pairs of feet sticking out of the water. I don't know whose they were. I watched them and tried to make my way over to where they were, but I couldn't buck the tide."

Lambert was now partly concealed from view himself by the ocean swells outside the Gate: the tide was carrying him out to sea. The feet remained above the surface "for a while. Then, a pair at a time, they disappeared, as if they were being sucked down.

"I closed my eyes for a moment, then looked up. About four feet away from me was the face of a man I knew. It was like paste. I tried to swim toward it. Then a terrible expression crossed the face. I knew the man was dead. He sank before my eyes."

A second later, Lambert saw Fred Dummatzen, twenty-four, floating in the water. "I swam toward him as fast as I could. For what seemed like hours, I supported his body. It was a dead weight. I was afraid he might be dead, but I couldn't take a chance. I thought if we were rescued quickly, they might be able to revive him."

On the bridge deck, some 200 men were now running around, shouting and swearing, pointing down in the water, trying to indicate where men were floating to the rescue boats, which had somehow not yet appeared. Among those watching from the bridge were two men, Peter Anderson and Hugh Hillen, who had brothers who had been working on the stripping platform. Anderson, working on the deck, had seen his brother Chris fall, while Hillen had known his brother Elbridge was in danger as soon as he heard the ropes start to snap.

As the net first began to rip, five warning blasts had been sounded on the horn on top of the field Engineering Office at Fort Point. Yet as the minutes ticked agonizingly by and the floating bodies and the net began to disappear from sight, rescue vessels from the Coast Guard station less than a quarter of a mile inside the Gate still had not yet appeared.

Casey, meanwhile, the surviving cement worker, was still hanging from the bottom of the bridge, both hands gripping one of the platform's broken roller casters. His pipe was still gripped tightly in his mouth. As men on the deck shouted to him to hang on, Todd, the painter, and William Foster, an inspector, lowered a rope down to him. Casey refused to let go even the one hand necessary to grab the rope. Instead, Foster rigged a loop in the end of the rope and maneuvered it between Casey's legs. "Then we told him to let go and ride on up. He did and we pulled him up, with that damn pipe still stuck in his face, and nothing between him and the bay but the rope."

Casey stood on the bridge deck, face white, legs trembling, pipe in mouth. Terrence Hallinan, the friend he'd come to work with that morning, was one of those who had gone "into the river." "Casey," recalls

Frenchy Gales, "walked down to the Field Office, picked up his time, and never came back."

Lambert, still supporting the stunned or dead Dummatzen, had been swept out the Gate. "My body felt like a block of ice. Blood was running into my eyes. All around me, I felt, were the dead bodies of my buddies, and I couldn't do anything about it."

A Coast Guard boat, arriving at last to pick up survivors, came within a hundred yards of Lambert, but the Coast Guardsman failed to spot him in the floating debris. The vessel turned back, while Lambert continued to drift out to sea.

Lambert was now drifting in the general direction of Point Bonita, the second to last headland on the northern approach to San Francisco Bay. "I was just about ready to give in," he admitted, "when a fishing boat came alongside."

The boat, coming in from crab fishing, was skippered by Mario Maryella of San Francisco. Maryella and his crew, with difficulty, wrestled Lambert and Dummatzen on board. Dummatzen was dead, but Lambert seemed in surprisingly good shape, apart from head lacerations, and the effects of shock and immersion. On the way in, Maryella and his crew, picking their way carefully among the debris, rescued another survivor. Oscar Osberg, a fifty-one-year-old carpenter, had clung to a piece of wreckage despite a broken leg, a broken hip, and severe internal injuries. Both he and Lambert, who, it turned out, had broken a shoulder, several ribs, and, in a lesser fracture undiagnosed until years later, several neck vertebrae, were taken to Mary's Help Hospital in San Francisco.

On the bridge, in the aftermath of the accident, some men were swearing and shouting while others stood silent and stunned. A few, like Peter Anderson, were weeping. From the Field Office came one long blast of the horn: the quitting signal, announcing at ten o'clock in the morning the end of work for the day.

Some two hours later, with the bridge empty of workmen and swarming with four different teams of investigators, the skeleton of the fallen platform slowly rose to the surface of the waters of the strait. "All the planking was gone," observed the *Chronicle,* "and the gaunt outlines resembled two huge picture frames."

A full riptide was then running, and the wreckage of the platform was swept out to sea, passing through what amounted to a flotilla of boats, now searching the strait for bodies, and, beneath, the dangling, windblown shrouds of Strauss's safety net.

CHAPTER

21

"I'M NOT ON TRIAL. I'M NOT GUILTY OF ANYTHING."

I

On the morning of the accident, Strauss had been scheduled to appear at the monthly meeting of the Bridge District Board of Directors, in the Hunter-Dulin Building at the corner of Sutter and Montgomery streets in downtown San Francisco. The meeting had already been called to order when the directors were notified that Strauss would be delayed because of trouble at the bridge. Knowing that something had gone wrong, uncertain exactly what it was, the directors turned to the first item on the agenda, a request by members of the Bridge Fiesta Committee for a postponement of the bridge opening ceremonies to coincide with the arrival, in San Francisco Bay, of 150 ships of the U.S. naval fleet.

It was noon before there was movement at the door, and the apprehensive directors looked up to see Strauss entering, holding a prepared "Preliminary Special Report." He read it to the board: "An accident to the stripping platform for the removal of the deck forms occurred at 9:30 A.M. today. It is too early to be able to determine the cause of the accident. It may be due to the failure of an aluminum casting in the stripping platform, but this can only be determined by investigations in process."

Strauss summarized the casualties, whose names were not yet all available. The only way to identify those who had actually died had been to take roll as the dismissed survivors had left the bridge for the day, and that list was not complete yet. So far, the only recovered corpse had been Dummatzen's, and the only survivors were Lambert, Osberg, and the dangling man, Casey. The Coast Guard, Strauss announced, was still searching for the remainder.

"I am taking immediate steps for a thorough investigation, and will report further as soon as possible. The accident," continued Strauss, visibly shaken, "is the more deplorable in view of the tremendous effort that has been made to conserve life."

The directors asked questions about the victims. "We don't know yet," Strauss answered, adding, "The Coast Guard has saved two." He begged off responding to further questions, pleading that his presence was required at the bridge.

In a single catastrophe, Strauss's proudest achievement at the Gate, the project's near-perfect safety record, had been obliterated. Because of the spectacular nature of the accident, the fact that he had cut, by two-thirds, the "acceptable average" of high-steel fatalities would be largely forgotten. The safety net, caught, as it ripped, in a spectacular photograph by a free-lance photographer named Joe Dearing, would forever be associated with the loss of ten lives, instead of the saving of nineteen. More ominously, the accident ushered in the association of the bridge and dramatic death, with its powerful attraction for potential suicides that has continued and grown up through the present day. To the bridge's embodiment of soaring aspirations had been added the dark, opposite lure of plummeting death.

The following day, four different agencies—the San Francisco Coroner's Office, the Bridge District, the State Industrial Accident Commission, and the Pacific Bridge Company—all announced investigations seeking to fix responsibility for the accident. Some were trying to fix it on each other. So far, the accusing fingers seemed to be pointing toward Pacific Bridge.

Frank MacDonald, a member of the State Industrial Accident Commission, announced to the press that the Pacific Bridge engineers had been warned twice that the scaffolding was unsafe and that the second warning had come on the morning of the accident. Commission inspectors, along with company engineers, said MacDonald, had been walking out on the bridge deck to inspect the scaffolding when it ripped loose and collapsed.

"If we had been able to get action a few hours sooner," MacDonald maintained, "all those lives would have been saved."

Phil Hart, Pacific Bridge's president, insisted, "The details of the construction and the scaffold itself were at all times visible and accessible to the inspectors of the Bridge District and the Industrial Accident Commission, and no objection was made to anyone of the scaffold."

San Francisco District Attorney Matthew Brady entered the controversy by offering to investigate any possibility of criminal negligence and dispatched one of his assistants to observe the coroner's inquest into the death of Dummatzen, whose was, as yet, the only recovered body.

Both Strauss and Cone were subpoenaed by the coroner. Cone testified that he had examined the scaffold when it was assembled and that the safety bolts were in place. However, according to Harold Fox, a cement worker who had left the scaffold fifteen minutes before it fell, "Those safety bolts weren't in our platform. I never saw any."

The coroner's jury found that Dummatzen "came to his death due to the spreading of the hooks used to support the scaffold on which the laboring crew were working. The spreading of these hooks was evidently through the failure to use certain safety equipment, namely, a bolt."

The verdict fixed blame on no one, and thus satisfied no one. The San Francisco Grand Jury announced that it would investigate the accident. Meanwhile, it was revealed that Strauss, a week before the collapse of the scaffold, had opposed the erection of permanent traveling scaffolds beneath the bridge. Charges and countercharges flew back and forth among Strauss, the State Industrial Accident Commission, and Local 261 of the Laborers Union, representing the deceased workmen.

On March 3, after several delays, the Bridge District opened its own inquiry. The proceedings, held at the district offices, were chaired by Strauss, who opened the hearing by announcing that he was postponing it. The State Industrial Accident Commission, said Strauss, had refused to permit its inspectors to testify, and the representatives of the laborers had not been officially designated. Under the circumstances, Strauss concluded, there was no point in proceeding with "half a hearing."

Elmer P. Delany, lawyer for the Laborers Union, rose to protest. The workers' witnesses were on hand, he insisted, and since the inspectors' testimony already existed in transcript, the investigation might just as well proceed. In a remark clearly pointed at Strauss, Delany argued, "We don't want people who may be concerned conducting this hearing."

Warren Shannon, the San Francisco supervisor who was also a bridge director, demanded to know why the State Industrial Accident Commission had refused to take part in the inquiry. The commission's letter was read aloud. Strauss, the letter pointed out, had been named defendant in a $100,000 damage suit arising from the accident. As chief engineer he was "perhaps responsible for all work on the bridge."

"It would appear," the commission spokesman had written Strauss, "that you as Inquisitor must have an interest in the inquiry you will conduct."

Strauss, though his physical energies may have flagged, retained his feisty risibility in the face of opposition. The chief engineer now objected, particularly to the inference that he was charged with blanket responsibility for the fitness of all work on the bridge. The commission's statements, Strauss protested, were "incorrect, misleading, and improper."

Delany, who also happened to be the attorney for the plaintiffs who had actually filed two $100,000 suits against Strauss, Cone, and "other un-known defendants," suggested that the chief engineer might be trying to use the hearings to whitewash the entire incident.

"I resent and object to that reference," Strauss now shouted. "I'm not on trial. I'm not guilty of anything whatever. There is no question about my integrity and unbiased honesty. My efforts alone installed the safety net which saved eleven lives and then saved two more in this accident. I pro-tected labor and no one else did."

Delany now rose to take exception to Strauss's remarks. As the lawyer spoke, Strauss shouted, "Not true!" Three or four of the directors broke in. Soon everyone was trying to talk at once. Strauss whammed his gavel on the table and declared: "I adjourn this meeting."

The directors refused to let Strauss cut the meeting off in the middle of someone else's speech and without their having a say. Then they discussed matters among themselves. Perhaps a postponement might be useful, allow-ing tempers to cool. "Then," suggested Director A. R. O'Brien, a Ukiah newspaper publisher, "we can approach the situation in a friendly spirit, instead of with meat axes."

2

While Strauss, the directors, the commissioners, and the various union and company lawyers continued to bicker and point fingers, work on the bridge quietly resumed. On the two side spans and the half of the remaining center span still protected by the net, carpenters and concrete workers continued to lay forms and pour concrete for the roadway. Along the cat-walks, ironworkers were wrapping the cables in a single layer of wound wire and adjusting and caulking the cable bands. On the cables, suspenders, and steel structure, painters, some of them hanging in bosun's chairs, were once again applying the final field coat of International Orange. The mood was more subdued now, with a minimum of joking and raillery: the idle center section, with its gaping emptiness of vanished net, fixed the after-image of what had happened in everyone's mind.

Two days after the accident, the torn section of net was located on the bottom of the strait and brought to the surface by a barge using a derrick and grappling hooks. One body was found caught in the net. Strauss had, almost immediately, ordered replacement of the original net, and the cleanup work preparatory to installing a new net had begun as early as March 3. The top priority was still getting the job done.

"Investigations, charges, and recriminations fill the air here," Reed wrote

Moisseiff on March 6, "but the whole truth is going to be brought out." Strauss and Cone were being sued by one of the widows, and similar suits were reported pending. "We have all been summoned to appear before the Grand Jury next week (criminal negligence), and the politicians are busy, as usual."

Strauss was privately consulting Ammann as to his legal liability, as chief engineer, for the accident. On March 23, he sent the New York Port Authority chief engineer a summary of his report, "from which it will be evident that the trouble was due to the fact that the aluminum side plates carrying the rollers on which the platform operated were over-stressed and consequently failed."

"Mr. Paine's figures on this," Strauss continued, "were confirmed by tests made at the University of California testing laboratory."

According to Strauss's summary, Pacific Bridge had failed to submit plans to or secure approval from Strauss or his staff before placing the scaffold in service. Strauss asked Ammann, in the light of "your long experience and extended practice," what he would consider proper engineering practice in this case, whether it was customary to make "the Engineer responsible for the work the Contractor is obligated to do, or for his defaults and mistakes.

"There is no doubt, of course, in my own mind," Strauss insisted, "that the Engineer is not responsible, and that no duty devolves on the Engineer to be on hand every instant of time to inspect scaffoldings." Still, the Engineer could use, at this point, the reassurance of another Engineer.

Ammann endorsed Strauss's position completely. "The provisions of the contract," he replied on March 29, "clearly put the responsibility fully and solely on the Contractor." Even if Pacific Bridge had submitted plans, "it was certainly not one of the duties of the Engineer to examine and investigate as to strength every detail of the scaffolds." This was not in accordance with general practice, or with Ammann's personal experience: "I do not recall any accident which resulted from failure of the Contractor's equipment, where an attempt was made to hold the Engineer responsible."

The grand jury, after investigating and hearing testimony, issued no indictment, thus cutting the ground from under Delany's civil suit. Responsibility for the accident was never legally fixed, leaving members of the Strauss & Paine staff and the Pacific Bridge engineers somewhat uneasily still at work with one another.

By March 31, all but the final few sections of the safety net had been restored, and the laying of the main span's concrete roadway had already been resumed. The wooden flooring of the east, or bay-side, catwalk had been removed, leaving just the wire support ropes flanking the cable, and work was scheduled to begin removing the walk on the west, or ocean, side. At the tops of the towers, ironworkers were ready to start centering the cable saddles on the tower tops and freezing the rollers in place with grout. The trestle that extended from the San Francisco shore out to the south pier would remain until sometime after construction was completed, as would the cofferdam around the base of the north, or Marin pier; yet the work of dismantling the mold in which the bridge had been formed had begun. The traveling scaffolds, one replaced and one repaired, had achieved a perverse sort of popularity: on March 29, the Pacific Bridge carpenters had gone on strike in a jurisdictional dispute with the ironworkers. Each claimed that the work of moving the scaffolds belonged to their respective unions.

It now appeared that, largely because of the speed of construction on the suspended span, the bridge would be ready for opening two months earlier than had been anticipated. Plans were already under way for a week of festivities that were not only to mark the completion of a monumental work but to serve as rites for San Francisco's passage from classic city to modern American metropolitan center. The rolling hills and cozy valleys of Marin County were to serve as the great green magnet, drawing the city's middle class, particularly its business and professional families, over the bridge, transforming Marin into a spacious, luxuriant bedroom community, connected by the passageway of the bridge to the sometimes crowded, sometimes vacant workshop of downtown San Francisco.

There was a pattern for bridge-opening ceremonies, established by the Brooklyn Bridge. "A friend I had known in New York sent a clipping from the New York *Times*," recalls Ted Huggins, "about how the Brooklyn Bridge had been reserved for pedestrians for a day before horse cars and electric trains were allowed over it." Huggins proposed a Pedestrians Day for the opening of the Golden Gate; the bridge management balked at it. The bridge had been built with private funds. It was intended to produce revenue, and cars, not people, generated revenue. Huggins, who had himself been working for Standard Oil when construction on the bridge started and had persuaded his boss to let him take photographs of the bridge to encourage driving and gasoline consumption, spoke with the conviction of the converted. One of his photographs, a lengthwise view of the bridge shot from the top of the Marin tower, was about to appear on the cover of *Life* magazine. The bridge management conceded: The pedestrians would have their day.

On the bridge itself, special pains had been taken to make the view as appealing as possible to the stroller's discerning eye. In addition to the 10½-foot-wide raised promenades facing the bay to the east, and the ocean to the west, Morrow had designed offset bays at intervals and large curving walkways around the towers to encourage people to pause and enjoy the view. The 3¼ miles of aluminum and steel hand railing, now being riveted into place, was not an obstacle, but a handsome balustrade, consistent with the bridge's angular, vertical motif. The space between posts was exaggerated so that walkers on the bridge would have an unbroken panorama of city, hills, water, and sky, and passing motorists a continuous, see-through view of the spectacular surroundings.

The concrete for the roadway, like that for the piers, pylons, and anchorages, was mixed in the batching plants on either shore. In all, enough concrete was poured for the bridge to lay a sidewalk five feet wide from San Francisco to New York. At the south side, concrete was trucked out on the trestle to the base of the San Francisco tower. From here it was derricked aloft to the level of the roadbed and dumped into a railroad-style hopper. Small gondola cars would be backed by the seven-ton locomotive beneath the hopper, which filled each car with concrete. The locomotive would then haul the cars, and the concrete, out to the pour.

The roadway was being poured in three long rows, each twenty feet wide. To maintain an even load on the cables and allow room for the rails, the two outside rows were poured first. Then, as the tracks were removed, the center patch was filled in. Every fifty feet, the rows and the finished deck came to a break, an intersection fitted with a copper expansion joint. Once the roadway was finished, whenever the steel structure of the bridge expanded or contracted with heat or cold, the roadway would be able to flex with it, instead of buckling under stress.

While the roadway was being poured by Pacific Bridge, the west lane of the deck was kept clear for Roebling workers who were removing the wooden sections and ropes from the west footwalk. At the same time, Bethlehem crews were busy on the deck with finish-up steelwork. The operations of all three contractors had to be carefully coordinated by Cone and his assistants to avoid confusion and accidents.

As fast as the form material for the roadway could be stripped away, it was placed in the space for sidewalks. Here the work of finishing cement proceeded as rapidly as enough experienced cement finishers could be found to do it. The demand for speed was so great that the residency requirement was waived for men in these skilled trades. "We worked with short crews, getting the catwalks down," Frenchy Gales recalls, "used pushers, supervisors, everyone."

With the beginning of pouring, the deck of the bridge had been cleared of much of the clutter of equipment and debris that had covered it. The great traveling cranes were now drawn in close to the towers and would soon be lowered out of sight themselves. The stacks of lumber for forms, the air compressors for the riveters' guns and oxyacetylene welders' tanks were gradually disappearing. For a few of the workmen on the bridge, the nearly completed roadway, with its great stretches of unblemished concrete, represented the same sort of attraction and challenge that a freshly cemented sidewalk does to a child. Frenchy Gales, who had put a penny with his initials on it in the last pour of the cement for the San Francisco pier, saw an opportunity to make a mark of another kind on the roadway.

"I had this old car," Gales recalled later, "and I had about five or six friends with me." Gales, who was living in Sausalito in a house shared by a group of ironworkers, admittedly was a man who liked to pull a cork in those days. "We started out from the Marin side and when we got up there near the San Francisco anchorage, there was a gap in the roadway. We got out and put down some planks and just drove on over.

"I stopped at the toll plaza," Gales added. "Made each of my friends pay a dime."

CHAPTER

22

"IT WAS NEVER JUST A JOB TO ME."

I

Bridges are among the most permanent of structures, and they are built by among the most transient of men. Like oil engineers, riggers, and drillers, bridge builders must go where the jobs are, commit themselves to protracted periods of high-risk work, and then, barely taking time to savor the satisfactions of completion, move on and begin again. For the past four years, there had been an uncommon concentration of major bridge building on San Francisco Bay, with two great bridges making the Bay Area the focus of international interest in long-span construction. Now that era was coming to an end, and hundreds of workmen, as well as more than a dozen contractors and subcontractors, were faced with the uncertainties of a locally saturated market for bridges and a Depression economy that was entering another slump.

At present, the best prospects seemed to be in the Pacific Northwest, where Pacific Bridge was one of the contractors scheduled to begin work on a new suspension bridge at the Tacoma Narrows. There had been an intense rivalry between the cities of Seattle and Tacoma as to which would be the site of the first bridge on Puget Sound, and Tacoma had won by supporting a design that would cost much less money. There were some misgivings about the strength of this new bridge, but the Golden Gate had expanded the possibilities of lightweight long-span construction in high-wind areas, encouraging the belief that the support structure could be made lighter, too. Both Ted Kuss, Cone's assistant resident engineer, and Ed Davenport, an inspector on Cone's staff, would eventually go to work for Pacific Bridge.

Cone, meanwhile, still overseeing finishing work on ten different bridge contracts, was also privately campaigning to replace Reed as general manager. Rumors were circulating that "the engineers actually responsible for bridge construction," which probably meant Strauss, Paine, and Cone, considered Reed "a pain in the neck." There was also disappointment, apparently among the directors, at Reed's inability to get action from federal, state, and city officials on the matter of highway approaches to the bridge. Reed, conscious of Cone's ambitions for his job, later claimed to bear him no malice "because he did not know that I had been planning to leave for over a year." Still, all was not as happy aboard ship as it might be as the four-year voyage of bridge building approached its destination.

In addition to exploring interest in a Richmond–San Rafael bridge, Strauss was also still waiting for his final payment as engineer on the Golden Gate. It appeared as though there would now be claims amounting to several hundred thousand dollars made against Strauss by some of the contractors. The Roebling lawyers, in particular, were insisting on being reimbursed for the cable bands and wire that had been rejected by Clifford Paine. The fact that it appeared Strauss would bring the bridge in under budget, with as much as a million extra dollars in the kitty, only heightened the contractors' hunger, while until matters were wound up financially there was little hope of Strauss or Paine undertaking another major bridge job.

Even the Bridge District itself was being considered for additional employment. As originally proposed to the voters of the six counties who signed petitions for it, the district was to be only a temporary organization. Its purpose was to raise money, build the bridge, then operate and maintain it until the tolls had paid off the bonds, at which point the bridge was to be turned over to the State Highway Department and run toll free, with maintenance paid for out of the state gas tax fund. The district, its Board of Directors, general manager, and staff would then be out of business.

Someone, apparently Cone, suggested a more "active" policy that "would be of greater benefit to the residents of the District. This policy would consist in making improvements to the roads and transportation throughout the entire District."

The money spent for these improvements would be spread throughout the six counties, thus tending to increase traffic and encourage use of the bridge. The model for such an organization was the Port of New York Authority, which had been originally organized for improving the shipping facilities of the port. Over the years, the authority had developed into a permanent district that built bridges and tunnels and had assumed responsibility for much of the public transportation of metropolitan New York.

The report quite accurately prophesied what actually happened to the Bridge District more than thirty years later. When the last of the bridge bonds was retired in 1967, the district, flush with revenue and under the pressure of a now too-heavily-traveled bridge, entered the transportation business, operating first commuter buses, which lost some money, and then, coming full circle from the district's birth struggles with the Southern

Pacific, a Marin ferry line, which continues to lose a lot. Instead of being out of existence, the Bridge District remains an organization of growing complexity and reach, requiring the attention of the same number of directors as when the bridge was built, and of a much larger full-time staff.

2

On April 19, the Pacific Bridge crews completed the concrete work for the bridge deck. Eight days later, with the official opening of the bridge still a month off, several hundred invited spectators gathered at the center of the main span for a ceremony marking the completion of construction.

The program began with the Sixth Army Band, followed by two companies of Sixth Coast Artillery, marching in crisp military formation up from the Presidio and out onto the bridge. As the marching soldiers approached the San Francisco side span, they were met by bridge engineers, frantically waving their arms and urging the army to stop. If the soldiers marched in step out onto the span, the engineers insisted, the vibrations sent up from their rhythm could "ruin" the bridge.

In fact, the effect of something as simple as a marching troop of men was one of the behavioral mysteries associated with suspension bridges. In 1829, the small Broughton suspension bridge at Manchester in Lancaster County, England, collapsed under the rhythmic impact of a marching party of troops. The bridge has sustained the weight of pedestrians and vehicles for more than two years, and its failure led to a general rule, persisting into modern times, requiring troops to break step when crossing bridges. With the comment that it must be a hell of a bridge that could be ruined by a couple of hundred marching men, the army captain at the head of the column ordered his men to shuffle out onto the bridge in the manner, reported the *Chronicle,* "of a bread line moving toward the soup."

At the center of the main span, the soldiers joined a gathering of various bridge and civic figures, ranging from Strauss and San Francisco Mayor Angelo J. Rossi to the warden of San Quentin prison. There were a number of speeches, including a brief address by Strauss, looking frail, with dark circles now rimming his eyes, comparing the occasion to the driving of the golden spike that in 1869 had completed the first transcontinental railroad. A man was introduced, Joseph Graham, who had been at that famous golden spike ceremony sixty-eight years before. The crowd was also presented with George Van Orden, who had come to California across the plains in 1846, and J.H.P. Gedge, who had sailed through this same Golden Gate in the Gold Rush year of 1849.

In keeping with the golden spike ceremony at Promontory, Utah, and the traditions of the Gold Rush, the bridging of the Golden Gate was to be completed by the driving of a golden rivet. Fashioned by Charles S. Segerstrom with gold from the historic Southern Mines around Sonora, the rivet gleamed in the bright light of a clear day, almost 100 percent pure gold.

"A lot of guys had brought knives and cutters with them out on the job that day," Frenchy Gales recalls. "Once that rivet was in, they were going to cut it out and take it home."

The rivet made of gold was much softer than the bridge's millions of other rivets of steel. As the riveter, Ed Stanley, attempted to hammer down the head of the rivet, while his buckerup Ed Murphy held the rivet in place, particles of fine gold were showered over the spectators. When Stanley took his gun away, the head of the golden rivet fell limply off and disappeared completely. Then the remainder of the rivet was punched out—and disappeared too. Of necessity, the superstructure of the bridge was then completed using the same material with which it had been begun: a simple Pottstown, Pennsylvania, steel rivet.

In the month that followed the completion of construction, the bridge and the city were made ready for an enormous celebration. Along with completing the highway link from Canada to Mexico, the opening of the bridge was to mark the return, temporarily at least, of San Francisco to preeminence among the cities of the West. No longer cut off by water, shrunken and humiliated by the explosive expansion of Los Angeles, San Francisco and its directly linked environs, which now included the universities at Berkeley and Stanford, the California wine country, the agricultural riches of the Central Valley, and the cathedral beauty of the redwoods, would assume their natural role as the centerpiece of the great banquet table of California. No longer would the "something that is wrong with San Francisco"—earthquakes, isolation, an antiquated and inadequate transportation system—serve to undermine the city's position as a metropolis of the first order.

To match the exuberance of these feelings, a week-long festival was planned to begin on May 27, scheduled as Pedestrians Day on the bridge. The festivities would include everything from historical pageants to beauty contests to fireworks displays; the entire U.S. fleet, including "battlewagons and airplane carriers" would sail beneath the bridge. Hundreds of carrier planes would fly in formation above its great towers. From Washington, President Roosevelt would touch the electric button that opened the entire West Coast to uninterrupted automobile traffic.

Because of its perfect proportions, it is easy to overlook the bridge's enormous scale. These people are standing in front of one lengthwise end of one leg of one of the towers.

It was to be Joseph Strauss's most long-awaited hour, and he took great pains to extract every ounce of satisfaction from it. His bridge would be so meticulously prepared that it would appear as near perfect as possible to the critical eye of the strolling taxpayer. In a final tidying up, men with steel wool and paint scrapers were put to work removing the last traces of cement and grout from the steel superstructure, while painters touched up scuffed and scratched spots with primer and a last coat of International Orange. Strauss held his suppliers accountable for this cosmetic work, under the terms of their contracts.

Near the toll plaza at the San Francisco end of the bridge, a bronze plaque was mounted honoring the memory of the ten men who had died when the safety net had broken. It also mentioned the fact that during the first forty-four months of construction there had been only one fatality. Strauss was determined not to be remembered as a man who had exacted the price of a great work in human lives.

On the east leg of the San Francisco tower was to be mounted the most prominent plaque, in honor of the officials of the district and the planners and engineers of the bridge. It was on this subject of official credit that Strauss was most touchy, and it got him into his most serious quarrel with his otherwise loyal subordinate and partner, Clifford Paine.

"Mr. Paine was responsible for so many things," says Ruth Natusch. "I

thought he had an awful lot to do with the design of the bridge. He and
Mr. Strauss had a row over the plaque on the base of the tower. Mr. Paine
wanted his title to be Assistant Chief Engineer. Mr. Strauss said no, Assistant *to* the Chief Engineer. He and Mr. Strauss argued. Mr. Strauss didn't
want anyone else to get credit. They had such an argument!"

Like many another man who has achieved a certain eminence in major
matters, Strauss remained dismayingly petty in minor ones. The plaque was
worded as he insisted.

"Mr. Strauss was jealous of other people. I think his size had a lot to do
with that."

Nowhere on the plaque, not in any of the news stories concerning completion of the bridge, nor in any of the press releases or promotional material
published by the Bridge District, then or since, was any mention made of
Charles Ellis.

3

Since the days of its original settlement in a rush of goldseekers from
South America, Europe, Australia, and Asia, San Francisco had been a
cosmopolitan community, with cultural ties in many ways closer to other
countries than to the rest of the United States. Over the years, to serve the
city's own cultural mix, as well as to function as a political, military, and
commercial listening post in the American West and on the Pacific rim,
San Francisco had become a diplomatic enclave, the site of more overseas
consulates than any other Western American city.

In the spirit of honoring the city's international ties, the committee
sponsoring the fiesta that was to celebrate the opening of the bridge had
arranged to have the parade route to the bridge decorated with suspended
banners representing the flags of other nations, among them, the German
swastika.

This was at the height of the Popular Front era, when a number of local
groups were actively involved in support for the Loyalist government in
Spain and vehemently opposed to German intervention. The idea of the
Nazi emblem being displayed on the streets of San Francisco, with local
officials and citizens parading beneath it, provoked protest to the mayor's
office from members of the powerful International Longshoremen's Association, representatives of the League Against War and Fascism, and the state
chairman of the Communist party. The Longshoremen's protest was accompanied by a threat to boycott the fiesta parades on streets where the Nazi
symbol was flown. Mayor Rossi refused to remove the German flags, insist-

ing that the festival was open to all nations and that the street decorations would reflect that policy.

On the evening of May 26, the night before the opening of the fiesta, a band of unidentified men carrying long poles attempted to tear down a swastika banner hanging in the heart of the downtown district, on Market Street near Grant Avenue. The men succeeded in ripping away a part of the banner before being chased away by the police. The incident, minor in itself, was received with great indignation in Berlin, coming as it did on the heels of recent anti-Nazi remarks by Mayor Fiorello La Guardia of New York and Cardinal Mundelein of Chicago. "New York, Chicago, San Francisco," said the *Donersenzeitung,* official organ of the German army, "the chain of serious provocations against the German people is apparently unending."

Under direct orders from Propaganda Minister Joseph Goebbels, the German press gave the incident headline treatment as an "insult to the German flag" and insisted that "satisfaction must be given." This diplomatic hot potato was dropped in the ample lap of Dr. August Ponschab, German consul-general in San Francisco and a long-suffering, non-Nazi German diplomat. Insisting publicly on behalf of his government that the persons accused of desecrating the German flag must be vigorously prosecuted, Ponschab was privately satisfied by Rossi's assurance that the mayor would remain "neutral" in the controversy. Although the anti-Nazi gesture had been made, the men responsible for it were never caught. All parties, with the exception of the German press, which continued to fume about "American democratic procedures," seemed to have been satisfied, and the incident, which also dramatized San Francisco's return to the status of a world city, was soon lost in the rush of festivities surrounding the opening of the bridge.

The following morning, Thursday, May 27, with local schools closed for the day and many offices and stores either on holiday or reduced to skeleton staff, people began heading toward the bridge from all over San Francisco and from the counties to the north. By 6 AM, the opening hour for Pedestrians Day on the bridge, it was estimated that nearly 18,000 people were massed at the San Francisco and Marin barriers. As the hour struck, there were great blasts from the foghorns mounted in the trusswork of the center span. The barriers were dropped, and the earliest, eagerest arrivals—most of them students of high school age—ran or walked out onto the bridge.

Donald Bryant, a San Francisco Junior College sprinter, was the first person across, wearing pants, sweater, and scarf against the early morning

chill. His was to be merely the beginning in a crescendo of firsts that was to turn Pedestrians Day into a personified version of the Guiness Book of Records. Carmen Perez and her sister Minnie were the first people to skate across; they had left their home on Bush Street at 3 AM and skated out to the bridge to get the jump on the competition. Florentine Calegari, a houseman on strike from the Palace Hotel, was the first person to cross over on stilts; he then turned around and crossed back. Two Balboa High School girls were the first twins to cross the bridge; they were disappointed at not receiving an expected prize. There was a uniformed Boy Scout troop that passed over, and two uniformed postmen; a man named John V. Royan and his daughter Betty, from Heckersville, Pennsylvania, carried twenty-five pounds of Schuylkill County anthracite across the bridge. There were people who tapdanced across, a man blowing a tuba, people on unicycles and playing harmonicas. People walked backward, balanced on lines, and walked dogs and cats. Six sprinters from the *Chronicle* crossed both ways barefoot. Henry Boder, a seventy-four-year-old San Francisco man, who had crossed the Brooklyn Bridge on its opening day in 1883, now performed a comparable feat on the West Coast. A woman, apparently in physical dis-

On Pedestrians' Day, 200,000 people walked, ran, climbed, danced, roller-skated, and stilt-walked across the Golden Gate Bridge.

tress, was stopped by police, who discovered that she wanted to be the first person across with her tongue out.

As the people took over their bridge, more conventional ceremonies took place at Crissy Field, the nearby airstrip in the San Francisco Presidio. Here the various floats, marching bands, and motoring dignitaries who had traveled the flag-decorated parade route through the city passed before a formal "Span of Gold" reviewing stand. Against the uncluttered backdrop of the full bridge, more imposing in its length and height, its warm red color and soaring grace than all the festivities that surrounded it, the miles of marching bands and mounted riders, pipers and motorcyclists, Scout troops and vehicles ranging from a Wells Fargo stagecoach to official limousines wound at last to their goal.

It was about 10 AM when Strauss arrived at the reviewing stand, a slight figure in a dark suit emerging from a long black car. There was a man on either side of him, as if to hold him up. From the stands there rose, reported the *Chronicle*, "a mighty ovation . . . for, as chief engineer, the task of planning the span was his." Standing here at last, perhaps two miles from the shoreside spot where he had built his Aeroscope twenty-two years before, Strauss was at last the man he had wished, then willed himself to be, the poet in steel, whose greatest work was now etched on the canvas stretching above and behind him.

Strauss was introduced, then took a place in the reviewing stand. There was no speech. Instead, Strauss had synthesized his thoughts about his bridge and himself in a commemorative ode, published the day before, along with his portrait, in the paper that had supported him the most faithfully, the San Francisco *News*.

The bridge, Strauss felt, had professionally and personally vindicated him.

> LAUNCHED 'MIDST A THOUSAND HOPES AND FEARS,
> DAMNED BY A THOUSAND HOSTILE SEERS,
> YET NE'ER ITS COURSE WAS STAYED;

"At last," Strauss wrote, as if summing up the spirit of his entire life, "the mighty task is done."

By early afternoon the crowd on the bridge, swollen by parade arrivals, had grown so large that the toll machines could no longer click people through the turnstiles a nickel at a time. Instead, pedestrians simply tossed a nickel into a bucket and strolled out onto the span. Under crowd pressure, some people on the bridge began displaying crowd behavior, crowd antics, crowd manners. Teenage boys began snatching people's hats and sailing them over the sides of the bridge. On the railings and on the towers, people scratched and gouged graffiti: initials, the date, pledges of affection. A few youngsters shinnied up the suspender ropes or climbed up on the cables, where they were chased down by police. About fifty people broke into the north and south towers, where they tried out the elevators, climbed the ladders, and got lost in the interior maze of cells. They also looted the towers of any and all loose nuts, bolts, and rivets. Souvenir hunters also snipped buttons off the uniform coats of two bridge guards. So much debris was dropped, jostled, or blown away that by the end of the day the safety net was littered with personal items: purses, hats, shoes, cameras. Two Marin County boys, short of the nickel pedestrian toll, scrambled onto the suspended structure girders and cross-beams and made their way into the safety net. They were about two hundred feet out and about to start trampolining above the water of the Gate when they were spotted by people walking on the roadway. Ordered out of the nets by the State Highway Patrol, they scrambled back the way they'd come without getting caught.

Few of the men who had worked on the bridge came out on opening day; most of them felt they had seen enough of the structure without having to fight crowds for another look at it. One of the men who did was Harold McClain. "I was in the parade, and I walked across the bridge. It was never just a job to me. I loved the work."

At about two o'clock the wind that rises in the Gate almost every afternoon began blowing through the bridge's great harp of towers, suspenders, and cables. In the absence of the usual steady burr of automobile traffic, another sound could be heard, and it brought the shuffling and larking of the crowd to a momentary halt. A man shouted for quiet and, holding up his hand, urged the people around him to listen. The hush spread. "Away down deep," reported the *Chronicle*, "there is a deep roar, like the bass notes of a piano. High up in the wires is a shrill sound that some gigantic violoncello might produce." From the towers came "a deep organ-like note, a series of different tones, changing, deepening, rising.

"They all blended into a splendid diapason, these different sounds, and those still crowds stood awe-stricken by one of the strangest symphonies the ear of man has ever heard."

4

The bridge's first day had belonged to people. Some 200,000 of them had walked across on Pedestrians Day. On its second, it was to be taken over by vehicles. At night, the bridge had been bathed in the glow of exploding rockets and falling flares; crowds had filled the grandstands at Crissy Field to watch a historical pageant on a specially built Redwood Grove Stage, featuring the songs of the baritone John Charles Thomas, a cast of 3,000, and a 100-piece orchestra. Then, with the bridge only a vague outline in the dusk, Irving Morrow's illumination gradually defined it: first the navigation lights, marking the channel; then the red airplane beacons atop the towers; then, to a deep, involuntary "oh" from the crowd, a ruddy band of sodium vapor, bridging in light what had already been bridged in steel, "strung," exulted the *Examiner*, "like a topaz necklace across the Golden Gate." Even at night, this was to be a bridge unlike any other.

The following day's official opening ceremonies began at the Marin entrance to the bridge, where, starting at 9:30 AM, Strauss, Mayor Rossi, California Governor Frank Merriam, and the governors of eleven other Western states, along with representatives of Canada and Mexico, gathered to watch three champion woodcutters race their way through a redwood log set across the roadway.

The party then moved to the Marin tower, which also marked the county line, where a chain symbolized the entrance to the bridge's main span. Frank Doyle, the Sonoma County banker who had called the first meeting of the Bridging the Golden Gate Association in Santa Rosa in 1923, was given the honor of holding the acetylene torch that cut the chain.

The official party now strolled across the span to the toll plaza at the San Francisco side and the last symbolic barrier: a line of flower-decked fiesta queens, representing all corners of California. They stepped aside as Strauss, apparently all but overcome with emotion, formally delivered the bridge to District President Filmer. His hands trembling, his voice barely audible, Strauss addressed the crowd with uncharacteristic diffidence:

"This bridge needs neither praise, eulogy, nor encomium. It speaks for itself."

At noon—3 PM Washington time, President Roosevelt pressed a telegraph key in the White House officially opening the bridge and setting off an enormous local din of sirens, bells, whistles, cannon, water spouts, car and fog horns. At the same time, the cacophony around the bay was joined by a low and gathering noise from the ocean. To the west, the sky darkened with droning squadrons of navy biplanes. Launched from the carriers *Lexington, Ranger,* and *Saratoga,* unseen and still at sea, some four hundred navy fighters, joined by seaplanes from the tenders *Wright* and *Langley* and

Pedestrians' Day: the squadron of tiny biplanes before the giant stepped-off structure, the crowd of tiny figures below . . . it suggests the climax of *King Kong.*

land-based patrol planes, headed in a mass flight like a thunderstorm of machines over the bridge.

Onto the span, led by an escort of Highway Patrol motorcycles, sirens screaming, rolled flag-flying official limousines, police cars, fire trucks, and military vehicles, forming, with the mass overflight of planes, a concentration of machinery likely to test the limits of Charles Ellis's estimate of the bridge's live-load capacity. Photographs from this time, with the tiny, swarming planes outlined against the great, dark Art Deco towers of the bridge, the people and vehicles on the span reduced to teeming figures far below, suggest the climax of the original *King Kong,* with Kong trapped on top of the Empire State Building, fighting off, like attacking mosquitoes, a squadron of strafing army planes. Only this was real. And it was bigger.

That afternoon, the largest fleet of American capital ships ever concentrated in a single harbor in peacetime sailed beneath the bridge into San Francisco Bay. In a demonstration designed, in part, to quell any remaining fears about the bridge as a defense or navigation hazard, some nineteen battleships and heavy cruisers, led by the U.S.S. *Pennsylvania,* steamed in formation beneath the great, striping shadow of the bridge, accompanied by twenty-three other major vessels, including all three of the country's operating aircraft carriers. It was a concentration of American naval strength that the attack on Pearl Harbor—only four years off—would make inconceivable in the future. And it fixed the bridge in the nation's imagination as the Pacific Ocean entrance to America.

As though riding the momentum of all the years of lobbying and litigation, the fiesta rambled on for five more days. There were parties and promotions, dances and contests, ceremonies and band concerts and appearances by movie stars. It was early June before the vast celebratory tide of planes and ships and dignitary-bearing limousines finally receded, leaving only the bridge behind. There was, surprisingly, no emotional letdown. The bridge, in and of itself, offered the senses satisfaction enough. It still does.

Most of its statistical achievements—longest, largest, fastest-spun— have been surpassed. Other structures, more difficult and challenging to engineers, have since been built. Yet this one retains its deep and powerful hold upon us. It is believed to be the most-photographed manmade structure in the world.

There is something beyond mere admiration in this, a suggestion of the possibility that mankind at its most practical, most ambitious, and most technologically advanced need not inevitably intrude upon and destroy nature but might complement, or even—as had happened here—enhance it.

In its natural state, before it was ever bridged, the Golden Gate was one of the inspiring sights of all the world. Artists painted its climactic convergence of mountains, bay, and ocean. Photographers took pictures of it, writers at least attempted poetry about it. Documentary evidence remains of the dramatic beauty of the unbridged Gate. It was a stirring sight, and these earlier pictures still move us emotionally. Yet they provoke no nostalgia, no sense of loss. There is no urge to return to a time before the bridge, nor even an itch to alter or improve upon its design. The feeling we are left with, upon seeing the unbridged Golden Gate, is not regret, but incompleteness, as if all this, beautiful as it was, was prelude, in waiting for the bridge itself to satisfy our hunger for fulfillment. To furnish the body of the work.

Of what other manmade structure, imposed upon so naturally beautiful a setting, can such a thing be said?

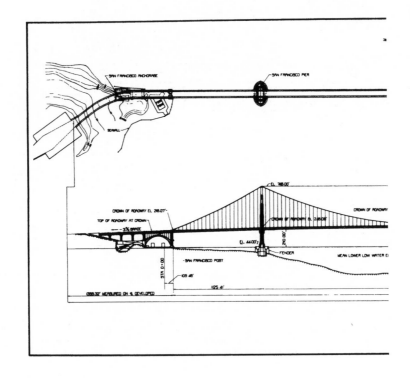

THREE

ENDURING

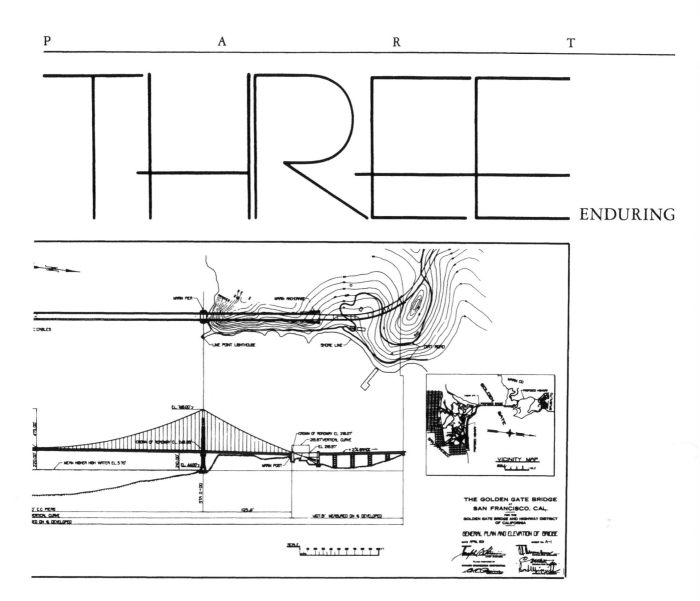

CHAPTER

"YOU SEE THE POSITION IN
WHICH HE IS NOW PLACED."

I

Finding themselves routed on the grounds of construction costs, politics, defense, navigation, aesthetics, and safety, the opponents of the bridge were reduced to taking consolation from the fact that, at first, the enterprise operated at a loss. The counties to the north, particularly Marin, were as yet underpopulated, and the bridge toll—fifty cents—still represented the price of a meal in 1937. In a last, desperate attempt to compete, the Southern Pacific–owned Golden Gate Ferries had cut its rates, making travel by boat between Marin and San Francisco considerably cheaper than on the bridge. Also, a significant number of vehicles were passing over the bridge toll free: as part of the price of its concession of land on either side of the strait, the army had wrung the right for all military personnel to cross without paying. Since there were army installations at both ends of the bridge, plus Hamilton Field and the Mare Island Navy Yard farther to the north, this amounted to an estimated monthly traffic of 35,000 vehicles that were producing no bridge revenue whatsoever. In addition, the rule about confining this privilege to military personnel who were actually on duty was all but impossible to enforce. Not surprisingly during these early, lean years of its operation, there was more than one attempt to turn the bridge over to the State Highway Department without waiting for the bonds to be paid off. Under these proposals, the tolls would be eliminated entirely and the bridge supported by the state gas tax fund. The Highway Department, however, found the proposition unattractive and turned it down, rejecting what would eventually prove to be an extremely lucrative source of revenue.

"Even the bridge people didn't think then that the bridge was ever going to make money," recalls Alfred Finnila. A former engineering student of Derleth's at Berkeley, Finnila had started as a riveting inspector on the suspended span. Now he found himself designing a parking lot and public restrooms and assisting on the design for a permanent traveling scaffold for painting.

"People didn't realize how much painting the bridge was going to require. We designed the scaffold on a tight budget. At one time, I was the youngest registered engineer in the state."

Eventually, there would be painters who would spend their entire careers working only on the bridge, traveling back and forth on the span and the cables and up and down on the suspenders and towers, measuring out lives in tens of thousands of gallons of International Orange.

Not all the workmen remaining on the bridge were painters. Frenchy Gales, for one, was kept on as part of the cleanup crew. "It was cumshaw," admits Gales, "for having been around so long." Gales and a partner were put to work hauling surplus items to a warehouse in a one-ton pickup truck.

"Somebody said it would be good to have a drink. I said I knew where to get some booze." Gales helped himself to a couple of rolls of cable wire, drove them out to a junkman in Marin, and sold them for $15. "I gave five bucks to the truck driver, who didn't drink, and bought two quarts of Old Grandad." Gales and the rest of the crew began drinking as they traveled back and forth between the bridge and the warehouse. "We took this big painting of the bridge. *Beautiful* painting." This must have been the original Maynard Dixon oil that had been used in promoting the bond issue in 1930. "One of the guys fell through the painting," Gales recalls. "We pushed it up against the wall, behind a lot of other stuff. We kept drinking. Some girls from the office came down. We went to their apartment. I stayed out all night.

"The next day my wife bawled me out. I said I didn't remember, which was bullshit—you always remember. I quit drinking then. That was 1937. I lost all my friends."

The painting has since been restored. It now hangs in the foyer of the Bridge District offices at the San Francisco toll plaza.

In July 1937, the first member of the management team that had built the bridge submitted his resignation. In quitting as general manager, James Reed announced that he was leaving "in justice to my family and my own future." He had anticipated that with the completion of construction and a tight operating budget he would be asked to take a smaller paycheck. Instead, he became president of a local steel-fabricating company.

Cone immediately stepped up his campaign to replace Reed as general manager, applying directly to the Bridge District Board of Directors. The Directors refused to act on the matter immediately, however, and it was not long before Cone realized that he had competition. Meanwhile Cone, who was also at this time helping Strauss prepare his final engineer's report, was appointed to the new job of maintenance engineer, at a salary of $7,500, or $2,500 less than he had been earning as resident engineer.

With the conclusion of construction, the nature of the bridge had changed from a work-in-progress to a political and financial entity. The directors, up to this point, had been forced to depend, to a degree that was sometimes uncomfortable, on the experience and knowledge of professional engineers. Now, some of the directors at least saw the vacant general manager's office as an opportunity to bring the management of the bridge in line with this new reality.

The man they proposed for the job, William H. Harrelson, was a former contractor, San Francisco supervisor, and, until recently, Bank of America vice-president in charge of administering bank-owned real estate. Harrelson, recently turned sixty-five and required to retire from the bank, had formally applied for Reed's job. The possibility of his appointment had been mentioned in the announcement of Reed's resignation.

Cone's supporters on the board, led by O'Brien, the Ukiah newspaper publisher, were indignant at having a sixty-five-year-old retired banker "thrust down our throats. We should take time," O'Brien insisted, "to examine qualifications before we appoint a man to manage this madhouse."

The bridge directorate immediately split into two camps, one favoring Harrelson and one Cone. Director Warren Shannon proposed that the maintenance engineer's and general manager's jobs be combined and given to Cone, thus saving an entire salary. When a ballot was taken at the July board meeting, Harrelson was one vote shy of confirmation, and the board fell into a deadlock, with ominous rumors of political infighting.

"The hue and cry against the appointment of Harrelson," observed Earl Behrens, the *Chronicle*'s "squire" of California political commentary, "came from a group of bridge directors closely aligned politically with former Sheriff Tom Finn and his organization."

For the next two months, Cone campaigned for the general manager's job, collecting endorsements from Moisseiff and Ammann as well as Frank Masters and other former employers. In the confusion, with nobody officially in charge of the bridge, concern arose about the care of the structure itself. "At present, the bridge is not being properly maintained," Strauss warned in September 1937. "It is being defaced and neglected. Work that should be done is not being done."

Strauss was knee deep in troubles of his own. In June the Board of Income Tax Appeals in Washington had ruled that Strauss and his wife were

$29,827 deficient in their income taxes for 1933 and 1934. Strauss had protested an earlier ruling that he owed nearly $60,000 in back taxes on his income from the Bridge District.

"I understand," confided Reed, eager to believe the worst of Strauss, in a gossipy letter to Ellis, "that some of this difficulty is of his own lawyer's making."

When the federal tax authorities had originally gone after Strauss for his 1933–34 taxes, Strauss's lawyer had pleaded that Strauss should be exempt on the grounds that he was an official of the Bridge District and therefore a state employee. However, in his earlier, and successful, suit against the district for the cost of materials inspection, Strauss's lawyer had established that Strauss was a contractor, serving the district for a fee, and that the additional inspection costs were "extras" under that contract, for which Strauss should be reimbursed.

"You see the position in which he is now placed," Reed wrote Ellis. "He attempted to evade the 1933–34 income taxes on the grounds that he was *not* a contractor, whereas in winning the 1936 suit for extras, he did so on the basis that he *was* a contractor.

"You can easily figure out," said Reed, savoring the thought of Strauss squirming, "that after paying for lawyers and the income taxes, which the government has now ruled that he must pay, it will probably cost him, all told, several hundred thousand dollars."

Strauss was indeed feeling beset on all sides: hounded by the federal tax authorities, his health shaky, his firm without prospects of another major bridge job. It seemed, he told a reporter from the *Examiner* in October, that he had built the Golden Gate Bridge "for nothing."

According to the *Examiner,* of the $1 million fee paid Strauss for designing and supervising the construction of the bridge, plus many thousands of dollars for extra work in designing approaches, supervising inspection, and making special engineering studies, virtually nothing remained.

Strauss blamed his financial difficulties on the delays caused by disputes over the legality of the bridge district. Added to this were the lawsuits by Meyers, seeking—and getting—a share of his fee, and, now, the government stepping in to claim a larger share of his income than he had estimated was owed.

"It is true," Strauss conceded, "that because my work extended over eight years instead of four years as I had anticipated I have practically no tangible financial gain to show for it."

He had, however, gained appreciably, in the public eye at least, in both authority and stature. Strauss's complaint that his bridge was not being properly looked after forced a resolution of the management dispute within the Board of Directors. At its September meeting, the board appointed Harrelson as general manager and confirmed Cone as the bridge's engineer, a compromise that resolved the immediate question of the bridge's maintenance, even as it promised to prolong the differences surrounding its management.

Strauss, having found little support for another bridge on San Francisco Bay, was now concentrating on coping with his tax problems, while overseeing the preparation of his final engineer's report. "I have been compelled to slow up very materially since the bridge was completed," he wrote Ammann on December 29, "and the doctors insist that I must take an extended rest."

By the first week of February, the report, eventually a bound volume of some 250 pages, including photographs and endpaper engineering drawings, was completed, and Strauss and his wife left for a two-month recuperative rest in Tucson. "I feel certain that when I come back," he wrote Ammann, "I will be okeh in every respect."

2

A few days later, on February 9, 1938, a storm bearing gale winds tore through the Gate, striking the bridge with gusts estimated as up to seventy-eight miles an hour. At 1 PM, with the wind roaring directly broadside to the bridge, Cone drove out on the roadway to examine firsthand the effect of such high winds on the span. The force of the wind was so strong that it was impossible to stand erect on the sidewalk or on the roadway.

"I drove to the San Francisco tower in a closed car," Cone later reported, "and was able to open the door on the leeward side and get out on the roadway." Only by crouching and standing in the lee of the west leg of the San Francisco tower was Cone able to cross the roadway.

Standing by the windward, or ocean, leg of the tower, Cone sighted along one of the offsets in the tower to the Marin shore, and saw "that the center of the bridge was deflected between eight and ten feet from its normal position, and was holding this deflected position." Cone also observed that the suspended structure of the bridge was "undulating vertically in a wave-like motion." These undulations were in the neighborhood of twenty to thirty vibrations per minute.

Because of the rapidity of the waves, Cone could only estimate their amplitude, "but it appeared to me that the stiffening truss was being disrupted as much as two feet vertically in 300 feet of bridge." The motion, Cone described as "a running wave similar to that made by cracking a whip."

To verify what he'd seen, Cone flagged down a passing pickup truck driven by a bridge electrician, F. L. Pinkham. Pinkham stopped his truck and tried to climb over the curb, but the wind blew him back over the curb and onto the roadway.

"He crawled over to where I was standing in the protection of the tower," Cone reported, "and I asked him to observe the movements and actions of the bridge, telling him I wanted a witness to substantiate what I had seen, since the oscillations and deflections of the bridge were so pronounced that they would seem unbelievable." Cone and Pinkham stood together for some time, observing the bridge.

Cone's report, submitted to the board, caused alarm among the directors and reinforced feelings among the board both for and against him. To Cone's supporters, like O'Brien and Shannon, Cone's candor arose from his dedication to the highest engineering standards and demonstrated the value of having a professional bridge engineer available for continuing study of the bridge. To the directors who had opposed Cone, however, the report, with its suggestion that the bridge's behavior in high winds required extensive investigation and perhaps structural modifications, smacked of opportunism and disloyalty. The report was kept secret, although it did not remain so for very long.

On March 28, as Strauss and his wife were returning to San Francisco from Arizona, Strauss suffered a coronary thrombosis. Upon his dismissal from the hospital, Strauss remained under a doctor's care at Annette Strauss's apartment in Los Angeles. The chief engineer's physical reserves, never strong, had at last been overdrawn. While he remained confined in Los Angeles, his two sons, Richard, a vice-president of Strauss & Paine in Chicago, and Ralph, an army lieutenant stationed at Fort Sill, Oklahoma, came to his bedside. On May 16, eleven days short of the first anniversary of the opening of his great bridge, Joseph Strauss died. He was sixty-eight.

The following day, tributes to Strauss's memory were made by California Governor Frank Merriam, San Francisco's Mayor Rossi, and by the San Francisco supervisors, who, after appointing a delegation to attend his funeral, adjourned "out of respect for his memory." Supervisor Warron Shannon, now president of the Golden Gate Bridge District Board of Directors, declared that "the entire credit for the success of the bridge belongs to him." The hagiography of Joseph Strauss had begun.

Strauss had the enviable fate of living a full and active life, yet dying at the summit of his reputation. There would be no further Strauss bridge required to suffer comparison to the Golden Gate, no further testing of Strauss's real abilities in the field of suspension bridge design; nor would there be a return to the less dramatic but more typical and sustaining work of turning out scores of easily reproduced rail and highway bascule bridges. These were all forgotten now, as were the associations with politicians and bagmen, the bitter quarrels with opponents and associates, the theft of credit for the work of others, particularly Ellis's, and even the original, ugly, and Strauss-designed Golden Gate Bridge.

Strauss's greatest conception, the idea that required his greatest application and his most arduous labors, was not the bridge itself but the Bridge District. Never before had a civic body been organized purely for the purpose of building a bridge, and it was Strauss's tireless campaigning, his speeches, press conferences, and official testimony, both to sell the idea and then to defend it, that established what has come to be a powerful and independent political body. No other bridge builder of stature, certainly not an Ammann or a Moisseiff or a Modjeski, could have sustained such an organizational effort, spread out over so many years, in the remote northern counties of California; but then stature, above all, was what Joseph Strauss had sought.

The district has remained faithful, to a fault, to Strauss's memory. Its official line has been that Strauss, with perhaps the technical assistance of Clifford Paine, who served as a consultant to the district into the 1970s, was the engineer who designed the bridge. In this there seems to be not so much a desire to mislead as the wish to avoid complexity: to simplify, and thus make more propagatable, the story of one man, Joseph Strauss, and his successful fight to build a great and beautiful bridge. He has become the classic Man Who Did What They Said Couldn't Be Done. The fact that, in this simplification, Ellis's unique conception has been completely obliterated, and Morrow's less profound but still significant architectural treatment largely obscured, becomes a regrettable but necessary concession to what is considered the greater good of satisfying explanation, happy endings, and heroes. Historians do not lie, V. S. Naipaul has informed us, they *elide*.

That the Golden Gate would have been bridged ultimately without Joseph Strauss there can be no doubt; but he was the only man with acceptable credentials who could have justified building it at the time it was built, within the budget that the voters accepted, and who could have brought the people he did to work on its design and construction. In this, perhaps, is his most enduring achievement.

Although some of the individuals Strauss had brought together in designing and building the bridge went on to other work of excellence and even eminence, there was never again an association with a project of the transcendent quality of that at the Golden Gate—a structure that reaches beyond engineering to the evocative realm of art.

O. H. Ammann, the most prominent of the engineers on the bridge consulting board, probably came the closest to duplicating the monumental achievement at the Golden Gate. Retiring as chief engineer of the Port of New York Authority in 1939, after directing planning and construction of the Triborough and Bronx-Whitestone bridges, Ammann operated his own consulting offices, specializing in bridge design, for the next seven years.

In 1946, at the age of sixty-seven, Ammann, in partnership with Charles S. Whitney, a specialist in reinforced concrete design, established the New York consulting engineering firm of Ammann & Whitney. By 1964, the firm had a staff of close to five hundred, and offices in Washington, Philadelphia, Paris, Athens, Teheran, Lahore, and Addis Ababa, and its projects included the lower level expansion of the George Washington Bridge, planning and supervision of the Walt Whitman Bridge in Philadelphia, the Throgs Neck Bridge in New York, Dulles International Airport, and structural design for three major buildings at Lincoln Center. The firm was also involved in projects in France, Spain, and Germany and is today one of the largest engineering organizations in the world, with responsibilities both in the United States and abroad, including the recent replacement of the original suspender ropes on the Golden Gate Bridge.

In 1964, the firm completed what Ammann regarded as his own greatest engineering work, the Verrazano Narrows Bridge across New York harbor between Brooklyn and Staten Island. Supported by four cables, with a suspension span sixty feet longer than the Golden Gate Bridge and a suspended load 75 percent greater, the Verrazano Narrows was the longest, heaviest suspension span in the world. Ammann, who believed the bridge represented both a technical and aesthetic milestone in long-span construction, died the year after it was completed. He was eighty-five.

Leon Moisseiff, who had suffered a heart attack in 1935, resumed his career as a consultant on long-span bridges, participating in the engineering design of the Triborough, East River, and Bronx-Whitestone bridges, the Tacoma Narrows Bridge, and the Mackinac Straits Bridge. His theories on the feasibility of lightweight, long-span construction in high-wind areas were taken to an extreme at the Tacoma Narrows, where, inspired by the success of the Golden Gate Bridge, cost-conscious builders eliminated the cross-bracing from the bridge towers not only above the roadbed but below it, as well as dispensing with the roadbed's stiffening truss, an economizing that was to prove disastrous and would call into question the safety of the design at the Golden Gate. Moisseiff, who continued to publish influential papers on bridge structural design, died in New York on September 3, 1943, at the age of seventy-one.

Charles Derleth, Jr., the Berkeley engineering dean who had been the bridge directors' man on the Engineering Board and who had been privy to the professional agonies endured by Ellis, had conflicting feelings about his experiences with Joseph Strauss. While expressing gratitude to Strauss for the opportunity of working with an engineer of his reputation on so monumental a project as the Golden Gate Bridge, Derleth also confided to Ammann and Moisseiff, not long after the bridge was opened, that "it is a regret that so many times in the past seven years it has been necessary for me to differ with the Chief Engineer."

Derleth remained chairman of the Department of Engineering at Berkeley until his retirement, on his seventieth birthday, in October 1944, with the title of professor emeritus. He continued to serve as a consultant on local and regional projects, including the Richmond-San Rafael Bridge, which was completed in 1957. Derleth died, at eighty-one, in 1956.

After Strauss's death, the management of the firm he founded was taken over by Clifford Paine. The name was changed to "Clifford E. Paine & Associates," with principal offices in Chicago, and the associates included Richard Strauss, Charles Clarahan, Dwight Wetherell, "and others of the engineering staff which Mr. Paine has headed for many years." While the organization continued to produce the highway and railroad bridge designs with which Strauss had been earlier identified, the stature achieved by Strauss's promotional drive at the Golden Gate could not be sustained. Clifford Paine & Associates would never serve as prime contractor on the design and construction of a major long-span bridge, and Paine's most prominent association remained that with the Golden Gate Bridge, where he was brought in from time to time by the directors as a consultant.

In June 1941, in the wake of the controversy over the Tacoma Narrows Bridge, Paine was requested by the president of the board to inspect the Golden Gate Bridge, which he did, finding it to be in "excellent condition." In December 1951, after a storm with winds so severe that the bridge had been closed, for the first time, to automobile traffic, Paine, who had now relocated in Fennville, Michigan, assured the directors that the bridge needed only a few minor repairs. This finding was somewhat at odds with Paine's later proposal for a major structural reinforcement of the bridge, which was to take more than a year to complete and cost more than $3 million.

In 1961 Paine, who through simple survival among other factors had taken on the aura of the original design engineer of the bridge, opposed the adaptation of the Golden Gate to rapid transit, insisting that "operation of rapid transit trains over the Golden Gate Bridge is not possible." Despite reports to the contrary by other engineers, Paine's moral authority as original design engineer carried the day, and the extension of the Bay Area Rapid Transit District to Marin was effectively barred. Over the years, the significance of Paine's role in the original conception of the bridge was to grow and grow with the fading of individual memories and the death, one by one, of the members of the Engineering Board. At his death, in July 1983, in Fennville, at the age of ninety-five, Paine was described by the Chicago *Tribune* as "designer and supervising engineer of the 4,200-foot Golden Gate Bridge in San Francisco."

Irving Morrow, whose use of light and color in the architectural treatment of the bridge took it beyond a work of engineering and into the category of sculpture, never worked on another project of anywhere near this significance, and his contributions on the bridge gradually were subsumed into the Strauss legend. Following his success on the bridge, there was an attempt to include Morrow on the board of architects who were to devise plans for the 1939–40 Golden Gate International Exposition, but Morrow was rejected on the ground of not having any large buildings among his architectural credits. "This fact is no reflection on his ability as an architect," Maynard Dixon wrote angrily to the San Francisco *News,* "but rather on his ability as a self-promoter. He is doing a distinguished piece of work on the Golden Gate Bridge and it is plain for all to see. What he does possess to an unusual degree is the faculty of visualization, of imagining things on a grand scale. And if there is any faculty needed in the planning of a great exposition, imagination is it."

Morrow, who as a writer for architectural publications had stepped on some local toes by criticizing the San Francisco obsession with Victorian-style architecture and architects' timidity in the use of color, found his contributions to the world's fair confined to participating in the design for the Alameda-Contra Costa County Building on Treasure Island. His practice returned to the character it had earlier, mostly the design of residential and commercial buildings, and he continued to work in partnership with his wife. Morrow, who had developed a heart condition, died of a heart attack while on a San Francisco city bus on October 28, 1952. He was sixty-eight.

Whether Charles Alton Ellis ever visited the bridge that once existed only in his mind, stood on the suspended structure whose dimensions he had specified, looked up at the cables and towers whose loads and wind stresses he had first calculated in a study that would eventually amount to ten volumes of precomputer higher mathematics—we do not know. He did visit San Francisco during the bridge's construction, but soon afterward came the war years, with extensive restrictions on civilian travel.

Ellis remained on the engineering faculty at Purdue until 1947, when he retired at the age of seventy-two, professor emeritus of the Division of Structural Engineering. He returned to his home at Evanston, Illinois, and died in Evanston Hospital on August 22, 1949. He never denied being the design engineer of the Golden Gate Bridge and never boasted of it. The Chicago *Tribune*, in his obituary, identified him as the bridge's designer, but then the *Tribune* had said the same thing earlier of Strauss. And would say it later of Paine.

In a way, it really doesn't matter: the bridge remained Ellis's in a way that he could carry with him wherever he went. The fact that other people didn't realize it only made it more so. The bridge was, and is, more his than anybody else's.

"THE MAN WHO BUILT THE BRIDGE"

On November 7, 1940, in the most spectacular failure in suspension bridge history, the Tacoma Narrows Bridge collapsed. The bridge, known as Galloping Gertie to the men who had built it, had been open for only four months: "It swayed so much men actually got seasick working on it," recalls Lefty Underkoffler. "When we saw the caissons being floated in, and realized how small they were, we made bets how soon the bridge would collapse. In six months, you'd have had a jackpot."

The Tacoma Narrows' main span had failed after being subjected to dynamic oscillation due to strong winds—a wave effect similar to although more pronounced than what Cone had observed in high winds at the Golden Gate. There were other disturbing parallels. The Tacoma Narrows had been, after the Golden Gate and George Washington bridges, the world's third longest single-span suspension bridge. Its relatively narrow roadway made its ratio of length to width between its cables greater than that of any similar structure; the Golden Gate Bridge was second. The collapse prompted demands within engineering circles and among the general public for an investigation of dynamic oscillation caused by winds in suspension bridges. Under the leadership of the Department of Commerce, an Advisory Board for the Investigation of Suspension Bridges was formed from "a group of competent engineers having an interest in the problem." Among the engineers chosen were Cone, Ammann, and Moisseiff, with Cone named secretary of the engineering board.

The selection stirred an understandable uneasiness among the Golden Gate Bridge directors about Cone and his earlier report. Should Cone's observations be made public as part of the current investigation, it could appear that the directors had suppressed information of vital importance to public safety. It was as if Cone himself, and not any possible structural shortcomings at the bridge, represented the real danger.

The start of the investigation coincided with a drive to economize on the operations of the bridge, which was still not producing the earnings, or the toll reductions, that had been anticipated. Reviewing the figures for 1940, Congressman Welch, still a director although now spending almost all his time in Washington, called for fewer board meetings, thus reducing the monthly stipends plus travel and hotel expenses for directors. The grand juries of the Bridge District counties, Welch suggested, might be invited to investigate the district in the interest of suggesting cuts and lowering tolls.

To avoid this sort of outside investigation, the bridge directors appointed their own economy committee, with O'Brien, the Ukiah newspaper publisher, as chairman. In January 1941, O'Brien submitted his committee's recommendations. Still smarting from the board's choice of Harrelson as general manager, O'Brien, besides urging fewer meetings, revived the earlier idea of combining the manager's and the engineer's jobs, with Cone the obvious dual choice. When this proposal was rejected, O'Brien, in retaliation, turned against his own recommendation of fewer meetings, and the directors concluded their economy drive by adopting no economy measures whatsoever.

Provoked by this atmosphere of continual feuding and rejected economies, the San Francisco Grand Jury announced its own investigation of the bridge administration, and summoned O'Brien to testify. On the day O'Brien was scheduled to appear, Cone offered to drive him to the San Francisco City Hall. When O'Brien finished his testimony, the grand jury foreman invited Cone to be sworn. In response to jury questions, which dealt mostly with insurance on the bridge, Cone said that, because of recent cancellations following the Tacoma Narrows collapse, the "insurable value" policy on the bridge had become so expensive and its coverage so reduced that the Bridge District would be better off dropping it. By carrying its own insurance and eliminating the mostly unused annual maintenance reserve, Cone said that the toll could be reduced to forty cents on the present bridge income.

The following week, while Cone was in St. Louis delivering a report to the engineering committee investigating the Tacoma Narrows disaster, his grand jury testimony came under attack at a meeting of the Bridge District Board of Directors.

"He went behind our backs," complained Director W. D. Hadeler of San Francisco, "and presumed to show he knows more than the directors or the finance committee about operation of the bridge."

A motion was proposed that Cone be dismissed. It passed by a vote of nine to four. "You are making a grave mistake," warned O'Brien, one of the dissenting directors. "He's the only man who knows every inch of this bridge."

The firing of Cone, who had just won the 1940 construction engineering prize of the American Society of Civil Engineers, was condemned as "reprehensible" by the grand jury, which threatened to reopen its investigation. It also left the bridge management without the services of a licensed, full-time bridge engineer. Speaking for the directors, Hadeler denied that Cone's removal had anything to do with his grand jury testimony and insisted it was strictly a measure of the district's concern for economy.

In a formal statement, two days after his firing, Cone defused the situation, expressing "no resentment that a majority of the directors have decided to dispense with my services. "I have never received any salary unless I was giving valued service," Cone maintained, adding his hope that his dismissal might help produce a toll reduction.

The last of the engineers responsible for the design or construction of the Golden Gate Bridge had left its regular employment.

At the June directors meeting, where it was proposed that Cone be replaced by a part-time engineer, O'Brien went public with Cone's eighteen-month-old report on the bridge's behavior in high winds. "Ever since that time," said O'Brien, "I have struggled to obtain for the district the engineering supervision this startling bridge performance demanded." He was revealing the details now, O'Brien insisted, because it was his only chance to have the bridge looked after "before it may be too late."

The directors, in response, turned to Clifford Paine. After making an examination of the structure "from cable housings to cable saddles," Paine reported that the bridge "is in good condition and that maintenance is likewise in a satisfactory condition." Together with Derleth, Paine recommended that, after the appointment of a part-time engineer, further aerodynamic studies of the bridge be made. A board of consultants, presumably with both Paine and Derleth on it, would be called in for "special problems."

In July, H. L. Nishkian was appointed engineer in charge of maintenance on the bridge, on a part-time basis, at a salary of $1,500 a year. Early in 1942, under Nishkian, the bridge was equipped with instruments to measure wind conditions automatically and to permit observation of bridge movements. In December 1951, a windstorm that forced closure of the bridge to vehicles rocked and damaged the structure in a way strongly reminiscent of Cone's report. Paine, who had at first insisted that the structure was in "magnificent" condition, was eventually awarded the year-long job of reinforcing the bridge floor with some 5,000 tons of lateral bracing steel. It was late, and unacknowledged, vindication of Cone's earlier professional concern.

The Tacoma Narrows Bridge, completely reconstructed after years of exhaustive study by the combined best bridge-engineering minds in the United States, was reopened in 1949. With its graceful twin towers, cross-braced below the roadway but open and ladderlike above, its long, slightly convex suspended span, its light but strong stiffening truss, and the soar and swoop of its cables, it bears a startling resemblance to Charles Ellis's design for the Golden Gate. The Tacoma Narrows remains, at thirty-seven, the graceful and successful younger sister of the more famous beauty to the south.

Following his firing, in the spring of 1941, Russell Cone formed a small company of fellow engineers hoping to sell themselves in the expanding market for war work. While in Washington, Cone encountered officials of the Silas Mason Company, an old-line engineering firm and prime contractor on the Grand Coulee Dam. They offered Cone the job of assistant superintendent, building an ammunition-manufacturing plant in Louisiana. As the company expanded with additional war work, Cone advanced with it, building ordnance plants in Illinois and elsewhere. By the end of the war the Mason Company was a major defense contractor, with several thousand employees, and Cone was its general manager.

The company now entered the manufacture of nuclear warheads, with Cone supervising projects like the Pantex plant in Amarillo, Texas. He had moved out of engineering into management and high-level defense contract negotiations and was involved in both the Nevada atomic tests and the pioneering rocketry installations at Cape Canaveral.

By the mid-1950s, Cone was vice-president and general manager, operating out of Manhattan, when an opportunity arose to return to bridge engineering. A second span had been proposed on the eastern arm of San Francisco Bay, at Carquinez Straits, paralleling the original bridge designed by Derleth in the 1920s. Here was a chance for Cone to return to his professional origins, make a good deal of money, show up the people who had dismissed him, and achieve reunion with the best part of himself. In collaboration with several partners, the Mason Company entered the project as a contractor, bidding on the foundations for the second Carquinez Straits Bridge. It was a particularly difficult place to build, over a deep, swift channel, where firm footings proved scarce. The Mason Company won the bid, but the partnership went sour. Work fell far behind schedule, and Cone was forced to abandon his other activities and supervise the project. The job was completed at a loss of more than $3 million. Cone, approaching sixty, would be forced to start over.

There was an opening, in San Francisco, that seemed ideal. Engineering work on the Bay Area Rapid Transit District, the area's greatest project since the building of the bridges, was about to get under way. Cone was proposed as chief engineer. He was about to be named to the job when he suffered a heart attack and had to withdraw from consideration. His active construction life had come to an end, in the shadow of his greatest accomplishment. He died, after a subsequent attack, in January 1961.

On May 28, 1941, just a few weeks after Cone's dismissal, a monument to Joseph Strauss was dedicated at a spectacular viewpoint just below the bridge toll plaza. An army band and color guard were present, and speeches were made by Strauss's lawyer and officials of the bridge district. Annette Strauss unveiled the monument: on a concrete pedestal 27½ feet high, a larger-than-life bronze statue of Joseph Strauss stood staring out toward the Pacific, a rolled set of plans clutched in one hand; behind him, blocking out the bay, were two concrete towers, modeled after the stepped-off design of the bridge towers. The inscription on the pedestal read: "Idealism and Action United To Form the Creative Power of This Achievement."

The monument has since been removed. It blocked the view—and the bridge is monument enough. The Strauss statue remains, near its original location, resting on a more modest pedestal, surrounded by flower beds, and beside a somewhat more intriguing cutaway of a section of bridge cable.

On the pedestal is a plaque, with Strauss's name, birth and death dates, followed by an inscription:

The Man Who Built the Bridge
Here at the Golden Gate is the eternal rainbow that he conceived and set to
form. A promise indeed that the race of man shall endure unto the ages.
Chief Engineer of the Golden Gate Bridge, 1929–1937

The only unchallengeable statement is the last.

Ted Huggins took this photo. The bridgeman is standing on top of an airplane beacon, which is on top of a cable, on top of a saddle, on top of the Marin tower.

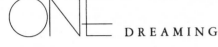
Chapter One: The First Bridge

17

"a man whom Albert Speer has revealed to us as a frustrated architect. . . ." Adolf Hitler's lifelong architectural yearning is one of the themes of Speer's memoir: Albert Speer, *Inside the Third Reich,* New York, Macmillan, 1970.

19

"Squat, bulky, unadorned": Factual information about the Fourth Street Bridge is from Judy Soto, Director's Office, Department of Public Works, City and County of San Francisco.

20

"the original City Hall took thirty years to complete": The most thorough account of the San Francisco earthquake and fire and the circumstances that surrounded it remains: William Bronson, *The Earth Shook, the Sky Burned,* Garden City, N.Y., Doubleday, 1969.

20

"At three o'clock in the afternoon of February 20, 1915, President Woodrow Wilson": Ben Macomber, *Panama-Pacific International Exposition: The Jewel City,* San Francisco: John H. Williams, 1915.

21

"Set in a 'City of Domes' spread over some three hundred landfill acres": John D. Barry, *The City of Domes,* San Francisco, John J. Newbegin, 1915.

21

"It was a happy time. The air was soft, and it was easy for a young man to believe in the inevitability of progress": As quoted in John P. Diggins, "The New Republic and Its Times," *The New Republic,* 70th anniversary issue, December 10, 1984.

21

"257 different engineering societies were scheduled to meet at the exposition": *Engineering News* 73, no. 7, February 18, 1915.

21

"Notable Engineering Works of the City Engineering Department of San Francisco": Ibid.

22

"Born in County Limerick, Ireland, on May 28, 1864, Michael Maurice O'Shaughnessy": *The National Cyclopedia of American Biography,* Vol. 24, New York, James T. White, 1941.

23

"Under the administration of a stooge mayor, Eugene Schmitz . . . and a wily political boss, Abe Ruef": The best account of the San Francisco of this era remains Lately Thomas's biography of Abe Ruef: *A Debonair Scoundrel,* New York, Holt, Rinehart and Winston, 1962.

23

"Public indignation . . . had led ultimately to the ouster and trial of both Schmitz and Ruef." Schmitz, known as "Handsome Gene," later ran successfully for the San Francisco Board of Supervisors. He was still around the city, a half-forgotten figure, in the 1920s. The author's father, then a young attorney, once served him with a subpoena.

24

"In the latter part of August, 1912, Mayor Rolph

wired me at San Diego": Michael M. O'Shaugh-nessy, *Hetch Hetchy, Its Origin and History,* San Fran-cisco, John J. Newbegin, 1934.

24

"He has not waited for the reports of his subordi-nates": San Francisco *Call,* April 22, 1912. From Michael O'Shaughnessy clip file, San Francisco *Ex-aminer.*

25

"the most significant structure of the project was named . . . after Michael O'Shaughnessy": San Francisco *Chronicle,* March 27, 1923.

26

"Hezekiah, . . . who stoppered the upper water-course of Gihon and brought it straight down . . . to the city of David": San Francisco *Ex-aminer,* July 27, 1919.

26

"he hiked through the territory he called Hobby-logland": Elfred Eddy, San Francisco *Call,* April 22, 1922.

26

"A chunky lump of a man": Ibid.

27

"Visitors rode to the fairgrounds on streetcars built in O'Shaughnessy's . . . shops": *Engineering News* 73, no. 7, February 18, 1915.

27

"the Aeroscope, the largest amusement ride at the fair": "The Bascule Bridge Turned Into an Amuse-ment Device": Ibid.

29

"Born in Cincinnati on January 7, 1870, Joseph Baermann Strauss": *Cyclopedia of American Biography,* vols. B, 27.

29

"we little children of the arts": From *Sherwood Anderson's Memoirs,* ed. Ray Lewis White, Chapel Hill, University of North Carolina Press, 1969.

30

"I recall distinctly the graduation exercises at the Old Pike Opera House": Alfred K. Nippert, "Joseph B. Strauss, Crusader—An Appreciation," from *The*

Record, SAE Fraternity Alumni magazine, undated, from Los Angeles *Times* Strauss file.

31

"Strauss was in fact neither a member of . . . nor a graduate of a formal college of engineering": Russ Cone, telephone conversations with Claire Young, Public Relations Office, University of Cincinnati.

31

"The University of Cincinnati did not establish its engineering college until 1901": Ibid.

31

"His degree was a liberal arts B.A., although he was awarded an honorary doctor of science degree in 1930": Ibid.

31

"He did . . . 'like to tinker in the machine shop a lot' ": Ibid.

32

"Strauss began his career in bridge engineering as a draftsman": Frank J. Taylor, "Strauss: Little Man Who Wanted To Build the Biggest Thing," Golden Gate Bridge Festival section, San Francisco *News,* May 26, 1937.

33

"He had also built a bridge—across the Neva River in Petrograd—that the Bolsheviks would storm over to seize the Tsar's Winter Palace": Jon Carroll, *California Magazine,* March 1981.

33

"In July of 1916 . . . Strauss . . . was physically attacked": Chicago *Tribune,* July 20, 1916.

34

"A man whose connections with the city engineer of Chicago, Alexander Murdock, would result in Murdock's firing": Chicago *Tribune,* May 19, 1923.

34

"On June 25, 1915, the Thomson Bridge Com-pany of San Francisco was awarded the contract for removing an old swing drawbridge": Soto, San Fran-cisco Department of Public Works.

Chapter Two: The Second Bridge

35

"It had become a kind of game with Michael O'-Shaughnessy": Taylor, San Francisco *News,* May 26, 1937.

35

"the city was the largest American metropolis still served primarily by ferry boats": *The Golden Gate Bridge at San Francisco, California,* Report of the Chief Engineer with Architectural Studies and Results of the Fact-Finding Investigation Conducted by the Board of Directors, sec. 3.

35

"San Francisco's expansion had fallen behind that of the city's own suburbs": *Engineer's Report,* sec. 3.

35

"Geologically, the Golden Gate is a gap in a mountain range": Andrew C. Lawson, "Report of the Consulting Geologist," *Engineer's Report,* sec. 5.

36

"The typical rough estimate that O'Shaughnessy received for the cost of constructing": Taylor, San Francisco *News,* May 26, 1937.

37

"Strauss . . . designed a number of portable searchlights . . . a disappearing tower . . . patented the first yielding traffic barrier": *Cyclopedia of American Biography,* vol. 27.

38

"When O'Shaughnessy suggested to Strauss that he address the problem of bridging the Gate": Taylor, San Francisco *News,* May 26, 1937.

38

"Strauss replied that he thought he could design and build the bridge": Allen Brown, *Golden Gate, Biography of a Bridge,* Garden City, N.Y.: Doubleday, 1965.

38

"The collaboration was to be an extension of that on the Fourth Street Bridge": Joseph B. Strauss and Michael M. O'Shaughnessy, *Bridging the Golden Gate,* San Francisco, privately printed, 1921.

39

"Federal experts believe it will be impossible": As quoted in Brown, *Golden Gate.*

39

"This information . . . O'Shaughnessy also sent to two other bridge builders." *The Golden Gate Bridge: Report of the Chief Engineer (1938).* sec. 3.

39

"According to the proposal prepared jointly in 1921 by Strauss and O'Shaughnessy": Strauss and O'Shaughnessy, *Bridging the Gate.*

42

"McMath never did, and it would be two years before Lindenthal would furnish his estimates": Stephen Cassady, *Spanning the Gate,* Mill Valley, Calif.: Squarebooks, Baron Wolman, 1979.

42

"In 1922 . . . Strauss turned up unexpectedly at a City Council meeting in Sausalito": Brown, *Golden Gate.*

42

"Strauss, whom Mayor Madden remembered as 'the world's worst speaker' ": Ibid.

43

"Strauss retained, as his 'public relations counsel,' Charles Duncan": Author's interview with Herbert T. Johnson, staff architect with Morrow & Morrow.

43

"the mayor of San Rafael . . . proclaimed that the bridge would 'open the gateway to thousands of homeseekers' ": Brown, *Golden Gate.*

43

"Frank Doyle called a mass meeting in the county seat of Santa Rosa": Ibid.

44

"It would be two and a half times larger than any similar bridge in the world": San Francisco *Chronicle,* March 18, 1923.

44

"I have conferred with the Secretary of War and several members of Congress": Ibid.

44

"San Francisco car owners . . . could drive clear to the Oregon border": Ibid.

45

"The Golden Gate could be bridged by 1927": Ibid., March 13, 1923.

45

" 'The engineering plans are fully solved,' he announced to the legislators": Brown, *Golden Gate*.

46

"O'Shaughnessy was included in some of the meetings and conferences . . . but only as a sort of junior partner": San Francisco *Journal*, San Francisco *Chronicle*, May 16, 1924.

46

" 'Dear San Francisco . . . in the interest of your own uniqueness, do not bridge the Golden Gate' ": Katherine Fullerton Gerould, *Harper's*, June 1924.

47

"Derleth . . . was known to have expressed the opinion that he considered an all-suspension bridge at the Golden Gate 'more feasible' than Strauss's combination design": Brown, *Golden Gate*.

47

"Derleth was born in New York City in 1877, graduated from Columbia": Biographical facts from the Papers of Charles E. Derleth, Jr., in the Water Resources Center Archives, University of California, Berkeley.

47

"Derleth was active in the Bohemian Club, whose Isle of Aves camp at the . . . Bohemian Grove was to prove extremely useful over the years": John van der Zee, *The Greatest Men's Party on Earth*, New York, Harcourt Brace, 1974.

47

"Richard J. Welch, who in November 1918 . . . had introduced the initial resolution in favor of building a bridge": San Francisco Board of Supervisors, resolution no 16,241 (new series), Nov. 12, 1918, in the Derleth Papers.

48

"On May 14, 1924, a more adventurous bridge proposal was submitted to the San Francisco supervisors": San Francisco *Chronicle*, May 16, 1924.

48

"Born in Parkman, Maine, in 1876, Ellis appears to have been an almost classic 'downeaster' ": *Purdue Engineer*, February 1938, June 1935.

49

"He was an active member of the American Society of Civil Engineers and contributed articles and papers to its journal and its meetings": *The Golden Gate Bridge: Chief Engineer's Report (1930)*, sec. 4.

49

"happiness cannot be found by merely seeking it. It is far more satisfactory than mere pleasure": *Purdue Engineer*, June 1935.

49

"He was, according to his former students, an inspiring and profoundly influential teacher": Author's interview with Professor Marion B. Scott of Purdue University (retired), July 23, 1984.

50

"Strauss made the most of Ellis's professional background and accomplishments": *Chief Engineer's Report* (1930); letters and telegrams from Strauss to Ammann and Moisseiff in Ammann & Whitney files.

50

"Although reserved, occasionally to the point of remoteness": Scott interview.

50

"He was what he was. . . . Competent, but never flouted it. Hard to convince, but could be won over": Ibid.

Chapter Three: "We will get this bridge, in the end."

52

" 'Engineer of Gate Bridge,' . . . City Engineer

M. M. O'Shaughnessy will assist Strauss": San Francisco *Journal,* San Francisco *Chronicle,* May 16, 1924.

52

"San Francisco has often done the impossible": Brown, *Golden Gate.*

53

"During the late War": Ibid.

53

"Some years earlier, O'Shaughnessy stated, he had been aboard the steamer *Alameda* when she had been wrecked": Ibid.

53

"In 1920, the city engineer had arranged to make a triumphant return visit to his mother's home in Limerick, Ireland": San Francisco *Examiner,* June 22, 1920.

54

"Even more gratifying was the ceremony, in May 1923": "Great Dam Named for O'Shaughnessy," San Francisco *Chronicle,* May 27, 1923.

54

"Among the critical eyes that O'Shaughnessy's string of successes had caught were those of William Randolph Hearst": "An Open Letter to Our City Engineers," San Francisco *Examiner,* December 10, 1923.

54

"There are indications that you and your assistants obstinately oppose municipal electricity": Ibid.

54

"Either to forget your opinions on matter of polity and finance": Ibid.

54

"O'Shaughnessy was defended by the other San Francisco papers": "O'Shaughnessy Is Not To Be Ruled," editorial, San Francisco *Bulletin,* December 10, 1923.

55

"In November 1924, a measure had been introduced . . . that would 'shear' O'Shaughnessy of his 'carte blanche' ": "Plan Made To Cut Power of City

Engineer," San Francisco *Examiner,* November 23, 1924.

55

"Strauss, perhaps sensing reservations about the appropriateness of his original design, made certain modifications in it": *Golden Gate Bridge, Perspective Conforming with U.S. Government Application,* Joseph B. Strauss, 1924, in the Derleth Papers.

56

"At last, on December 20, 1924, the idea of a bridge received its first nod of approval from Washington": "Weeks O.K. on Gate Span Received; Plan Speeded," San Francisco *Chronicle,* December 21, 1924.

57

"in March of 1925, the Chicago *Tribune* described Strauss as": "Wins Frisco Contract; Sells Chicago Home," Chicago *Tribune,* March 8, 1925.

57

"Strauss had arranged for Professor George F. Swain of Harvard . . . and Leon S. Moisseiff": "Experts To Aid on Gate Bridge," San Francisco *Examiner,* February 7, 1925.

57

"Strauss also requested Moisseiff to prepare a report on a comparative design for a suspension bridge": Moisseiff to Strauss, November 15, 1925, in the Golden Gate Bridge file, Ammann & Whitney, Inc., 2 World Trade Center, New York.

57

"Moisseiff's report . . . suggests a bridge similar in many ways to the one that was eventually built": Leon S. Moisseiff, *Report on a Comparative Design of a Stiffened Suspension Bridge Over the Golden Gate at San Francisco,"* New York, November 1925, in Ammann & Whitney Golden Gate Bridge file.

58

"Under the terms of the 1923 enabling act, the Northern California counties that agreed to form a bridge district would be able to float bond issues, take out loans, and charge bridge tolls" Brown, *Golden Gate.*

59

"The wheelhorse of the campaign was Joseph Strauss": Ibid.

59

"If you had followed me today you would be tired": "Strauss Has Busy Day on Bridge Plan," San Francisco *Chronicle*, March 15, 1923.

59

"You are bottled up now, with a wonderful undeveloped country lying to the north": Ibid.

59

"The supervisors of Mendocino County . . . voted to join the Bridge District on January 7, 1925": Brown, *Golden Gate*.

59

"Strauss gave a speech in Sausalito before . . . approximately 25 percent of the town's population": Ibid.

60

"Responding to the most emphatic, if not the most numerous, of their constituents, the Humboldt County Board of Supervisors declined": Ibid.

60

"When the Lake County Board of Supervisors received the petition . . . the board members promptly voted against it": Ibid.

61

"In San Francisco the Board of Supervisors . . . now put the district-favoring ordinance in committee": Ibid.

61

"Welch told the people that they had been 'slapped in the face' ": San Francisco *Examiner*, January 27, 1925.

61

"McLeran found himself, following his day in the sun as acting mayor, attacked editorially in the *Examiner*": San Francisco *Examiner* editorial, January 27, 1925.

62

"Frank L. Coombs . . . who had introduced the enabling legislation, now warned the San Francisco supervisors": Brown, *Golden Gate*.

62

"At a meeting in the city of Napa on April 5, 1925, a compromise was ironed out": Ibid.

62

"When, on August 24, tiny Del Norte County . . . voted in favor of the district": Ibid.

62

"In Mendocino County . . . there was now an intense reaction against the idea": Ibid.

63

"The Mendocino counterpetition . . . caused the licensing secretary of state . . . to refuse to certify": Ibid.

63

"they argued that . . . one's signature on a petition was like the casting of a ballot and could not be withdrawn": Ibid.

65

"the Joint Council of Engineering Societies held a general discussion of the Golden Gate Bridge project": *Engineering News-Record* 96, no. 12, March 25, 1926.

65

"The resolution . . . was a direct slap at Joseph Strauss, the outsider, the promoter-engineer": "Promotion Methods in Golden Gate Bridge Project Disapproved," Ibid.

66

"In the six counties that had agreed to form the district, hearings were now called for": "California Counties Debate Benefits of Golden Gate Bridge," *Engineering News-Record* 99, no. 22, December 1, 1927.

66

"In opposition to Strauss's figures, a group of Sonoma County taxpayers presented another report": Ibid.

66

"The proposed center span . . . 'is probably not practicable' ": Ibid.

67

"it would afford no saving in cost and would probably increase the total weight on foundations": Ibid.

67

"The actual cost of the bridge . . . the engineers estimated at $112,344,778": Ibid.

67

"Such a bridge . . . would start its long deficit period": Ibid.

68

"No competent engineer has said that the Golden Gate cannot be bridged": Editorial, ibid.

68

"It is natural . . . that such a bridge scheme . . . should be opposed by the engineers of San Francisco": Ibid.

69

"A 'mortification to the proud Golden State,' . . . 'a misfortune to the engineering profession' ": Ibid.

69

"No scheme for bridging the Golden Gate that meets all or any of these requirements has yet been proposed": Ibid.

69

"At the time Strauss had first proposed his design, the largest cable ever constructed was for the Manhattan Bridge over the East River in New York": Charles A. Ellis, Address to the National Academy of Scientists, Berkeley, California, September 18, 1930.

70

"Launched midst a thousand hopes and fears": From Joseph B. Strauss, "The Mighty Task Is Done," San Francisco News, Bridge Fiesta section, May 26, 1937.

70

"His marriage, strained by his absences and travel": Author's interview with Ruth Natusch.

71

"At the next feasibility hearing . . . Strauss brought in his own support troops": "Golden Gate Span Fight Opens Today," San Francisco Chronicle, February 15, 1928.

71

"an official-sounding firm calling itself the San Francisco Bureau of Governmental Research": "Gate Bridge Study Urged by Bureau," San Francisco Chronicle, February 22, 1928.

71

"it will not even be known whether the bridge is possible": Ibid.

71

"the supervisors appointed three of their own members to represent San Francisco on the bridge board": Brown, Golden Gate.

72

"What voter wouldn't be gun-shy": Ibid.

72

"Born in Poland, Hirschberg had emigrated to America at fourteen": Hirschberg obituary, San Francisco Chronicle, January 15, 1959.

73

"Havenner . . . resigned from the bridge board and introduced a resolution . . . calling on the other supervisors to do the same": Brown, Golden Gate.

73

"On January 16, 1929, an irked . . . assemblyman named R. R. Ingels introduced a bill": Ibid.

73

"Harlan . . . made 'a spectacularly successful presentation' in support of the bridge": Ibid.

Chapter Four: "I can have Professor Ellis adjust matters."

74

"in September of 1926 . . . O'Shaughnessy had been so stung . . . that he was contemplating resigning": "O'Shaughnessy Resignation Probable, Says Mayor Rolph," San Francisco Examiner, September 25, 1926.

74

"I thought it was a private conversation between

the Mayor and myself": "O'Shaughnessy Denies Intention To Quit City Post," ibid., September 26, 1926.

74

"Fremont Older, the respected San Francisco *Call* editor . . . had launched a campaign to get him out." Thomas, *A Debonair Scoundrel*, p. 392.

75

"O'Shaughnessy made a public statement that he had not been consulted in the choice": "O'Shaughnessy Ignored in Wilcox Hiring, He Declares," San Francisco *Chronicle*, October 2, 1927.

75

"A green grocer, a boilermaker and a retired military officer": "O'Shaughnessy Barbs Stir City Hall," San Francisco *Chronicle*, April 8, 1928.

75

"In 1929, however, O'Shaughnessy was still a man of enormous power and prestige in San Francisco": There is reason to believe that Dashiell Hammett, living and writing in San Francisco at this time, took the name of his manipulative villainess in *The Maltese Falcon*, Brigid O'Shaughnessy, from that of San Francisco's prominent city engineer.

76

"He had heard the rumors that Strauss had imported an advance agent, one H. H. Meyers": Meyers later sued Strauss for his fee, and Strauss settled.

77

"On April 10, 1929, Alan MacDonald . . . was appointed manager of the Golden Gate Bridge and Highway District": San Francisco *Chronicle*, April 11, 1929.

77

"The morning after the announcement of MacDonald's new job, Derleth wrote him a formal letter": Charles Derleth to Alan MacDonald, April 12, 1929, in the Derleth Papers.

78

"a number of other bridge builders . . . were approached by the district during the spring and summer of 1929": Cassady, *Spanning the Gate;* Brown, *Golden Gate.*

78

"According to Congressman Welch, the new survey would determine whether the two rock shoals . . . extended far enough": San Francisco *Chronicle,* April 23, 1929.

79

" 'a 13 years' war . . . a long and tortuous march'": Joseph Strauss, Golden Gate Bridge Dedication Speech, Fairmont Hotel, San Francisco, February 25, 1933, in the Derleth Papers.

80

"Welch and Filmer returned from New York riding a tide of lifted optimism and lowered estimates": San Francisco *Chronicle,* June 13, 1929.

80

"Among the engineers they had met with . . . were men handling 'the great New York tube and bridge projects' ": San Francisco *Examiner,* June 13, 1929.

80

"The new bridge over the Hudson River": This was the George Washington Bridge: Ibid.

80

"The directors approved a budget of $300,000 for preliminary work": Ibid., July 25, 1929.

80

"In February 1929 lawyers representing nearly one thousand taxpayers filed suit": San Francisco *Chronicle,* April 1, 1929.

80

"Derleth had been in contact with the Bridge District manager at least two or three times a week": Derleth, Golden Gate Bridge diary, in the Derleth Papers.

81

"O. H. Ammann, Swiss-born, was to the practical implementation of engineering design what Moisseiff was to theory": *National Cyclopedia of American Biography,* vols. D, 52.

81

"Derleth . . . was willing to work as an engineer on the bridge, in just about any capacity": Derleth to MacDonald, April 12, 1929.

82

"Strauss . . . had once sued the city of Chicago for infringing on one of his bridge patents": Chicago *Tribune,* November 27, 1924.

82

"In March of 1929, working through Ellis, Strauss approached Moisseiff and Ammann": Ammann to Strauss, March 15, 1929, in Ammann & Whitney Golden Gate Bridge file.

82

"Ammann was more reluctant": Ammann telegram to Strauss: "Unable to accept your offer," April 3, 1929, in Ammann & Whitney Golden Gate Bridge File.

82

"Ammann was personally determined not to be put in the position of competing with other engineers": Ibid.

83

"On March 21, Ammann received a letter from Filmer": Filmer to Ammann, March 21, 1929, in Ammann & Whitney Golden Gate Bridge file.

83

"confirm our verbal agreement that you will act in an advisory capacity to me": Strauss to Ammann, March 29, 1929, in Ammann & Whitney Golden Gate Bridge file.

83

"This is a new situation not before realized": Ammann telegram to Strauss, April 3, 1929.

83

"As I understand it you had already accepted offer": Strauss telegram to Ammann, April 3, 1929, in Ammann & Whitney Golden Gate Bridge file.

83

"I have already named you and Moisseiff": Ibid.

84

"I can have Professor Ellis come on and adjust

matters with you": Strauss telegram to Ammann, April 4, 1929 in Ammann & Whitney Golden Gate Bridge file.

84

"Ammann wrote to Strauss . . . explaining . . . his position": Ammann to Strauss, April 6, 1929, in Ammann & Whitney Golden Gate Bridge file.

84

"Ammann wrote Filmer": Ammann to Filmer, April 6, 1929, in Ammann & Whitney Golden Gate Bridge file.

84

"On August 13, Ammann received a telegram from Filmer, on behalf of the board": Filmer telegram to Ammann, August 13, 1929, in Ammann & Whitney Golden Gate Bridge file.

84

"On July 24, MacDonald made a tentative offer to Derleth": MacDonald telegram to Derleth, July 24, 1929, Derleth diary.

84

"The following week, at the Bohemian Grove, Derleth visited with Sproul and left a letter for MacDonald at his camp": Ibid.

84

"reservations concerning the appointment of Derleth": San Francisco *Examiner,* August 16, 1929.

85

"On August 7, Filmer made another 'tentative' offer": Derleth diary.

85

"one of 'four engineers of national . . . reputation' ": Oakland *Tribune,* August 15, 1929; San Francisco *Chronicle,* San Francisco *Examiner,* August 16, 1929.

85

"Derleth . . . wrote a diplomatic letter to Strauss": Derleth to Strauss, August 16, 1929, in the Derleth Papers.

85

"the terms of Strauss's contract were 'the lowest

ever written for a bridge job of such magnitude' ": San Francisco *Chronicle,* August 16, 1929.

86

"On August 20, Moisseiff . . . wired Strauss that he was withdrawing his acceptance": Moisseiff telegram to Strauss, August 20, 1929, In Ammann & Whitney Golden Gate Bridge file.

86

"Filmer wired both Moisseiff and Ammann, urging them to come to San Francisco anyway": Filmer telegram to Ammann, August 20, 1929, in Ammann & Whitney Golden Gate Bridge file.

86

"At Moisseiff's and Ammann's insistence, their contracts were modified to specify that they were being appointed by the Bridge District" Ammann to Moisseiff, November 9, 1929, in Ammann & Whitney Golden Gate Bridge file.

86

"O'Shaughnessy . . . was quoted as saying . . . in his opinion . . . its costs would be in excess of $100 million": San Francisco *Chronicle,* September 10, 1929.

86

"Knowing as I do the disposition of my friend": Ibid.

87

"In Marin County . . . the local county real estate board . . . passed a resolution": Ibid., September 15, 1929.

Chapter Five: The Third Bridge

89

"In one of those rare photographs": Photo of first meeting of the Engineering Board, from the Papers of Russell Cone (Russ Cone, Jr.).

90

"Now, some eleven days later the three consultants, along with Ellis . . . were making a three-day reconnaissance": San Francisco *Chronicle,* August 29, 1929.

91

"Born in Riga, Latvia, in 1872, Leon Solomon Moisseiff": *Who's Who in America; Who Was Who in American History* (Science and Technology).

92

"Moisseiff had developed a theory to distribute wind stresses": Leon S. Moisseiff and Frederick Lienhard, "Suspension Bridges Under the Action of Lateral Forces," American Society of Civil Engineers Proceedings, no. 1849, 1932.

92

"Moisseiff would begin by guessing what portion of the total wind load was carried by the truss": This is Ellis's explanation of Moisseiff's theory. Charles Ellis, address to National Academy of Scientists, Berkeley, Calif., September 18, 1930, in the Derleth Papers.

92

"While four different integrations were required . . . for the stiffening truss, only two were required for the cable": Ibid.

92

"According to this theory . . . as much as half of the pressures caused by winds could be absorbed by the main cables themselves": Ibid.

93

"Joseph Strauss was now willing to concede the desirability of building a suspension bridge at the Golden Gate": Ibid.

93

"The second choice, a suspension span with straight backstays . . . Strauss eliminated for two reasons": Ibid.

94

"If, however, as Moisseiff's new theories suggested, a suspension span of much greater length could safely be built at this site": Ibid.

95

"On August 28, Strauss, along with Moisseiff and Ammann, departed for the East": San Francisco *Chronicle.* August, 29, 1929.

95

"Meanwhile, Ellis would be his personal representative in San Francisco": Ibid.

95

"Conferring regularly with Derleth, the Berkeley dean": Derleth diary.

96

"On September 7, it was announced that . . . work crews under Ellis": San Francisco *Chronicle,* September 7, 1929.

96

"on September 11, Ellis . . . asked the directors . . . to call for bids": Ibid., September 12, 1929.

96

"Major General Charles Jadwin . . . stated emphatically . . . that he was opposed to the construction of the bridge": Ibid., October 19, 1929.

96

"Welch . . . sent a telegram to Filmer": Ibid.

97

"Calling upon California Senator Samuel J. Shortridge on October 15": Filmer telegram to R. H. Felt, October 15, 1929, in the Derleth Papers.

97

"Filmer wired Welch the results that afternoon": Filmer telegram to Welch, October 16, 1929, in the Derleth Papers.

97

"On October 23 their attorneys filed an appeal before the U.S. Supreme Court": San Francisco *Chronicle,* October 23, 1929.

97

"On November 25, 1929, the first diamond drill bit into the rocky soil": Ibid., November 26, 1929.

Chapter Six: "Mr. Strauss gave me some pencils . . ."

98

"on December 9, the day the test drillings . . . began . . . more than a thousand spectators": Ibid., December 9, 1929.

98

"On the shore at Fort Point": Ibid.

99

"Strauss's presence is intermittent": Derleth diary.

99

"The man who was there on the job every day was Ellis": Ibid.

100

"Lawson asked Ellis what would happen if one of the anchorages should instantly slip six inches": Ellis, Berkeley speech.

100

" 'Supposing I have a fish,' Ellis improvised in explanation": Ibid.

100

"Lawson . . . next asked Ellis what would happen if one of the main piers dropped instantly six inches": Ibid.

100

"Professor Lawson proposed . . . What would happen if one of the piers were to be moved sideways": Ibid.

101

"If I knew that there was to be an earthquake in San Francisco tomorrow": Ibid.

101

"On Feburary 3, 1930, Strauss proudly announced . . . that a rock formation so hard": San Francisco *Call-Bulletin,* February 4, 1930.

102

"Privately, the conclusions . . . were somewhat less enthusiastic . . . particularly those of Professor Lawson": A. C. Lawson, *Geology Report,* April 7, 1930, Berkeley, Calif., in the Derleth Papers.

102

"He had found and demonstrated to Lawson that the rock could sustain a load of 32 tons per square foot,": Ibid.

102

"It was the south *anchorage* of the bridge . . . that represented the greatest problem to Lawson": Ibid.

102

"tho it faces possible destruction": Ibid.

102

"More surely destructive . . . would be the fact that the life of a steel bridge near salt water is limited": Ibid.

103

"On March 7 . . . Lawson abruptly resigned": Derleth diary.

103

"Derleth . . . spent two hours talking to the geologist. . . . Then he called Ellis": Ibid.

103

"The following day, Derleth met with Lawson": Ibid.

103

"On March 3, the Supreme Court dismissed the appeal of the four counties' taxpayers": *Engineering News-Record* 104, no. 11, March 13, 1930.

103

"the finest kind of rock": San Francisco *Chronicle,* February 26, 1930.

103

"Strauss . . . is out of the city and will not return until Monday": Ibid.

104

"It was to be an all-suspension bridge, since that was the most economical kind to build": Ellis, Berkeley speech.

104

"In the interests of lightness and economy, the rail lines . . . would be dropped": Ibid.

104

"The roadbed would be made of concrete slabs": Ibid.

104

"At this point, Mr. Strauss gave me some pencils and a pad of paper": Ibid.

Chapter Seven: "Some very mysterious things . . ."

105

"This was something new. Moisseiff understood it, and Ellis, and that was about all": Derleth, Remarks before the Commonwealth Club of California, October 16, 1930, in the Derleth Papers.

105

"Ellis was now going to jump across it—go beyond the state of the art": Scott interview.

105

"According to Ellis, any young engineer who has mastered the fundamental theories . . . can design a girder or truss": Ellis, Berkeley speech.

105

"an engineer 'will find himself in totally undiscovered country' ": Ibid.

106

"some very mysterious things": Ibid.

106

"with a suspension bridge, this is reversed": Ibid.

106

"Another mystery concerns the stiffening truss": Ibid.

107

"Now it takes a genius to discover a new idea": Ibid.

107

"Ellis segmented the stiffening truss into parts": Ibid.

107

"What he was doing was devising new algebraic formulas (he came up with thirty-three in all), working with as few as six and as many as thirty unknown quantities": Brown, *Golden Gate.*

107

"In considering any type of bridge . . . the real test is whether it . . . can be shipped to the site and there be put together": Ellis, Berkeley speech.

108

"Ellis based his calculations on a live load of 4,000 pounds per linear foot": Ibid.

108

"allowances had to be made for bending, both lengthwise and sideways": Ibid.

108

"The towers would also have to bend sideways, because of wind load": Ibid.

109

"Ellis then considered what the stresses would be if all four of these conditions were combined": Ibid.

109

"In this design, however, we have assumed that all these four extreme conditions may happen simultaneously": Ibid.

109

"On March 1, Ellis had sent copies of his preliminary design specifications to the members of the Engineering Board": Ellis, Preliminary Design Specifications, Golden Gate Bridge, in Ammann and Whitney Golden Gate Bridge file.

109

"Before I return West . . . sufficient progress should have been made": Strauss to Ammann, March 6, 1930, in Ammann and Whitney Golden Gate Bridge file.

109

"What do you consider maximum allowable deflection": Ellis telegram to Moisseiff, March 28, 1930, in Ammann and Whitney Golden Gate Bridge file.

110

"Deflection side span six thousand": Moisseiff telegram to Ellis, March 28, 1930, in Ammann and Whitney Golden Gate Bridge file.

110

"The pressure was constantly being brought to bear to rush the work": Ellis to Ammann and Moisseiff, October 3, 1932, in Derleth Papers.

110

"I always explained . . . as clearly as I could that it was impossible": Ibid.

110

"Although I have made many sacrifices": Strauss to Moisseiff, April 16, 1930, in Ammann and Whitney Golden Gate Bridge file.

110

"an able chap, but this was, to the best of my knowledge, his first introduction": Ellis to Ammann and Moisseiff, October 3, 1932, in Derleth Papers.

110

" 'You or I . . . would not have to look at them ten seconds": Ibid.

111

"The meetings of the Engineering Board continued through the spring and summer of 1930": Meetings of the Engineering Board, in Ammann and Whitney Golden Gate Bridge file.

111

"the geologist was recommending that the pier base go down twenty-five feet below": Strauss to Ammann and Moisseiff, April 21, 1930, in Ammann and Whitney Golden Gate Bridge file.

111

"the work at Chicago is progressing at full speed . . . on or about the 20th of June": Strauss to Ammann, May 1, 1930, in Ammann and Whitney Golden Gate Bridge file.

113

"the Philadelphia-Camden towers were characterized by three great steel-cell cross-members": Alfred Bendiner, *Bendiner's Philadelphia*, New York, A.S. Barnes, 1964.

113

"these slender, simplified structures gave the bridge": *Pennsylvania, a Guide to the Keystone State*, WPA Writers' Program, New York, Oxford University Press, 1940.

113

"On the Ambassador Bridge at Detroit": Comparative drawing from *The George Washington Bridge Over the Hudson River at New York*, Bethlehem, Penn., McLintic-Marshall Corporation, 1932.

113

"Where the giant cross-members had been . . . at Philadelphia, there would now be great open rectangles": *Engineering News-Record* 112, no. 3, January 25, 1934.

114

"I have in mind the stepped-off type of architecture": Strauss to Ammann, March 6, 1930, in Ammann and Whitney Golden Gate Bridge file.

114

"Eberson was an electrical contractor who became a self-taught architect": *Cyclopedia of American Biography*, vol. 40.

114

"He was particularly concerned with the architecture on the towers and anchorages": *Chief Engineer's Report (1930)*, sec. 55.

115

"Using figures from the George Washington Bridge job, which Ellis felt were too high": Ellis to Derleth, August 9, 1930, in Ammann and Whitney Golden Gate Bridge file.

115

"I have not yet received from Mr. Ellis the final total cost": Strauss to Ammann, July 28, 1930, in Ammann and Whitney Golden Gate Bridge file.

115

"It was about this time that Strauss received another surprise . . . from the federal authorities": Ellis to Moisseiff and Ammann, August 9, 1930, in Ammann and Whitney Golden Gate Bridge file.

115

"Strauss, in a rush . . . decided to hire Irving F. Morrow": Strauss to Moisseiff and Ammann, August 15, 1930, in Ammann and Whitney Golden Gate Bridge file.

115

"Eberson had submitted an estimate . . . which Strauss considered too high": Ibid.

116

"His opinion carries great weight here": Ibid.

116

"Born in Oakland, California, . . . Irving Foster Morrow": Biographical data from Morrow Papers, Mrs. Eleanor Morrow Mead, Tucson, Arizona.

116

"On the further shore, the long peninsula": Irving F. Morrow, *Monastery of the Visitation of the Blessed Virgin Mary*, San Francisco: John Henry Nash,

1919. Original is in Bancroft Library, University of California, Berkeley.

117

"Strauss . . . asked Charles Duncan, his public relations counsel, to recommend someone": Author's interview with Herbert T. Johnson.

117

"Morrow's original plans called for two elaborate plazas": Architectural drawings, *Chief Engineer's Report (1930)*.

118

"Derleth, who . . . was keeping the directors informed of developments within the Engineering Board": Author's interview with Frank Stahl, vice-president Ammann & Whitney, May 1985.

118

"On July 9, Welch assured the directors": Minutes of the Directors, Golden Gate Bridge District, July 9, 1930, in the San Francisco District Office.

118

"on August 21 . . . a call was issued for a special meeting": Minutes of the Directors, Golden Gate Bridge District.

118

"The report . . . runs to some 285 pages": *The Golden Gate Bridge at San Francisco, California*, Report of the Chief Engineer with Architectural Studies, San Francisco, the District, 1930.

119

"According to John Hughes, publicist for the NYPA, Strauss was not involved with the George Washington Bridge at all": Russ Cone, telephone conversation with John Hughes, Publicity Department, New York Port Authority.

119

"has never heard of Strauss being connected with that (the Bayonne Arch) or any of our other projects": Leon Katz of the New York Port Authority, ibid.

121

"The first version was addressed to William Filmer": Moisseiff to Filmer, August 19, 1930, in Ammann and Whitney Golden Gate Bridge file.

121

"The second version of the letter": Moisseiff to Strauss, Ammann, and Derleth, August 19, 1930, in Ammann and Whitney Golden Gate Bridge file.

Chapter Eight: "It is not Mr. Moisseiff and Mr. Ellis . . ."

123

"In May 1930, the powerful Pacific American Steamship Association had issued a statement": Brown, *Golden Gate*.

123

"The San Francisco Chamber of Commerce reacted": Ibid.

125

"On September 18, the first conference of the National Academy of Scientists ever held on the Pacific Coast": San Francisco *Examiner,* September 18, 1930.

125

"Ellis's speech ran the better part of an hour": Ellis, Address to the National Academy of Scientists, Berkeley, September 18, 1930, in the Derleth Papers.

125

"At this point, Mr. Strauss gave me some pencils and a pad of paper": Ibid.

126

"On October 16, before another important audience": Derleth remarks before Commonwealth Club of California, October 16, 1930, in the Derleth Papers.

127

"Oh, I met so many of them and they made good friends for me": Ibid.

127

"That is much too much, 200 is enough": Ibid.

127

"we have made two designs for the anchorage": Ibid.

128

"The original design . . . was made in Chicago under Mr. Strauss's direction and ably assisted by . . . Charles A. Ellis": Ibid.

128

"Mr. Ellis . . . was in charge of that first design": Ibid.

128

"it is not Mr. Leon Moisseiff and Mr. Ellis, even if Mr. Strauss and Mr. Ammann and I do not know anything about the subject": Ibid.

129

"Taxpayers! Before you vote": Newspaper ad, Proposition No. 37, in the Derleth Papers.

129

"Net income for that bridge . . . had been less than one fifth": Ibid.

130

"The *Examiner* . . . had tagged him 'More Millions' ": San Francisco *Examiner,* October 11, 1930.

130

"serious investigations of the 'Czar of City Hall' were either threatened or pending": Ibid., May 21, 1931.

130

"the touchy city engineer responded by filing a formal complaint": Ibid.

130

"The Golden Gate is One of Nature's Perfect Pictures": Election ad, San Francisco *Chronicle,* November 3, 1930.

131

"The . . . *Call-Bulletin* rendered the Old Guard as a giant, potbellied ogre": San Francisco *Call-Bulletin,* October 27, 1930.

131

"To the San Francisco *News,* the Old Guard was a military force": San Francisco *News,* undated editorial cartoon, in the Derleth Papers.

131

"A giant social-realist muscleman, wearing a block cap tagged 'Labor' ": Editorial cartoon, in the Derleth Papers.

131

"On November 4, the bridge bonds were carried by a three-to-one margin": Brown, *Golden Gate.*

Chapter Nine: "The structure was nothing unusual . . ."

132

"Eventually, there would have to be 130 different plans drawn": Attorney Harlan to District Manager Reed, November 1936, in the Ammann & Whitney Golden Gate Bridge file.

132

"In February of 1931 . . . Ellis responded with another gem": Strauss to Ammann and Moisseiff, February 9, 1931, in the Ammann & Whitney Golden Gate Bridge file.

133

"To allay voters' fears of an unlimited power to tax . . . the directors would have to present a final bill in advance": Ibid.

133

"the directors received a tempting offer of . . . a buy-out": Brown, *Golden Gate.*

133

"when the bids were opened on July 17, it was found that twenty-seven different firms had submitted more than forty bids": Ibid.

134

"A series of rumors . . . maintained that the floor of the Gate could not possibly support": Ibid.

134

"I have been unable . . . to get him to modify his position": Strauss to Derleth, May 7, 1930, in the Ammann & Whitney Golden Gate Bridge file.

134

"Lawson's statement . . . I think a gratuitous contribution to a geological report": Derleth to Strauss, April 30, 1930, in the Ammann & Whitney Golden Gate Bridge file.

135

"the Board initiated a completely new survey of the foundations": Brown, *Golden Gate.*

135

"Kinzie made it known that he would render an unfavorable report": Joseph Strauss, *Supplementary Report of the Chief Engineer of the San Francisco Pier Site,* Golden Gate Bridge file no. 1226, June 10, 1931, in the Derleth Papers.

135

"a 'rotten rock foundation' that resembled 'plum pudding' ": Ibid.

135

"Strauss submitted his own report, in rebuttal": Ibid.

135

"Mr. Kinzie himself . . . directed the location of the first hole": Ibid.

135

"It is the opinion of . . . all except Mr. Kinzie": Ibid.

136

"Mr. Kinzie 'seems terror-stricken by the very word serpentine' ": Ibid.

136

"This means that he is willing to surrender the opinions": Ibid.

136

"The Kinzie report . . . has already taken its place in the limbo of disregarded things": Editorial, San Francisco *News,* June 18, 1931.

136

"The bond buyers . . . announced that they would boycott the offering": Brown, *Golden Gate.*

137

"only one offer was received, from the brokerage house of Dean Witter": Ibid.

137

"the district secretary announced that he was declining to sign the bonds": Ibid.

137

The best accounts of the Southern Pacific's machinations in California politics remain George E. Mowry, *The California Progressives,* Berkeley, University of California Press, 1951; and Thomas, *A Debonair Scoundrel.*

138

"Such opposition did develop": *Engineer's Report (1938)*.

138

"On August 15, 1931, Warren Olney, Jr. . . . entered the Bridge District's case": Brown, *Golden Gate*.

139

"The lawsuit was now expanded": Ibid.

139

"The district, in fact, had just about gone broke": Ibid.

140

"The Southern Pacific-Golden Gate Ferries . . . once more attacked the validity": San Francisco *Examiner*, November 29, 1931.

140

"An outcry against railroad interference now arose": Brown, *Golden Gate*.

140

"Shoup . . . claimed the Southern Pacific, as only one of many stockholders": Ibid.

141

"Relations between Ellis . . . and Strauss . . . had been brought to a certain pressure point": Ellis to Ammann and Moisseiff, October 3, 1932, in the Derleth Papers.

141

"Beginning in October 1931, Strauss . . . began sending Ellis letters": Ibid.

142

"By November, Strauss . . . had become extremely critical": Ibid.

142

"In late November, Strauss wrote Ellis urging him . . . to start on his . . . vacation": Ibid.

142

"On December 5, 1931, Charles Ellis . . . left the offices of the Strauss Engineering Corporation for . . . the last time": Ellis to Derleth, March 23, 1932, in the Derleth Papers.

142

"When I left the office": Ibid.

142

"the structure was nothing unusual and did not require all the time" Ellis to Ammann and Moisseiff, October 3, 1932.

142

"an 'indefinite vacation without pay' ": Ibid.

142

"The surrogate struggle between the Bridge District and the Southern Pacific had intensified into direct conflict" Brown, *Golden Gate*.

143

"Not all Mr. Eastman's words": Ibid.

143

"The San Francisco Supervisors now called attention to the fact that the franchise": Ibid.

143

"On January 1, 1932, the most direct threat yet": Ibid.

144

"Ellis's firing had come as a shock not only to himself but also to Derleth and Moisseiff": Derleth to Moisseiff, September 12, 1932, in the Derleth Papers.

145

"In all my interviews, I have simply expressed a desire to return to university work": Ellis to Derleth, March 23, 1932.

145

"I should of course say nothing": Ibid.

145

"In my previous computations": Ibid.

146

"To speak very frankly, I just do not know what to do": Ibid.

146

"If any criticism of my work be made": Ibid.

146

"On July 16, 1932, an opinion on the legality of the organization of the Bridge District": Brown, *Golden Gate*.

146

"On August 9, in separate statements": Ibid.

Chapter Ten: "The responsibility would be placed at my door . . ."

148

"It is caressed by breezes from the blue bay throughout the long golden afternoon": Morrow, *Monastery of the Visitation of the Blessed Virgin Mary.*

149

"He sought out and absorbed the feelings of people like himself": Correspondence, Morrow Papers.

150

"Who would do this today?," Jon Carroll, *California Magazine,* March 1981.

150

"Morrow himself was considering painting the bridge in a variety of tones": Irving F. Morrow, "Color and Lighting for the Golden Gate Bridge, San Francisco," undated but probably fall 1931, Morrow Papers.

151

"This hard bargaining . . . was justified . . . as . . . the kind of treatment the Bridge District had been receiving": Brown, *Golden Gate.*

152

"On July 19 . . . Filmer . . . wrote to . . . the Reconstruction Finance Corporation": Ibid.

152

"Now it was the Bridge District's turn to be the target": Ibid.

152

"The district opened three legitimate bids": Ibid.

153

"On September 2 . . . the bridge directors voted to withdraw the application . . . and to rescind the announced tax": Ibid.

153

"with nothing in particular to do and with the thought of the condition of the towers constantly on my mind": Ellis to Ammann and Moisseiff, October 3, 1932.

153

"the responsibility for any errors or delay on account of them would be placed at my door": Ibid.

154

"in a position to speak with some assurance and authority concerning the design": Ibid.

154

"Mr. Ellis realizes that he may be placing me in an embarrassing position": Derleth to Moisseiff, September 12, 1932, Derleth Papers.

154

"My relations with him are very friendly . . . and I desire that such a relation continue": Ibid.

155

"These grave differences . . . may not really exist": Ibid.

155

"I have worked three years and more on this job . . . and I cannot face the possibility": Ellis to Ammann and Moisseiff, October 3, 1932.

155

"I do not wish to put the consultants in an embarrassing position": Ibid.

155

"I am much vexed by the actions of Mr. Ellis": Moisseiff to Derleth, December 7, 1932, in the Ammann & Whitney Golden Gate Bridge file.

156

"I sympathize deeply with Mr. Ellis": Derleth to Moisseiff, September 12, 1932.

156

"Paine, who had spent some twenty years in railway and bridge construction": *Engineer's Report (1930),* sec. 4.

Chapter Eleven: "We'll take the bonds."

157

"the Bank's president, Will F. Morrish, was op-

posed to the continued presence of MacDonald":
Brown, *Golden Gate.*

157

"Finn was a survivor of what now seemed the
Pleistocene era of local politics": Henry S. Peters,
Tom Finn, A Biography, San Francisco *Chronicle,* five-
part series, January 1938.

157

"through Hirschberg . . . Finn was now rumored
to have acquired a monopoly on the insurance":
Brown, *Golden Gate.*

158

"some of the early construction bids . . . had
given off a peculiar odor": Ibid.

158

"Strauss and the directors decided to appeal directly to the Bank of America's legendary chairman,
A. P. Giannini": Ibid.

159

"Giannini . . . introduced, defended, and established the idea of branch banking in America": Julian Dana, *A. P. Giannini, Giant in the West,* New
York, Prentice-Hall, 1947.

159

"When they met, Strauss and Giannini were both
sixty-two years old": Ibid.

160

"We'll take the bonds . . . we need the bridge":
Ibid.

160

"There were charges . . . of cost overruns on the
Hetch Hetchy project": San Francisco *Chronicle,*
April 15, 1931.

161

"In December 1931, the investment prop-
erty . . . that he had leased . . . was raided by federal agents": San Francisco *Examiner,* December 5,
1931.

162

"Morrish . . . wrote a letter to the Bridge
Board . . . demanding the dismissal . . . of Mac-
Donald": Brown, *Golden Gate.*

162

"According to the . . . *Examiner,* MacDonald,
along with Shannon and Stanton . . . represented
the Finn interest within the board": Ibid.

162

"Shannon attempted to board a return ferry
and . . . fell into the bay": San Francisco *Examiner,*
January 26, 1933.

162

"The bands didn't play as scheduled . . . the . . .
salute was never fired": Ibid., March 1, 1933.

163

"The time has come . . . to break loose from
'Murphy' Hirschberg's snake-charming": Brown,
Golden Gate.

164

"He began his address with a quote from Tennyson": Strauss, Golden Gate Bridge Dedication
Speech, Fairmont Hotel, February 25, 1933, Derleth Papers.

164

"Although accompanied constantly by a man rumored to be a physician": Natusch interview.

165

"In these giant arms of steel": Strauss dedication
speech.

165

"The ideal and the altruistic march side by
side with the material and the commercial":
Ibid.

165

"Oh, the glory of all together": Ibid.

P   A   R   T

BUILDING

168

"You couldn't help know it": Author's interview
with Harold McClain, June 9, 1985.

Chapter Twelve: "Beginning to feel like my old self . . ."

169

"In the winter of 1932–33 . . . Strauss concentrated his considerable powers of persuasion": Author interview with Izetta Cone and Russ Cone, May 16, 1985.

169

"Born in Ottumwa, Iowa, . . . Russell G. Cone": Russ Cone interview.

170

"the film director Frank Borzage . . . drew, from his personality, traits incorporated into a film character later played by Spencer Tracy." The film, *Big City,* with Tracy and Luise Rainer, in which Tracy plays the head of a group of independent taxi drivers, was released in 1937.

170

"Starting as early as 1931, Cone began receiving first inquiries": Izetta Cone and Russ Cone interview.

170

"Ellis was something of a hero to Cone": Ibid.

170

"To . . . work for Strauss . . . would . . . be a serious loss of prestige": Ibid.

170

"Clement Chase . . . saw an advantage to Cone's going West": Ibid.

171

"To close the deal, Strauss . . . sent Clifford Paine to New Jersey with $2,000 in cash": Ibid.

171

"The water of the strait was so consistently rough that the use of floating equipment . . . was out of the question": Russell G. Cone, "Battling Storm and Tide in Founding Golden Gate Pier," *Engineering News-Record,* August 22, 1935.

172

"Based on my own experience . . . I do not concur with . . . the Chief Engineer": Reed to Ammann, February 6, 1933, in Ammann & Whitney Golden Gate Bridge file.

172

"I am pleased to note what you say": Ammann to Derleth, February 7, 1933, in Ammann & Whitney Golden Gate Bridge file.

173

"Reed, described by people who knew him as '100 percent navy' ": Izetta Cone and Russ Cone interview.

173

"the bridge could not be built because there was no plant large enough to fabricate the steel": Natusch interview.

173

"The steel . . . had to be rolled in four different forms, in four different plants": Strauss, *Chief Engineer's Final Report (1938),* book 5.

174

"Fabrication of the steel into sections . . . took place in two . . . shops": Ibid.

174

"It towered over the shed": Ibid.

175

"What the opposition . . . would have . . . been pleased to call a crevasse": Derleth to Reed, June 13, 1933, in Ammann & Whitney Golden Gate Bridge file.

175

"There were no steady work crews": Author interview with Alfred "Frenchy" Gales, May 11, 1985.

175

"There were other guys standing by the office waiting to go to work": Ibid.

176

"I had the caulker's job": Ibid.

176

"Everybody dropped everything and went to grab some fish": Ibid.

176

"To seat the framework for the trestle": Cone, "Battling Storm."

177

"When the elder of the Graham brothers found out, . . . 'he nearly jumped out the window' " Natusch interview.

177

"the consulting engineers agreed to have a model made and tested": *Engineering News-Record,* January 25, 1934.

178

"the chief engineer had disappeared from sight entirely": San Francisco *Examiner,* June 21, 1933.

178

"Strauss had suffered a nervous breakdown": Ibid.

178

"If Mr. Strauss is well enough to be interviewed": Ibid., June 29, 1933.

178

"Doctor Bloomfield . . . said that he had been unable to determine": Ibid., June 21, 1933.

178

"the directors adopted a motion urging the chief engineer to return . . . and 'get busy' " Ibid., June 29, 1933.

178

"On June 30 Strauss wired Filmer": Ibid., June 30, 1933.

179

"in 1939 Meyers, under indictment . . . for selling fraudulent oil leases" J. S. Swenson, Inspector, U.S. Post Office Department, Tacoma, Washington, to Ammann, May 17, 1939, in Ammann & Whitney Golden Gate Bridge file.

179

"a very wealthy financier and businessman": Ibid.

179

"It was an association that Moisseiff and Ammann quickly and vehemently denied": Ammann to Swenson, May 22, 1939; Moisseiff, letter to Swenson, May 19, 1939, both in Ammann & Whitney Golden Gate Bridge file.

179

"On June 26, 1933, Strauss . . . was married": Strauss marriage license, Prince Georges County Clerk's Office.

179

"She was a much younger woman": Izetta Cone and Russ Cone interview.

180

"In 1935, Annette Strauss sang in recital": Ibid.

180

"Strauss's own 1932 poem, 'The Redwoods,' was published as a song": Strauss entry, *National Cyclopedia of American Biography,* vol. 27.

181

"In 1848 a man named Charles Ellet, Jr." Freidrich Bleich, C. B. McCullough, Richard Rosecrans, and George S. Vincent, *Vibration in Suspension Bridges,* Washington, D.C.: Department of Commerce, U.S. Bureau of Public Roads, 1950.

182

"I worked on a night pour at one of the anchorages": Gales interview.

182

"On top of the base block, there was poured an anchor block": Cassady, *Spanning the Gate;* Brown, *Golden Gate.*

182

"On June 29, well within its budget, . . . the Marin pier was presented": Brown, *Golden Gate.*

183

"The method of blasting at the pier site was an expansion of the . . . technique": A. E. Graham, "Rock Excavation for San Francisco Pier, Golden Gate Bridge," *Architect & Engineer,* December 1933.

183

"They'd put charges in, a lot of little ones at first": Author interview with George W. Albin, June 5, 1985.

183

"The bomb would be driven into the blast hole by as many as a dozen of the smaller bombs": Graham, "Rock Excavation."

183

"You'd see a lot of bubbles": Albin interview.

184

"The operator of the bucket could not see what he was digging": Graham, "Rock Excavation."

184

"By early August, the trestle reached nearly all the way out to the pier site": Cone, "Battling Storm."

184

"As the pressure area known as the Pacific High moves north with the summer sun": Harold Gilliam, *Weather of the San Francisco Bay Region,* Berkeley, University of California Press, 1962.

185

"on August 14, 1933 . . . the McCormick Line freighter *Sidney M. Hauptman":* Brown, *Golden Gate;* Cone, "Battling Storm."

Chapter Thirteen: "I'll never forget that day . . ."

186

"Looking back on it all afterward, Russell Cone ranked the building of the San Francisco pier": Cone, radio interview, "San Francisco Progress Radio Program," March 7, 1937.

186

"On August 23, . . . the Engineering Board revised the specifications for the cement": Brown, *Golden Gate.*

186

"Eleven directors of the Golden Gate Bridge . . . voted yesterday to give their fellow director": San Francisco *Examiner,* September 14, 1933.

187

"Once again, . . . the bridge was being subjected to 'such wild, exaggerated, and absurd criticisms' ": Strauss to Bridge Board of Directors, September 15, 1933, in Ammann & Whitney file 1226.

187

"a 'selfish interest . . . heedless of what is best for the Bridge District' ": Ibid.

187

"I am astonished . . . that reputable concerns would be so unethical": Ibid.

187

"is a big belly laugh to me": Brown, *Golden Gate.*

187

"according to the *Examiner,* . . . the pink cement 'would be in danger of not only probable but practically certain cracking' ": Ibid.

188

"Unanticipated swells, . . . coming out of nowhere in otherwise calm weather": Cone, "Battling Storm."

188

"By October 19, the trestle had been extended out to the location": Ibid.

189

"Finally the tower was set in position": Ibid.

189

"I'll never forget that day": Cone, *San Francisco Progress* interview, Cone Papers, undated.

190

"By November 22, Strauss was able to report to the directors": Joseph Strauss, *Interim Report of the Chief Engineer,* November 22, 1933, in Ammann & Whitney Golden Gate Bridge file.

190

"It tore the remainder of the trestle from its foundation": Cone, *San Francisco Progress* interview.

190

"Reed accused the management of Barrett & Hilp . . . of transferring 'your principal activities from the San Francisco side' ": Reed to Barrett & Hilp, November 20, 1933, in Ammann & Whitney Golden Gate Bridge file.

191

"In October, Clifford Paine . . . was met by Reed, and the two men then visited the Princeton offices of Professor Beggs": Strauss, *Interim Report,* October 25, 1933, in Ammann & Whitney Golden Gate Bridge file.

191

"There had, in the past, been models made of

bridge towers": *Engineering News-Record,* January 25, 1934.

191

"On January 20, 1934, a group of 200 engineers": Ibid.

191

"An ingenious method . . . had been worked out by C. A. Ellis": Ibid.

192

"An article, by Clifford Paine, . . . would make no mention of Ellis": Clifford E. Paine, "Designing Bridge Towers 700-ft High," *Engineering News-Record,* October 8, 1936.

192

"It wasn't until the autumn of 1934 that Ellis found another position": *Purdue Engineer,* June 1935.

192

"When he was at Purdue . . . Professor Ellis told his staff and students that he was the designer of the Golden Gate Bridge": Scott interview.

192

"I designed every stick of steel . . ." Charles Kring, interview with Russ Cone, May, 1986.

193

"I helped set the base plates for the Marin tower": Author interview with Walter "Peanuts" Coble, May 31, 1985.

193

"At the base of the Marin pier, an Erector-Set-like derrick 85 feet high": Cassady, *Spanning the Gate.*

193

"Each crane had a carrier that boomed out with a load block on it": Albin interview.

193

"We had a state inspector . . . with a feeler gauge": Ibid.

193

"A connecting gang . . . would work the joints": Cassady, *Spanning the Gate.*

194

"As the tower rose . . . the traveler was 'jumped up' ": Ibid.

194

"There is a lot of coming and going when you are jumping hammerhead cranes": Albin interview.

194

"men working in teams . . . bonded the sections together with lengths of hot steel": Cassady, *Spanning the Gate.*

194

"You'd have a pipe in one end": Albin interview.

194

"I had a hammer to see if a rivet was loose": Author interview with Alfred Finnila, May 15, 1985.

195

"In a raising gang . . . everybody likes to get to the top first": Coble interview.

195

"It was my job to straddle the cables up there": Ibid.

195

"The red lead would come up like out of a smokestack": Ibid.

195

"Lead which enters the body through the respiratory tract is the most toxic": L. K. Reinhardt, California Department of Industrial Relations, to James Reed, February 14, 1934, in Ammann & Whitney Golden Gate Bridge file.

196

"I didn't wear a mask . . . You could suffocate": Albin interview.

196

"Guys from New York, Pennsylvania, Louisiana, Oklahoma, Texas—the best ironworkers in the country": Author interview with Al Zampa, May 21, 1985.

196

"Building buildings is almost like doing piecework in a factory. . . . But bridgework is always changing": Author interview with Harold McClain, June 9, 1985.

196

"They were crying for good hands": Ibid.

Chapter Fourteen: "Pudding stone."
198

"Strauss informed the directors, the plans for the San Francisco pier . . . would have to be completely revised": Brown, *Golden Gate.*
198

"Three weeks ago outsiders told me what was happening": Ibid.
199

"There may be what is known as subconscious ducking": Moisseiff to Derleth, January 20, 1934, in Ammann & Whitney Golden Gate Bridge file.
199

"The contractor's proposal . . . contains a clause binding him to work a twenty-four-hour day,'" Joseph Strauss, *Special Report on Explorations, San Francisco Pier and Fender,* February 2, 1934, in Ammann & Whitney Golden Gate Bridge file.
199

"This time it was decided to base the trestle not on steel sections but on rounded timber pilings": Cone, "Battling Storm."
199

"a Pacific Bridge diver . . . fell off a steep bank; . . . another diver attempting the same work" Derleth to Ammann, January 4, 1934, in Ammann & Whitney Golden Gate Bridge file.
199

"On November 29 . . . the directors had elected to rescind the change in specifications": Brown, *Golden Gate.*
200

"Judge Maurice Dooling . . . a man chosen . . . in hope of geographical impartiality" Ibid.
200

"On March 22, the first concrete had been poured in the fender": Cone, "Battling Storm."
200

"The plan now required building a considerably larger surface of framework underwater": Ibid.

200

"The bottom form . . . included the legs of the guide tower": Ibid.
201

"From forty feet below on up, the boxes were replaced by skeleton steel frames": Ibid.
202

"In early April, the assistant engineer of the state of California in charge of administering Public Works Administration funds, . . . received a . . . letter and memorandum from Dr. Bailey Willis": Reed to Ammann, April 11, 1934, in Ammann & Whitney Golden Gate Bridge file).
202

"At first [it was] assumed . . . Willis was fronting for the same interests": Ibid.
202

"No reputable geologist would consider it seriously": Professor Andrew Lawson, memorandum to Reed, April 10, 1934, in Ammann & Whitney Golden Gate Bridge file.
203

"utterly incompetent . . . puerile . . . cannot be taken seriously": Derleth to Reed, April 30, 1934, in Ammann & Whitney Golden Gate Bridge file.
203

"The mere fact that Professor Willis received the degree of Civil Engineer . . . more than fifty years ago": Ammann to Strauss, September 29, 1934, in Ammann & Whitney Golden Gate Bridge file.
203

"Willis limited his appearance to a conversation with Reed": Notes of conversation, Professor Bailey Willis, Reed, Francis V. Keesling, April 7, 1934, in Amman & Whitney Golden Gate Bridge file.
203

"his 'happy faculty of reaching his conclusions without taking the trouble to get the facts' ": Strauss to Ammann, October 3, 1934, in Ammann & Whitney Golden Gate Bridge file.
204

"The first day I worked out there . . . I was

warned how cold it would be, but I didn't believe it": McClain interview.

204

"I'd never been so cold in my life. I don't think I've ever been as cold since": Ibid.

204

"The fog was wetter than on the Bay Bridge. . . . It was miserable": Zampa interview.

205

"We'd be in sunshine on the top, and you couldn't see the water because of the fog": Ibid.

205

"On May 4, 1934, Cone, Reed, and the McLintic-Marshall engineers . . . rode the elevator up": Brown, *Golden Gate;* Cassady, *Spanning the Gate.*

206

"Ammann . . . favored a field coat of aluminum paint, like that used on the George Washington and the San Francisco-Oakland Bay bridges": Minutes of Engineering Board, July 16, 1934, in Ammann & Whitney Golden Gate Bridge file.

206

"it was decided to establish a test for the final field coat at the site" Strauss, memorandum to Reed, June 25, 1934, in Ammann & Whitney Golden Gate Bridge file.

207

"The painters used to hang off of the cable": Coble interview.

207

"The change in architects . . . was a very happy one indeed": Strauss to Moisseiff, July 20, 1934, in Ammann & Whitney Golden Gate Bridge file.

207

"the stepped-off construction of the towers . . . Strauss informed Paine . . . 'originated with me, and is covered . . . by . . . patents issued to me' ": Ibid.

207

"Strauss . . . now proposed that the architect be in charge of lighting design . . . toward which Ammann and Moisseiff both expressed strong reserva-

tions": Reed, memorandum to Strauss, September 14, 1933, in Ammann & Whitney Golden Gate Bridge file.

208

"I am convinced that any discussion with you would serve no good purpose": Strauss to Professor Bailey Willis, September 21, 1934, in Ammann & Whitney Golden Gate Bridge file.

208

"rank discourtesy and unethical conduct": Ibid.

208

"we are not impressed with anything that you might say now": Ibid.

208

"In my previous letters . . . I stated my opinion that you were not qualified to pass on engineering matters": Ibid.

208

"We are having court reporters make a record of all his statements": Strauss to Ammann, October 3, 1934, in Ammann & Whitney Golden Gate Bridge file.

209

"I remember him and Mrs. O'Shaughnessy dancing together": Izetta Cone interview.

209

" 'God bless you all,' O'Shaughnessy told his family": San Francisco *Examiner,* October 13, 1934.

Chapter Fifteen: "We descended in a bucket . . . to the ocean floor."

210

"According to the specifications written by Charles Ellis in 1931": Charles A. Ellis, "Pneumatic Caisson," April 1931, in Ammann & Whitney Golden Gate Bridge file.

210

"The caisson, 185 feet long and 90 feet wide": Brown, *Golden Gate.*

210

"As the plan now went, the fender for the pier

would be built with one end left open": Cone, "Battling Storm."

211

"on October 8, at four-thirty in the morning of a calm, still, Indian summer day": Brown, *Golden Gate*.

211

"That night . . . another . . . storm . . . began . . . tipping the caisson crazily": Joseph B. Strauss, "Special Report on Change of Construction Method of South Pier," November 19, 1934, in Amman & Whitney Golden Gate Bridge file.

211

"Jack Graham and Phil Hart . . . called Cone, who came out": Cone, *San Francisco Progress* interview.

211

"I went out to the scene at 5:00 AM": Ibid.

212

"Hart and Graham insisted that the only safe way to handle the situation": Ibid.

212

"Caisson battering fender due to extreme swells": Strauss, telegram to Ammann, October 9, 1934, in Ammann & Whitney Golden Gate Bridge file.

212

"Only a week before, Professor Willis had charged": Bailey Willis, "Golden Gate Bridge," memorandum on positioning of caisson, October 5, 1934, in Ammann & Whitney Golden Gate Bridge file.

212

"Eliminating the caisson . . . involved merely the extension of these wells": Strauss, "Special Report on Change of Construction."

212

"The change of plan involving junking of the great caisson . . . does not remove the threat": Bailey Willis to Charles Blyth, Blyth & Co., October 11, 1934, in Ammann & Whitney Golden Gate Bridge file.

213

"Bailey Willis, Geologist, Member Earthquake Safety Committee and Junior Chamber of Commerce": Ibid.

213

"Strauss and Paine are not at present playing ball with us": Derleth, telegram to Moisseiff, October 29, 1934, in Ammann & Whitney Golden Gate Bridge file.

213

"The decision . . . had already been made": Strauss to Ammann and Moisseiff, October 27, 1934, in Ammann & Whitney Golden Gate Bridge file.

213

"It appears that I am like the private": Strauss to Derleth, November 5, 1934, in Ammann & Whitney Golden Gate Bridge file.

214

"I hope . . . that luck will be with us": Strauss to Derleth, November 2, 1934, in Ammann & Whitney Golden Gate Bridge file.

214

"By October 28 the fender ring was completed, and by November 4 the blanket of concrete poured within it . . . had reached sixty-five feet below the level of the strait": Cone, "Battling Storm."

214

"On November 27 the pier area inside the fender wall was pumped out": Brown, *Golden Gate*.

215

"On November 8 Strauss . . . submitted a detailed report . . . challenging Willis on just about every point": Strauss report to Building Committee, November 8, 1934, in Ammann & Whitney Golden Gate Bridge file.

215

"Willis had plotted an alleged fault . . . in three different locations": Ibid.

215

"faulty hypotheses, false assumptions, erroneous measurements": Ibid.

349

216

"On the evening of December 3, three men . . . were bolted inside an airlock at the top of one of the inspection wells": Cone report to Strauss, December 4, 1934, in Ammann & Whitney Golden Gate Bridge file.

216

"We descended in a bucket. . . . The foundation bed was exposed and dry": Ibid.

216

"Cone and Graham were grinning and pounding one another on the back": San Francisco *News*, December 5, 1934.

216

"I walked about and carefully examined all of the area": Cone report to Strauss.

217

"my opinion is that the bottom had been satisfactorily cleaned": Ibid.

217

"Cone, Graham, and Hansen rode the bucket up. . . . Here they waited thirty-five minutes": San Francisco *News*, December 5, 1934.

217

"I was pregnant and it was a bitter night . . ." Izetta Cone, interview.

"He was a fool . . . ," Ibid.

217

"On December 7, Professor Andrew Lawson . . . was lowered . . . to the bottom": Lawson to Strauss, December 8, 1943, in Ammann & Whitney Golden Gate Bridge file.

217

"When struck with a hammer, it rings like steel": Ibid.

217

"at the west end of the pier diamond drilling was taken down to 250 feet": Lawson report to Strauss, December 24, 1934, in Ammann & Whitney Golden Gate Bridge file.

218

"The west shaft . . . has been concreted":
Strauss, *Interim Report*, December 19, 1934, in Ammann & Whitney Golden Gate Bridge file.

218

"I have been watching very closely the progress of the towers": Beniamino Bufano to Morrow, January 12, 1935, Morrow Papers.

219

"I have watched the Marin tower from the ferry . . . in almost every kind of weather and light": Robert B. Howard to Morrow, February 23, 1935, Morrow Papers.

219

"The tone is beautiful under all light conditions": E P. Meinecke to Morrow, February 25, 1935, Morrow Papers.

219

"The magnitude and . . . its position . . . suggest that it should be emphasized rather than played down": Morrow report to Strauss, December 10, 1934, in Ammann & Whitney Golden Gate Bridge file.

220

"When the bridge is completed, and has been observed through an entire . . . year": Ibid.

220

"In December, Morrow had submitted . . . a detailed plan for illuminating the bridge": Ibid.

220

"Uniform distribution . . . would seem too artificial to be real": Ibid.

220

"I am much impressed by Mr. Morrow's presentation . . . and am inclined to agree with him": Moisseiff to Strauss, March 6, 1935.

220

"On the San Francisco tower, they lined up the sections": Gales interview.

222

"People who weren't good didn't last": McClain interview.

222

"When construction on the San Francisco pier had

fallen behind . . . the bridge directors had ordered McLintic-Marshall to stop fabricating steel": Brown, *Golden Gate.*

222

"It was a stick-up, and Strauss and the directors knew it": Ibid.

222

"the fellow who puts you up against a wall and takes money out of your pocket": Ibid.

222

"Hugo D. Newhouse . . . pointed out that . . . the board . . . would be saving $3,100 a day": Ibid.

223

"Most accidents were guys hit with flying stuff": Gales interview.

223

"If somebody did get seriously hurt, Cone went to visit him": Russ Cone interview.

223

"I was on the corner like a point on the very top": Coble interview.

224

"I was up on the tower when the earthquake hit": Gales interview.

224

"The following day there was an aftershock that set the tower swaying again": Gales interview; Cassady, *Spanning the Gate.*

Chapter Sixteen: "Can he deliver?"

225

"Strauss had filed suit against the Bridge District": San Francisco *Examiner,* September 2, 1934.

225

"Strauss, as chief engineer, had written into the contract the very terms whose vagueness": Ibid., September 21, 1934.

226

"On March 19, 1935, the mysterious H. H. 'Doc' Meyers . . . filed suit against the chief engineer": Ibid., March 20, 1935.

226

"The suit threatened to blow the lid off the various personal and political arrangements": Ibid.

226

"this man Meyers has been around here saying that he can deliver": Ibid, March 29, 1935.

226

"The chief engineer had already made payments to Meyers": San Francisco *Chronicle,* April 17, 1935.

226

"there were rumors that Strauss's first wife had been brought to San Francisco and would be subpoenaed": San Francisco *Examiner,* March 29, 1935.

227

"Strauss it is understood . . . will carry out . . . the mystery agreement": San Francisco *Chronicle,* April 17, 1935.

227

"frankly . . . have never been satisfied that the contract . . . will secure that high class of work which such an outstanding structure deserves": Ammann, memorandum to Reed, March 3, 1933, in Ammann & Whitney Golden Gate Bridge file.

228

"In a board meeting on January 24, 1934, this was what Strauss . . . threatened to do": San Francisco *Examiner,* January 25, 1934.

228

"matters have been put into the minutes . . . which are radically at variance with what happened": Ammann, note to Derleth on letter to Strauss, October 1935, in Ammann & Whitney Golden Gate Bridge file.

229

"Your husband . . . is my right arm": Izetta Cone interview.

229

"Not long after Cone had arrived in San Francisco, Clement Chase . . . had . . . suffered a heart attack": Ibid.

229

"Cone . . . might well be able to do what amounted to the same thing here": Ibid.

229

"Reed began to suspect that the much younger . . . resident engineer was trying to undermine him": Reed to Ellis, December 22, 1937, Derleth Papers.

230

"we have repainted the cables and suspenders of the George Washington Bridge with aluminum paint": Ammann to Strauss, June 17, 1935, in Ammann & Whitney Golden Gate Bridge file.

230

"Derleth questioned the use of a red color": Derleth to Strauss, March 4, 1935, in Ammann & Whitney Golden Gate Bridge file.

230

"As Evan Connell has pointed out": Evan S. Connell, Jr., *Holiday,* April 1961.

231

"Modjeski . . . had rejected Strauss's first idea for a cheaper . . . bascule bridge": Cassady, *Spanning the Gate.*

231

"This . . . rivalry extended to the ironworkers . . . whose two locals . . . historically did not get along": Author interview with "Lefty" Underkoffler, June 7, 1985.

231

"The . . . International Association . . . had been riven with dissension since . . . 1911": Ibid; Clarence Darrow, *Attorney for the Damned,* New York: Simon & Schuster, 1957.

232

"Bridge, bridge . . . that's only a trestle": Natusch interview.

232

"The huge junked caisson . . . was now . . . hauled by two tugs . . . out to sea": Brown, *Golden Gate.*

Chapter Seventeen: "Roebling brought class to the job."

233

"On the Brooklyn Bridge, John Roebling . . . and his son Washington Roebling . . . built, on site, cables of steel wire": "A Century Old, The Wonderful Brooklyn Bridge", *National Geographic,* May 1983.

233

"This basic method . . . had been refined and improved over the years, chiefly by Roebling Company engineers": Charles M. Jones, "Largest and Longest Bridge Cables Spun at New Record Rate," *Engineering News-Record* 116, no. 18, April 30, 1936.

234

"To keep the Golden Gate Bridge on schedule, the existing cable-spinning speed record would have to be multiplied": Ibid.

234

"Roebling brought class to the job": Albin interview.

234

"The Roebling people were the most experienced": McClain interview.

234

"The ironworkers were strongly unionized": Gales interview.

234

"In July 1935, the Roebling engineers . . . put in place the company's first contribution": Cassady, *Spanning the Gate.*

235

"As the cable changes length . . . the bridge itself moves and deflects": Cone, *San Francisco Progress* interview.

235

"On August 2, 1935 . . . the entrance to San Francisco Bay was closed to all shipping": Brown, *Golden Gate.*

236

"For the shore-to-tower backspans, the flooring . . . would be pulled up in hundred-foot lengths": Cassady, *Spanning the Gate.*

236

"Rigging up ahead of time was dirty work": Albin interview.

237

"I helped tie down a cable at night": Gales interview.

237

"There was this guy from Washington with me. I pushed him ahead of me": Author interview with Ted Huggins, May 8, 1985.

238

"The target selected was the *Shensu Maru*": Gales and Albin interviews.

239

"overhead tramways were now installed, cross-braced . . . against the wind": Jones, "Largest and Longest Bridge Cables."

239

"After the tramway and hauling ropes were in place, guide wires . . . were fixed to the Marin and San Francisco anchorages": Albin interview.

239

"a complex of men and mechanisms, suggesting the interior of an enormous pocketwatch": Jones, "Largest and Longest Bridge Cables."

241

"You couldn't help know it": McClain interview.

241

"I stayed away from the places where the bridgemen drank": Coble interview.

241

"The bars in Sausalito made a fortune": Gales interview.

241

"There was a first aid station. . . . He kept a bottle of Canadian whisky there": Ibid.

241

"On . . . November 11 . . . the first wire of the west cable": Strauss, *Interim Report,* November 13,

1935, in Ammann & Whitney Golden Gate Bridge file.

241

"The top or 'live' wire was spun out continuously; the bottom or 'dead' wire . . . would be adjusted from point to point": Albin interview.

241

"You'd get to a certain point. . . . The guy at the tower top had a come-along machine": Ibid.

241

"You're out on the catwalk with a good view of the wire": McClain interview.

241

"It was like splicing a piece of rod": Ibid.

242

"For the first time in cable spinning, the number of wires per strand would be varied": Jones, "Largest and Longest Bridge Cables."

243

"Each strand of wire was tested at a certain time at night": Gales interview.

243

"You could see the expansion and contraction in the main cable": Albin interview.

243

"We worked between the strands . . . even though we weren't supposed to": Gales interview.

243

"We used to splice each line by hand": Ibid.

Chapter Eighteen: "I have felt all along that the organization was not as it should be."

244

"My other stenographers wouldn't work for Mr. Strauss": Natusch interview.

244

"He was a nice person, sort of a hanger-on, with a lot of braggadocio": Ibid.

245

"Clifford Paine was all over that bridge": Albin interview.

246

"his boy friend Clifford": Reed to Ellis, December 22, 1937, Derleth Papers.

246

"I remember Morrow's face when he returned from the meeting": Author interview with Herbert T. Johnson, June 7, 1985.

247

"On December 18 Derleth . . . questioned . . . some of the foundation footings": Derleth to Strauss, December 21, 1935, in Ammann & Whitney Golden Gate Bridge file.

247

"I regret that you did not call me on the phone and discuss the matter": Strauss to Derleth, January 4, 1936, in Ammann & Whitney Golden Gate Bridge file.

247

"I believe . . . that they must be guided in technical matters": Ibid.

250

"Chief Engineer Strauss has not informed me in response to my inquiries": Reed to Ammann, Moisseiff, and Derleth, January 13, 1936, in Ammann & Whitney Golden Gate Bridge file.

250

"I have felt all along that the organization . . . was not as it should be": Ammann to Reed, February 7, 1936, in Ammann & Whitney Golden Gate Bridge file.

250

"The situation is such": Moisseiff to Ammann, January 17, 1936, in Ammann & Whitney Golden Gate Bridge file.

250

"The question of a replacement . . . was referred to the Finance and Auditing Committee": Reed to Ammann, February 11, 1936, in Ammann & Whitney Golden Gate Bridge file.

251

"It occurred to C. C. Sunderland, chief engineer of Roebling's bridge department": Jones, "Largest and Longest Bridge Cables."

251

"It wasn't something you could pick up quickly": Albin interview.

251

"You'd have five men on a gang, including a foreman": Ibid.

252

"the cable-spinning total had jumped to an average of 271 tons a day, or more than four times the peak output": Jones, "Largest and Longest Bridge Cables."

252

"The cable-spinning crews were now completing one 'set' . . . every twenty-four working hours": Ibid.

253

"The squeezing machine that compacted the cables had so much pressure": Albin interview.

253

"It is unnecessary for me to repeat here the exceedingly unsatisfactory history of the castings": Paine to W. E. Joyce, June 29, 1936, in Ammann & Whitney Golden Gate Bridge file.

253

"Nor is it necessary to remind you of my warnings": Ibid.

253

" 'It was our understanding . . . that your inspecting engineer would accept' ": Charles Jones to Directors, July 1936, in Ammann & Whitney Golden Gate Bridge file.

254

"You may be interested in knowing": Ibid.

254

"On August 10 . . . Jones requested an extension of the Roebling contract, as well as payment of some $250,000 in compensation": Jones to Reed and Board of Directors, August 10, 1936, in Ammann & Whitney Golden Gate Bridge file.

254

"As you know . . . our work has been practically shut down since June 15": Ibid.

254

"Dear Joe,": Paine to Strauss, August 13, 1936, in Ammann & Whitney Golden Gate Bridge file.

254

"It is very disgusting to have to work with such an organization": Joseph B. Strauss, *Report on Cable Bands,* August 19, 1936, in Amman & Whitney Golden Gate Bridge file.

255

"In my opinion . . . the criticism directed against the engineer was totally unwarranted": Ibid.

255

"The firm rented Paradise Cove, a park on a quiet Marin County shore behind Tiburon, and threw a barbecue for the bridgemen": Gales interview.

255

"We had a tug-of-war on the beach": Ibid.

Chapter Nineteen: "You have to have a little bit of fear."

256

"In those days, a man's life wasn't worth a nickel, anyway": Underkoffler interview.

257

"On the Brooklyn Bridge . . . Washington Roebling had hired sailors": *National Geographic,* May 1983.

258

"I worked on the Delaware River Bridge": Underkoffler interview.

258

"You couldn't visit or bullshit": Zampa interview.

258

"I was walking the catwalk . . . and I slipped my watch into my pocket": Author interview with Alfred Finnila, May 15, 1985.

258

"People freeze up there,": Zampa interview.

258

"You have to have a little bit of fear": Ibid.

258

"I was boarding with this widow lady in Sausalito": Coble interview.

259

"the ironworkers' union had been fighting for safety nets for years": Underkoffler interview.

259

"The manufacture of the net itself was contracted through Roebling": A. F. McLane to Billings Wilson, N.Y. Port Authority, February 16, 1937, in Ammann & Whitney Golden Gate Bridge file.

260

"In ample time . . . I warned the Roebling Company": Strauss, *Special Report,* August 19, 1936.

260

"On a bridge, particularly of this magnitude, patchwork should not be permitted": Ibid.

260

"Roebling's president, on August 10, sent Reed and the district a bill": W. Hammond to Reed, August 10, 1936, in Ammann & Whitney Golden Gate Bridge file.

261

"Some 254 bands were now clamped around the smoothly rounded cable and bolted into place": Cassady, *Spanning the Gate.*

261

"To keep the load . . . evenly distributed . . . work on one side . . . would have to be slowed or halted": Albin interview.

261

"Once again, the steel would be lifted up on the tower by cranes": Cassady, *Spanning the Gate.*

262

"a little further each day, out over the stretch of water the bridgemen . . . referred to simply as 'the river' ": Zampa interview.

262

"We were connectors. . . . I was the right size": Ibid.

263

"You had to be surefooted like a mountain goat, and hang on like a monkey": Ibid.

263

"The punks would get a kick out of jumping into the net": Underkoffler interview.

264

"I flipped three times and fell into the net": Zampa interview.

264

"From a distance, in the fall of 1936, the bridge was perhaps at its most interesting": Bridge District photograph, October–November 1936.

264

"On October 6 . . . O. H. Ammann . . . formally proposed that . . . 'it . . . approve of the final coat' ": Minutes, Engineering Board Meeting, October 6, 1936, in Ammann & Whitney Golden Gate Bridge file.

265

"It was the color he had always wanted": Johnson interview.

266

"The crews on these jobs ranged from as few as 4 to as many as 257 men": Golden Gate Bridge, *Daily Construction Report*, May 4, 1937, Russell Cone (Russ Cone, Jr.) Papers.

266

"He was just doggone good. . . . He knew his work": Davenport interview.

266

"the dinner given by Filmer in February 1935": Derleth diary.

266

"shortly after the bridge was finished Cone's marriage would end": Russ Cone interview.

267

"There was a lot of stuff dangling out there": Albin interview.

267

"On October 21,1936, the project's safety record was overtaken by the actuarial odds": Brown, *Golden Gate.*

267

"Jack's brother-in-law just froze": Underkoffler interview.

268

"In December the veterans . . . formed the 'Halfway to Hell Club' ": Zampa interview.

268

"In November 1936, Dean Kinter, a riveting inspector": Brown, *Golden Gate.*

268

"Less than two months later, a three-man crew was at work dismantling a work tower": Brown, *Golden Gate;* Cassady, *Spanning the Gate.*

268

"Later the same month, a seven-ton locomotive": Brown, *Golden Gate.*

Chapter Twenty: "I felt a funny shudder . . ."

269

"Strauss's lawyer . . . wrote a letter to Filmer": John McNab to William Filmer, November 18, 1936, in Ammann & Whitney Golden Gate Bridge file.

269

"Under the terms of Strauss's original contract, it had been assumed that the detailed plans . . . would be worked out . . . one unit at a time": Ibid.

270

"At the first meeting of the Bridge District board . . . the directors had passed a resolution": Ibid.

270

"With the start of actual construction . . . these engineering plans had . . . to be redrawn": Ibid.

270

"Strauss's lawyer maintained that . . . these changes had required the work of Clifford Paine and thirty-two other engineers": Ibid.

270

"I make my living playing bridge": *Purdue Engineer,* June 1935.

271

"His students called him 'Uncle Charlie' ": Scott interview.

271

"At some point he entered into a correspondence with Reed": Reed to Ellis, December 22, 1937, Derleth Papers.

272

"There was some skepticism about whether the communities around the bay": Ibid.

272

"There was a small ceremony": Cassady, *Spanning the Gate.*

273

"In the spring of 1936 . . . Morrow submitted to Strauss his recommendations on lighting": Morrow, *Report on Color and Lighting,* April 6, 1935, Morrow Papers.

273

"The object is to reveal aspects of a great monument which are unsuspected under the conditions of natural . . . lighting": Ibid.

273

"Threading through all this . . . would be 'the continuous horizontal line of diffusion' ": Ibid.

274

"sodium-vapor lights . . . bathing the bridge and everything on it, in a strange, transforming orange glow": The author's sister, as a child, was reduced to tears by finding her prized red coat transformed, by the light of the bridge, to a drab brown one.

275

"As work . . . moved toward its conclusion, the complexities . . . increased." Derleth report to Francis V. Keesling, July 23, 1936, in Ammann & Whitney Golden Gate Bridge file.

276

"Strauss pointed out that the platform . . . would infringe upon the strict navigational clearances": Reed to W. H. Ringe, January 8, 1937, in Ammann & Whitney Golden Gate Bridge file.

277

"We made some tests recently . . . and found that the . . . rope was still as good": A. F. McLane to Billings Wilson, February 16, 1937, in Ammann & Whitney Golden Gate Bridge file.

277

"I felt the stripper give a funny shudder": San Francisco *Chronicle,* February 18, 1937.

277

"The whole mechanism . . . now started downward": Ibid.

277

"The net cracked like a machine gun . . . it ripped like a picket fence splintering": Ibid.

277

"The net sagged slowly . . . like a slow-motion picture": Ibid.

278

"I felt the scaffold start to go out from under me": San Francisco *Examiner,* February 18, 1937.

279

"I got so excited, I kneeled on the bridge floor": San Francisco *Chronicle,* February 18, 1937.

279

"the scene was framed like a picture": Ibid.

279

"I could hear faint babylike cries": Ibid.

279

"All of us . . . were scared stiff": San Francisco *Examiner,* February 18, 1937.

279

"As I was falling, a piece of lumber fell on my head": San Francisco *Chronicle,* February 18, 1937.

280

"I swam toward him as fast as I could": Ibid.

280

"Foster rigged a loop . . . and maneuvered it between Casey's legs": San Francisco *Examiner,* February 18, 1937.

280

"Casey picked up his time, and never came back": Gales interview.

281

"My body felt like a block of ice": San Francisco *Chronicle,* February 18, 1937.

281

"The boat, coming in from crab fishing, was skippered by Mario Maryella": San Francisco *Examiner,* February 18, 1937.

281

"the gaunt outlines resembled two picture frames": San Francisco *Chronicle,* February 18, 1937.

Chapter Twenty-One: "I'm not on trial. I'm not guilty of anything."

282

"The meeting had already been called to order when the directors were notified that Strauss would be delayed": San Francisco *Examiner,* February 18, 1937.

282

"An accident to the stripping platform for the removal of deck forms": Joseph B. Strauss, *Preliminary Special Report,* February 17, 1937, in Ammann & Whitney Golden Gate Bridge file.

283

"The following day, four different agencies": San Francisco *Examiner,* February 19, 1937.

283

"Pacific Bridge engineers had been warned twice": San Francisco *Chronicle,* February 18, 1937.

284

"Those safety bolts weren't in our platform": Brown, *Golden Gate.*

284

"Strauss . . . opened the hearing by announcing that he was postponing it": San Francisco *Examiner,* March 4, 1937.

285

"We don't want people who may be concerned": Ibid.

285

"I resent and object to that reference": Ibid.

285

"instead of with meat axes": Ibid.

285

"Two days after the accident . . . One body was found caught in the net": Brown, *Golden Gate.*

286

"Investigations, charges, and recriminations fill the air here": Reed to Moisseiff, March 6, 1937, in Ammann & Whitney Golden Gate Bridge file.

286

"the trouble was due to the fact": Strauss to Ammann, March 23, 1937, in Ammann & Whitney.

286

"There is no doubt, of course, in my own mind": Ibid.

286

"The provisions of the contract . . . clearly put the responsibility . . . on the contractor": Ammann to Strauss, March 29, 1937, in Ammann & Whitney Golden Gate Bridge file.

287

"The Pacific Bridge carpenters had gone on strike": Strauss, *Interim Report,* March 31, 1937, in Ammann & Whitney Golden Gate Bridge file.

287

"A friend . . . in New York sent a clipping . . . about how the Brooklyn Bridge had been reserved for pedestrians": Huggins interview.

288

"One of his photographs . . . of the bridge . . . was about to appear on the cover of *Life* magazine": Ibid.

288

"The roadway was being poured in three long rows": Cassady, *Spanning the Gate.*

289

"We worked with short crews . . . used pushers, supervisors": Gales interview.

289

"I had this old car": Ibid.

Chapter Twenty-Two: "It was never just a job to me."

290

"There had been an intense rivalry between the

cities of Seattle and Tacoma": Underkoffler interview.

290

" 'the engineers actually responsible for bridge construction' . . . considered Reed 'a pain in the neck' ": San Francisco *News,* editorial, July 15, 1937.

291

"It appeared as though there would now be claims . . . made against Strauss": Reed to Ellis, December 22, 1937, Derleth Papers.

291

"Someone . . . suggested a more 'active' policy": Paper, June 2, 1937, unsigned copy, Cone Papers.

292

"a ceremony marking the completion of construction": San Francisco *Examiner,* April 28, 1937.

292

"In 1829, the small Broughton suspension bridge . . . collapsed under the rhythmic impact": Bleich, McCullough, Rosecrans, and Vincent, *Mathematical Theory of Vibration in Suspension Bridges.*

292

"a bread line moving toward the soup": San Francisco *Chronicle,* April 28, 1937.

292

"Strauss . . . comparing the occasion to the driving of the golden spike": Ibid.

293

"Once that rivet was in, they were going to cut it out and take it home": Gales interview.

294

"In a final tidying up, men with steel wool and paint scrapers": Richard Dillon, Thomas Moulin, and Don DeNevi, *High Steel: Building the Bridges Across San Francisco Bay,* Millbrae, Calif.: Celestial Arts, 1979.

295

"He and Mr. Strauss had a row over the plaque": Natusch interview.

295

"Mr. Strauss was jealous of other people": Ibid.

295

"The idea of the Nazi emblem being displayed on the streets of San Francisco": San Francisco *Examiner,* May 27, 1937.

296

"insult to the German flag": Ibid.

296

"By 6 AM . . . it was estimated that 18,000 people": San Francisco *Chronicle,* May 28, 1937.

297

"Donald Bryant, a San Francisco Junior College sprinter": Ibid.

297

"Carmen Perez and her sister Minnie . . . Florentine Calegari . . . John V. Royan and his daughter Betty . . . Henry Boder . . . A woman apparently in physical distress": San Francisco *Chronicle,* San Francisco *Examiner,* May 28, 1937.

298

"a mighty ovation . . . for . . . the task of planning the span was his": San Francisco *Chronicle,* May 28, 1937.

298

"Launched 'midst a thousand hopes and fears": Joseph Strauss, "The Mighty Task Is Done," San Francisco *News,* May 26, 1937.

299

"crowd behavior . . . crowd antics, crowd manners": San Francisco *Chronicle,* May 28, 1937.

299

"I was in the parade, and I walked across the bridge": McClain interview.

299

"Away down deep . . . there is a deep roar": San Francisco *Chronicle,* May 28, 1937.

300

"Strung . . . like a topaz necklace": Ibid.

300

"This bridge needs neither praise, eulogy, nor encomium": San Francisco *Examiner,* May 29, 1937.

300

"an enormous local din of sirens, bells, whistles": Ibid.

302

"nineteen battleships and heavy cruisers . . . accompanied by twenty-three other major vessels": Ibid.

P   A   R   T

 ENDURING

Chapter Twenty-Three: "You see the position in which he is now placed."

306

"In a last desperate attempt . . . Golden Gate Ferries had cut its rates": Strauss to Ammann, February 3, 1938, in Ammann & Whitney Golden Gate Bridge file.

306

"there was more than one attempt to turn the bridge over to the State Highway Department" Brown, *Golden Gate.*

306

"Even the Bridge people didn't think then that the bridge was ever going to make money": Finnila interview.

307

"It was cumshaw . . . for having been around so long": Gales interview.

307

"Reed announced that he was leaving 'in justice to my family and my own future' ": Brown, *Golden Gate.*

308

"Cone . . . at this time helping Strauss prepare his final engineer's report": Cone to Frank M. Masters, July 2, 1937, Cone Papers.

308

"Cone . . . was appointed to the new job of maintenance engineer, at a salary . . . $2,500 less than

he had been earning": Reed to Ellis, December 22, 1937.

308

"a sixty-five-year-old retired banker 'thrust down our throats' ": Brown, *Golden Gate.*

308

"The hue and cry against the appointment of Harrelson": San Francisco *Chronicle,* July 31, 1937.

308

"At present, the bridge is not being properly maintained": Brown, *Golden Gate.*

308

"the Board of Income Tax Appeals . . . had ruled that Strauss and his wife were . . . deficient": San Francisco *Examiner,* June 10, 1937.

309

"I understand . . . that some of this difficulty": Reed to Ellis, December 22, 1937.

309

"You see the position in which he is now placed": Ibid.

309

"he had built the Golden Gate Bridge 'for nothing' ": San Francisco *Examiner,* October 22, 1937.

309

"It is true . . . I have practically no tangible financial gain": Ibid.

310

"I have been compelled to slow up very materially": Strauss to Ammann, December 29, 1937, in Ammann & Whitney Golden Gate Bridge file.

310

"I will be okeh in every respect": Strauss to Ammann, February 3, 1938, in Ammann & Whitney Golden Gate Bridge file.

310

"I drove to the San Francisco tower in a closed car": Cone report, San Francisco *Chronicle,* June 12, 1941.

310

"the center of the bridge was deflected": Ibid.

311

"He crawled over to where I was standing in the protection of the tower": Ibid.

311

"On March 28 . . . Strauss suffered a coronary thrombosis": San Francisco *Examiner*, May 17, 1938.

311

"the entire credit for the success of the bridge belongs to him": Ibid.

313

"O. H. Ammann . . . probably came the closest to duplicating the monumental achievement": *National Cyclopedia of American Biography*, vol. 52.

313

"Leon Moisseiff . . . resumed his career": *Who Was Who*, "Science and Technology."

314

"it is a regret that so many times": Derleth to Ammann and Moisseiff, July 27, 1937, in Ammann & Whitney Golden Gate Bridge file.

314

"Derleth remained chairman of the Department of Engineering at Berkeley": Derleth obituary, San Francisco *Examiner*, June 14, 1956.

314

"excellent condition": Ibid., June 13, 1941.

315

"only a few minor repairs": Ibid., December 8, 1951.

315

"operation of rapid transit trains . . . is not possible": Ibid., August 1, 1961.

315

"designer and supervising engineer": Chicago *Tribune*, July 15, 1983.

315

"the faculty . . . of imagining things on a grand scale": San Francisco *News* editorial, June 28, 1935.

315

"His practice returned to the character it had earlier": Mrs. Eleanor Mead to author, May 11, 1985; Johnson interview.

316

"Ellis remained on the engineering faculty at Purdue until . . . the age of seventy-two": Chicago *Tribune* obituary, August 23, 1949.

Chapter Twenty-Four: "The Man Who Built the Bridge"

317

"On November 7, 1940 . . . the Tacoma Narrows Bridge collapsed": Bleich, McCullough, Rosecrans, and Vincent, *Mathematical Theory of Vibration in Suspension Bridges*, p. 8.

317

"a group of competent engineers having an interest in the problem": Ibid.

317

"The start of the investigation coincided with a drive to economize": Brown, *Golden Gate*.

318

"Cone said that . . . the 'insurable value' policy on the bridge": *Engineering News-Record*, March 6, 1941.

318

"He went behind our backs": Brown, *Golden Gate*.

318

"You are making a grave mistake": San Francisco *Chronicle*, March 6, 1941.

319

"Cone . . . had just won the 1940 construction engineering prize": Ibid., January 10, 1941.

319

"I have never received any salary unless I was giving valued service": Ibid., March 7, 1941.

319

"Ever since that time . . . I have struggled": Ibid., June 12, 1941.

319

"the bridge 'is in good condition' ": San Francisco *Examiner*, June 13, 1941.

319

"In July, H. L. Nishkian was appointed engineer

in charge": Hugo D. Newhouse to Ammann, July 30, 1941, in Ammann & Whitney Golden Gate Bridge file.

319

"The bridge was equipped with instruments to measure wind": *Western Construction News,* April 1942.

320

" 'magnificent' condition": San Francisco *Examiner,* December 15, 1951.

320

"reinforcing the bridge floor with 5,000 tons of lateral bracing steel": Ibid., September 20, 1954.

320

"Following his firing . . . Russell Cone formed a small company": Russ Cone interview.

320

"The company now entered the manufacture of nuclear warheads": Ibid.

320

"an opportunity arose to return to bridge engineering": Ibid.

321

"Cone was proposed as chief engineer": Ibid.

321

"a monument to Joseph Strauss was dedicated": *Engineering News-Record,* June 1941.

BIBLIOGRAPHY

BOOKS

Anderson, Sherwood. *Sherwood Anderson's Memoirs.* Edited by Ray Lewis White. Chapel Hill, University of North Carolina Press, 1969.

Barry, John D. *The City of Domes.* San Francisco, John J. Newbegin, 1915.

Bendiner, Alfred. *Bendiner's Philadelphia.* New York, A.S. Barnes, 1964.

Bleich, Freidrich, C. B. McCullough, Richard Rosecrans, and George S. Vincent. *Vibrations in Suspension Bridges.* Washington, D.C., U.S. Department of Commerce, Bureau of Public Roads, 1950.

Bronson, William. *The Earth Shook, the Sky Burned,* Garden City, N.Y., Doubleday, 1969.

Brown, Allen. *Golden Gate, Biography of a Bridge,* Garden City, N.Y., Doubleday, 1965.

Cassady, Stephen. *Spanning the Gate.* Mill Valley, Calif., Squarebooks, Baron Wolman, 1979.

Dana, Julian. *A. P. Giannini, Giant in the West.* New York, Prentice-Hall, 1947.

Darrow, Clarence. *Attorney for the Damned.* New York, Simon & Schuster, 1957.

Dillon, Richard, Thomas Moulin, and Don DeNevi. *High Steel: Building the Bridges Across San Francisco Bay.* Millbrae, Calif., Celestial Arts, 1979.

The George Washington Bridge Over the Hudson River at New York. Bethlehem, Penn., McLintic-Marshall, 1932.

Gilliam, Harold. *Weather of the San Francisco Bay Region.* Berkeley, University of California Press, 1962.

Macomber, Ben. *Panama-Pacific International Exposition: The Jewel City.* San Francisco, John H. Williams, 1915.

Moisseiff, Leon S., and Frederick Lienhard. *Suspension Bridges Under the Action of Lateral Forces.* American Society of Civil Engineers *Proceedings,* no. 1849, 1932.

Morrow, Irving F. *Monastery of the Visitation of the Blessed Virgin Mary.* San Francisco, John Henry Nash, 1919.

Mowry, George E. *The California Progressives.* Berkeley, University of California Press, 1951.

National Cyclopedia of American Biography. New York, James T. White, 1941–.

O'Shaughnessy, Michael M. *Hetch Hetchy, Its Origin and History.* San Francisco, John J. Newbegin, 1934.

O'Shaughnessy, Michael M., and Joseph B. Strauss. *Bridging the Golden Gate.* San Francisco, privately printed, 1921.

Pennsylvania, A Guide to the Keystone State. WPA Writers Program. New York, Oxford University Press, 1940.

Speer, Albert. *Inside the Third Reich.* New York, Macmillan, 1970.

Strauss, Joseph B. *The Golden Gate Bridge at San Francisco, California.* Report of the Chief Engineer with Architectural Studies and Results of the Fact-Finding Investigation. San Francisco, Bridge District Board of Directors, 1930.

Strauss, Joseph B. *The Golden Gate Bridge: The Chief Engineer's Final Report.* San Francisco, Bridge District Board of Directors, 1938.

Thomas, Lately. *A Debonair Scoundrel.* New York, Holt, Rinehart and Winston, 1962.

van der Zee, John. *The Greatest Men's Party on Earth.* New York, Harcourt Brace, 1974.

Who Was Who in American History—Science and Technology. A component of *Who's Who in American History.* Bicentennial edition. Chicago, Marquis Who's Who, 1976.

ARTICLES

Cone, Russell G. "Battling Storm and Tide in Founding Golden Gate Pier." *Engineering News-Record* 115, no. 8, August 22, 1935.

Graham, A. E., "Rock Excavation for San Francisco Pier, Golden Gate Bridge." *Architect & Engineer,* December 1933.

Holway, Donal F., and John G. Morris. "100 Years Old, the Wonderful Brooklyn Bridge." *National Geographic,* May 1983.

"Introducing John E. Pearson, Prof. C. A. Ellis." *The Purdue Engineer,* June 1935.

Jones, Charles M. "Largest and Longest Bridge Cables Spun at New Record Rate." *Engineering News-Record* 116, no. 18, April 30, 1936.

Markwart, A. H. "Engineering Problems of the Panama Pacific Exposition," *Engineering News* 73, no. 7, February 18, 1915.

Meier, R. A. "Purdue Personalities." *The Purdue Engineer,* February 1938.

Moisseiff, Leon S. "Report on a Comparative Design of a Stiffened Suspension Bridge Over the Golden Gate at San Francisco." New York, November 1925, Ammann & Whitney Golden Gate Bridge file.

"The Municipal Engineering Works of the City of San Francisco." *Engineering News* 73, no. 7, February 18, 1915.

Paine, Clifford E. "Designing Bridge Towers 700-ft. High." *Engineering News-Record* 117, no. 15, October 8, 1936.

Peters, Henry S. "Tom Finn, A Biography." San Francisco *Chronicle,* 5-part series, January 1938.

Strauss, Joseph B. "The Mighty Task Is Done" (poem). San Francisco *News,* May 26, 1937.

Taylor, Frank J. "Strauss: Little Man Who Wanted to Build the Biggest Thing." San Francisco *News,* Golden Gate Bridge Festival section, May 26, 1937.

PAPERS

Cone, Russell G. Papers of Russell Cone. Correspondence, proposals, engineer's daily reports, photographs, news clippings, provided by Russell Cone, Jr.

Derleth, Charles E., Jr. Papers of Charles E. Derleth, Jr. Seven Golden Gate Bridge volumes covering 1929–1941, including a Golden Gate Bridge diary, correspondence, speeches, reports, proposals, and news clippings. Water Resources Center Archives, University of California, Berkeley.

Golden Gate Bridge File, Amman & Whitney, Inc. Twenty-eight volumes covering 1929–1941; letters, weekly chief engineer's reports, memoranda, proposals, telegrams. Ammann & Whitney, Inc., 2 World Trade Center, New York, New York.

Morrow, Irving Foster. Papers of Irving F. Morrow. Correspondence, proposals, drawings, news clippings. Provided by Mrs. Eleanor Morrow Mead, Tucson, Arizona.

NEWSPAPER FILES

O. H. Ammann file. San Francisco *Examiner*. Microfiche.

Charles E. Derleth, Jr., file. San Francisco *Examiner*. Microfiche.

Leon Moisseiff file. San Francisco *Examiner*. Microfiche.

Michael O'Shaughnessy file. San Francisco *Examiner*. Incorporates material from San Francisco *Call*, San Francisco *Bulletin*, San Francisco *News*, and San Francisco *Chronicle*. Microfiche.

Clifford Paine file. San Francisco *Examiner*. Microfiche.

Joseph Strauss file. Chicago *Tribune*. Newspaper Room, Doe Library, University of California, Berkeley.

Joseph Strauss file. Los Angeles *Times*. Photocopy.

Joseph Strauss file. San Francisco *Examiner*. Microfiche.

INDEX

ABOUT THE AUTHOR

Born in San Francisco, John van der Zee lives with his wife, son and daughter within view of the Golden Gate Bridge. *The Gate* is his tenth book.

More than two years before the start of construction, consulting architect Irving Morrow caught the spirit of the finished Golden Gate Bridge in a series of boldly imaginative charcoal drawings. This is the first time Morrow's drawings have been reproduced.

0-595-09429-5

Made in the USA
Middletown, DE
03 March 2016